T0329568

PRACTICAL APPROACHES TO METHOD VALIDATION AND ESSENTIAL INSTRUMENT QUALIFICATION

PRACTICAL APPROACHES TO METHOD VALIDATION AND ESSENTIAL INSTRUMENT QUALIFICATION

Edited by

CHUNG CHOW CHAN
CCC Consulting, Mississauga, Ontario, Canada

HERMAN LAM
Wild Crane Horizon Inc., Scarborough, Ontario, Canada

XUE MING ZHANG
Apotex, Inc., Richmond Hill, Ontario, Canada

WILEY

A JOHN WILEY & SONS, INC., PUBLICATION

Published by John Wiley & Sons, Inc., Hoboken, New Jersey.
Published simultaneously in Canada.

For general information on our other products and services or for technical support, please contact our Customer Care Department within the United States at (800) 762-2974, outside the United States at (317) 572-3993 or fax (317) 572-4002.

Wiley also publishes its books in a variety of electronic formats. Some content that appears in print may not be available in electronic formats. For more information about Wiley products, visit our web site at www.wiley.com.

Library of Congress Cataloging-in-Publication Data:

Practical approaches to method validation and essential instrument qualification / edited by Chung Chow Chan, Herman Lam, Xue Ming Zhang.
    p. ; cm.
   Complement to: Method validation and instrument performance verification / edited by Chung Chow Chan ... [et al.]. c2004.
   Includes bibliographical references and index.
    ISBN 978-0-470-12194-8 (hardback)
  1. Drugs—Analysis—Methodology—Evaluation. 2. Laboratories—Equipment and supplies— Evaluation. 3. Laboratories—Instruments—Evaluation. I. Chan, Chung Chow. II. Lam, Herman. III. Zhang, Xue Ming. IV. Method validation and instrument performance verification.
   [DNLM: 1. Chemistry, Pharmaceutical—instrumentation. 2. Chemistry, Pharmaceutical— methods. 3. Clinical Laboratory Techniques—standards. 4. Technology, Pharmaceutical— methods. QV 744 M5915 2010]
   RS189.M425 2010
   615'.1901—dc22

                 2009054243

10 9 8 7 6 5 4 3 2 1

CONTENTS

Contributors   **xi**

Preface   **xiii**

**1   Overview of Risk-Based Approach to Phase Appropriate
     Validation and Instrument Qualification**   **1**
*Chung Chow Chan, Herman Lam, Xue Ming Zhang, Stephan Jansen,
Paul Larson, Charles T. Manfredi, William H. Wilson, and
Wolfgang Winter*

   1   Risk-Based Approach to Pharmaceutical Development, 1

   2   Regulatory Requirements for Performance Verification of
       Instruments, 3

   3   General Approach to Instrument Performance Qualification, 4

       References, 10

2   Phase Appropriate Method Validation   **11**
Chung Chow Chan and Pramod Saraswat

   1   Introduction, 11

   2   Parameters for Qualification and Validation, 14

   3   Qualification and Validation Practices, 15

   4   Common Problems and Solutions, 25

       References, 25

**3   Analytical Method Verification, Method Revalidation,
     and Method Transfer**   **27**
Chung Chow Chan and Pramod Saraswat

     1   Introduction, 27

     2   Cycle of Analytical Methods, 28

     3   Method Verification Practices, 28

     4   Method Revalidation, 36

     5   Method Transfer, 40

     6   Common Problems and Solutions, 42

            References, 43

4   Validation of Process Analytical Technology Applications   **45**
Alison C. E. Harrington

     1   Introduction, 45

     2   Parameters for Qualification and Validation, 48

     3   Qualification, Validation, and Verification Practices, 56

     4   Common Problems and Solutions, 69

            References, 73

**5   Validation of Near-Infrared Systems for Raw Material
     Identification**   **75**
Lenny Dass

     1   Introduction, 75

     2   Validation of an NIR System, 77

     3   Validation Plan, 78

            References, 90

6   Cleaning Validation   **93**
Xue Ming Zhang, Chung Chow Chan, and Anthony Qu

     1   Introduction, 93

     2   Scope of the Chapter, 94

     3   Strategies and Validation Parameters, 94

     4   Analytical Methods in Cleaning Validation, 100

5   Sampling Techniques, 102

6   Acceptance Criteria of Limits, 104

7   Campaign Cleaning Validation, 107

8   Common Problems and Solutions, 108

    References, 110

**7   Risk-Based Validation of Laboratory Information
     Management Systems**   **111**
R. D. McDowall

1   Introduction, 111

2   LIMS and the LIMS Environment, 114

3   Understanding and Simplifying Laboratory Processes, 117

4   GAMP Software Categories and System Life Cycle for a
    LIMS, 121

5   Validation Roles and Responsibilities for a LIMS Project, 124

6   System Life-Cycle Detail and Documented Evidence, 126

7   Maintaining the Validated Status, 147

8   Summary, 151

    References, 152

8   Performance Qualification and Verification of Balance   **155**
Chung Chow Chan, Herman Lam, Arthur Reichmuth, and Ian Ciesniewski

1   Introduction, 155

2   Performance Qualification, 159

3   Common Problems and Solutions, 161

    References, 169

    Appendix, 170

9   Performance Verification of NIR Spectrophotometers   **177**
Herman Lam and Shauna Rotman

1   Introduction, 177

2   Performance Attributes, 179

3   Practical Tips in NIR Performance Verification, 192

    References, 195

Appendix 1, 197

Appendix 2, 199

**10 Operational Qualification in Practice for Gas
    Chromatography Instruments**   **201**
*Wolfgang Winter, Stephan Jansen, Paul Larson, Charles T. Manfredi,
William H. Wilson, and Herman Lam*

1   Introduction, 201

2   Parameters for Qualification, 202

3   Operational Qualification, 208

4   Preventive Maintenance, 219

5   Common Problems and Solutions, 220

    References, 229

**11 Performance Verification on Refractive Index, Fluorescence,
    and Evaporative Light-Scattering Detection**   **231**
Richard W. Andrews

1   Introduction, 231

2   Qualification of Differential Refractive Index Detectors, 238

3   Qualification of Fluorescence Detectors, 242

4   Qualification of Evaporative Light-Scattering Detectors, 250

    Reference, 253

**12 Instrument Qualification and Performance Verification
    for Particle Size Instruments**   **255**
Alan F. Rawle

1   Introduction, 255

2   Setting the Scene, 257

3   Particle Counting Techniques, 258

4   Particle Size Analysis and Distribution, 259

5   Instrument Qualification for Particle Size, 260

6   Qualification of Instruments Used in Particle Sizing, 272

7   Method Development, 279

8   Verification: Particle Size Distribution Checklist, 291

9   Common Problems and Solutions, 292

10   Conclusions, 295

        References, 295

        Appendix, 297

**13 Method Validation, Qualification, and Performance
     Verification for Total Organic Carbon Analyzers**   **299**
José E. Martínez-Rosa

        1   Introduction, 299

        2   TOC Methodologies, 302

        3   Parameters for Method Validation, Qualification, and
               Verification, 307

        4   Qualification, Validation, and Verification Practices, 310

        5   Common Problems and Solutions, 321

        References, 323

14 Instrument Performance Verification: Micropipettes   **327**
George Rodrigues and Richard Curtis

        1   Introduction, 327

        2   Scope of the Chapter, 328

        3   Verification Practices: Volume Settings, Number of
               Replicates, and Tips, 338

        4   Parameters: Accuracy, Precision, and Uncertainty, 342

        5   Summary, 345

        References, 346

**15 Instrument Qualification and Performance Verification
     for Automated Liquid-Handling Systems**   **347**
John Thomas Bradshaw and Keith J. Albert

        1   Introduction, 347

        2   Commonalities Between Volume Verification Methods for
               Performance Evaluation, 354

        3   Volume Verification Methods, 356

        4   Importance of Standardization, 371

        5   Summary, 372

        References, 373

**16 Performance Qualification and Verification in Powder X-Ray
    Diffraction**   **377**

Aniceta Skowron

1   Introduction, 377

2   Basics of X-Ray Diffraction, 378

3   Performance Qualification, 381

4   Performance Verification: Calibration Practice, 387

        References, 389

Index   **391**

CONTRIBUTORS

Keith J. Albert, Artel, Inc., Westbrook, Maine, USA

Richard W. Andrews, Waters Corporations, Milford, Massachusetts, USA

John Thomas Bradshaw, Artel, Inc., Westbrook, Maine, USA

Chung Chow Chan, CCC Consulting, Mississauga, Ontario, Canada

Ian Ciesniewski, Mettler Toledo Inc., Columbus, Ohio, USA

Richard Curtis, Artel, Inc., Westbrook, Maine, USA

Lenny Dass, GlaxoSmithKline Canada Inc., Mississauga, Ontario, Canada

Alison C. E. Harrington, ABB Ltd., Daresbury, United Kingdom

Stephan Jansen, Agilent Technologies Inc., Amstelveen, The Netherlands

Herman Lam, Wild Crane Horizon Inc., Scarborough, Ontario, Canada

Paul Larson, Agilent Technologies Inc., Wilmington, Delaware, USA

Charles T. Manfredi, Agilent Technologies Inc., Wilmington, Delaware, USA

José E. Martínez-Rosa, JEM Consulting Services Inc., Caguas, Puerto Rico

R. D. McDowall, McDowall Consulting, Bromley, Kent, United Kingdom

Anthony Qu, Patheon Inc., Cincinnati, Ohio, USA

Alan F. Rawle, Malvern Instruments Inc., Westborough, Massachusetts, USA

Arthur Reichmuth, Mettler Toledo GmbH, Greifensee, Switzerland

George Rodrigues, Artel, Inc., Westbrook, Maine, USA

Shauna Rotman, Wild Crane Horizon Inc., Scarborough, Ontario, Canada

Pramod Saraswat, Azopharma Product Development Group, Hollywood, Florida, USA

Aniceta Skowron, Activation Laboratories Ltd., Ancaster, Ontario, Canada

William H. Wilson, Agilent Technologies Inc., Wilmington, Delaware, USA

Wolfgang Winter, Matthias Hohner AG, Karlsruhe, Germany

Xue Ming Zhang, Apotex Inc., Richmond Hill, Ontario, Canada

PREFACE

This book is a complement to our first book, *Method Validation and Instrument Performance Verification*. As stated there, for pharmaceutical manufacturers to achieve commercial production of safe and effective medications requires the generation of a vast amount of reliable data during the development of each product. To ensure that reliable data are generated in compliance with current good manufacturing practices (cGMPs), all analytical activities involved in the process need to follow good analytical practices (GAPs). GAPs can be considered as the culmination of a three-pronged approach to data generation and management: method validation, calibrated instrumentation, and training.

The chapters are written with a unique practical approach to method validation and instrument performance verification. Each chapter begins with general requirements and is followed by strategies and steps taken to perform these activities. The chapters end with the authors sharing important practical problems and their solutions with the reader. I encourage you to share your experience with us, too. If you have any observations or solutions to a problem, please do not hesitate to email it to me at chung_chow_chan@cvg.ca.

The method validation section focus on the strategies and requirements for early-phase drug development, the validation of specific techniques and functions [e.g. process analytical technology (PAT)], cleaning, and laboratory information management systems (LIMSs). Chapter 1 is an overview of the regulatory requirements on quality by design in early pharmaceutical development and instrument performance verification. Instrument *performance verification* and *performance qualification* are used as synonyms in this book. Chapter 2 is an overview of the strategies of phase 1 and 2 development from the analytical perspective. Chapter 3 provides guidance on compendial method verification, analytical revalidation, and analytical method transfer. Discussed are strategies for an equivalent analytical method and how that can be achieved. Chapters 4 and 5 cover method validation of specific techniques in PAT and near-infrared identification. Chapters 6 and 7 give guidance on cleaning validation and LIMS validation.

The instrument performance verification section (Chapters 8 to 16) provides unbiased information on the principles involved in verifying the performance of instruments that are used for the generation of reliable data in compliance with cGXPs (all current good practices). Guidance is given on some common and specialized small instruments and on several approaches to the successful performance verification of instrument performance. The choice of which approach to implement is left to the reader, based on the needs of the laboratory. Chapter 8 provides background information on the most fundamental and common but most important analytical instrument used in any laboratory, the balance. A generic protocol template for the performance verification of the balance is also included to assist young scientists in developing a feel for writing GXP protocol. Performance verification requirements for near-infrared, gas chromatographic, and high-performance liquid chromatographic detectors are described in Chapters 9, 10, and 11. Chapter 12 gives guidance on performance verification of particle size, which is very challenging for its concept. Chapter 13 covers the requirements needed for the specialized technique of total organic content. Performance verification of small equipment used in pipettes and liquid-handling systems is discussed in Chapters 14 and 15. Chapter 16 provides an overview of x-ray diffraction technique and performance verification of this instrument.

The authors of this book come from a broad cultural and geographical base—pharmaceutical companies, vendor and contract research organizations—and offer a broad perspective to the topics. I want to thank all the authors, coeditors, and reviewers who contributed to the preparation of the book.

CHUNG CHOW CHAN

CCC Consulting
Mississauga, Ontario, Canada

1

OVERVIEW OF RISK-BASED APPROACH TO PHASE APPROPRIATE VALIDATION AND INSTRUMENT QUALIFICATION

CHUNG CHOW CHAN
CCC Consulting

HERMAN LAM
Wild Crane Horizon Inc.

XUE MING ZHANG
Apotex, Inc.

STEPHAN JANSEN, PAUL LARSON, CHARLES T. MANFREDI, AND WILLIAM H. WILSON
Agilent Technologies Inc.

WOLFGANG WINTER
Matthias Hohner AG

1 RISK-BASED APPROACH TO PHARMACEUTICAL DEVELOPMENT

In the United States, the U.S. Food and Drug Administration (FDA) ensures the quality of drug products using a two-pronged approach involving review of information submitted in applications as well as inspection of manufacturing facilities for conformance to requirements for current good manufacturing practice

Practical Approaches to Method Validation and Essential Instrument Qualification,
Edited by Chung Chow Chan, Herman Lam, and Xue Ming Zhang
Copyright © 2010 John Wiley & Sons, Inc.

(cGMP). In 2002, the FDA, together with the global community, implemented a new initiative, "Pharmaceutical Quality for the 21st Century: A Risk-Based Approach" to evaluate and update current programs based on the following goals:

- The most up-to-date concepts of risk management and quality system approaches are incorporated while continuing to ensure product quality.
- The latest scientific advances in pharmaceutical manufacturing and technology are encouraged.
- The submission review program and the inspection program operate in a coordinated and synergistic manner.
- Regulatory and manufacturing standards are applied consistently.
- FDA resources are used most effectively and efficiently to address the most significant issues.

In the area of analytical method validation and instrument performance qualification, principles and risk-based orientation, and science-based policies and standards, are the ultimate driving forces in a risk-based approach to these activities.

1. *Risk-based orientation*. To comply with the new guiding regulatory principle to provide the most effective public health protection, regulatory agencies and pharmaceutical companies must match their level of effort against the magnitude of risk. Resource limitations prevent uniform intensive coverage of all pharmaceutical products and production.
2. *Science-based policies and standards*. Significant advances in the pharmaceutical sciences and in manufacturing technologies have occurred over the last two decades. Although this knowledge has been incorporated in an ongoing manner, the fundamental nature of the changes dictates a thorough evaluation of the science base to ensure that product quality regulation not only incorporates up-to-date science but also encourages further advances in technology. Recent science can also contribute significantly to assessment of risk.

Related directly or indirectly to implementation of the risk-based approach to pharmaceutical quality, the following guidance affecting the analytical method and instrument qualification had been either initiated or implemented.

FDA 21 Code of Federal Regulations (CFR) Part 11: Electronic Records Requirements. The final guidance for industry Part 11, Electronic Records, Electronic Signatures: Scope and Application, clarifies the scope and application of the Part 11 regulation and provides for enforcement discretion in certain areas. The guidance explains the goals of this initiative, removes barriers to scientific and technological advances, and encourages the use of risk-based approaches.

ICH (International Conference on Harmonization) Q9: Risk Management. The goal of the guidance is to manage risk to patients, based on science, from information on the product, process, and facility. The level of oversight required is commensurate with the level of risk to patients and the depth of product and process understanding.

FDA Guidance for Industry PAT: A Framework for Innovative Pharmaceutical Manufacturing and Quality Assurance. This guidance is intended to encourage the voluntary development and implementation of innovative pharmaceutical manufacturing and quality assurance technologies. The scientific, risk-based framework outlined in this guidance, process analytical technology (PAT), helps pharmaceutical manufacturers design, develop, and implement new and efficient tools for use during product manufacture and quality assurance while maintaining or improving the current level of product quality assurance. It also alleviates any concerns that manufacturers may have regarding the introduction and implementation of new manufacturing technologies.

FDA Guidance for Industry: Quality Systems Approach to Pharmaceutical cGMP Regulations. One of the objectives of this guidance is to provide a framework for implementing quality by design, continual improvement, and risk management in the drug manufacturing process.

FDA Guidance for Industry INDs: cGMP for Phase 1 Investigational Drugs. This guidance recommended that sponsors and producers of phase 1 material consider carefully risks in the production environment that might adversely affect the resulting quality of an investigational drug product.

Implementation of a risk-based approach to analytical method validation and performance verification should be done simultaneously and not in isolation. It is only through a well-thought-out plan on the overall laboratory system of instrument performance verification that quality data for analytical method validation will be obtained. The laboratory will subsequently be able to support the manufacture of either clinical trial materials or pharmaceutical products for patients. Details of risk-based approaches to phase appropriate analytical method validation and performance verification are presented in subsequent chapters.

2 REGULATORY REQUIREMENTS FOR PERFORMANCE VERIFICATION OF INSTRUMENTS

System validation requirements are specified in many different sources, including 21 CFR Part 58 [good laboratory practice (GLP)], 21 CFR Parts 210 and 211 (cGMP) [1], and more recently, in the GAMP 4 guide [2]. GLP, and GMP/cGMP are often summarized using the acronym GXP. Current GXP regulations require that analytical instruments be qualified to demonstrate suitability for the intended use. Despite the fact that instrument qualification is not a new concept and regulated firms invest a lot of effort, qualification-related deviations are frequently cited in inspectional observations and in warning letters by regulatory agencies such as the FDA and its equivalents in other countries. In common terms, the objective of qualification is to establish documented evidence that a system has been designed and installed according to specifications and operates in such a way that it fulfills its intended purpose.

GLP makes the following provisions in 21 CFR 58.63 about maintaining, calibrating, and testing equipment:

- Equipment is to be adequately inspected, cleaned, maintained, calibrated, and tested.
- Written standard operating procedures (SOPs) are required for testing, calibration, and maintenance.
- Written records are to be maintained for all inspection, maintenance, calibration, and testing.

cGMP makes the following provisions in 21 CFR 211.68(a):

- Automatic equipment, including computers, that will perform a function satisfactorily may be used.
- Equipment is to be calibrated, inspected, or checked routinely according to a written program designed to assure proper performance.
- Written records of calibration checks and inspections are to be maintained.

Many validation professionals in regulated firms are not sure what exactly to qualify or requalify, test, and document. How much testing is enough? Unlike analytical method validation, there were no clear standards for equipment qualification. The *United States Pharmacopeia* (USP) has addressed this issue by publishing General Chapter ⟨1058⟩ on analytical instrument qualification (AIQ) [3,4]. The USP establishes AIQ as the basis for data quality and defines the relationship to analytical method validation, system suitability testing, and quality control checks. Similar to analytical method validation, the intent of AIQ is to ensure the quality of an instrument before conducting any tests. In contrast, system suitability and quality control checks ensure the quality of analytical results right before or during sample analyses.

3   GENERAL APPROACH TO INSTRUMENT PERFORMANCE QUALIFICATION

Testing is one of the most important analytical measures for system developers and system users when verifying that a system fulfills the defined system requirements and is fit for the intended purpose. Generally, the fitness of systems for the intended purpose (i.e., their quality) needs to be ensured through constructive and analytical measures. Constructive measures are defined in terms of recognized professional engineering practices and include formal design methodologies that typically follow a life-cycle approach. System qualification follows a structured approach that uses test cases and test parameters based on a scientific and risk-based analysis. Defining and executing these tests typically require the use of metrology.

Other analytical measures include trending analysis of metrics such as error rates, formal methods of failure analysis, and formal reviews and inspections. Testing and the associated collection of documented evidence on the system test activities are key tasks of quality assurance. The documented evidence comprises test planning, test execution, test cases, and test results, all of which must be traceable to the requirements documented in various levels of specification documents (i.e., user requirements specification, functional specifications, design specifications, test specifications, etc.).

3.1 Definition of Terms

Many different definitions are used for the relevant terms in the area of equipment qualification. Not all of them are identical. For the sake of this chapter, we use the terms *design qualification* (DQ), *installation qualification* (IQ), *operational qualification* (OQ), and *performance qualification* (PQ), in line with the definitions originally published by the Valid Analytical Measurement Instrument Working Group (see Figure 1). Similar system qualification approaches are discussed thoroughly in GAMP (Good Automated Manufacturing Practice) Forum publications and in USP General Chapter ⟨1058⟩. DQ, IQ, OQ, and PQ constitute important phases that result in key deliverables during the overall validation activities necessary over a system's life cycle (see Figure 2).

Design Qualification During DQ, the functional and operational specifications of an instrument need to be defined and documented. DQ is an important decision-making tool for selecting the best system and supplier. The right type of equipment is selected for specific tasks, and the supplier's ability to meet and reproduce these performance criteria consistently through appropriate quality processes in design, development, manufacturing, and support is crucial for efficacy and risk mitigation. DQ is primarily the user's responsibility, because this is the only logical place to define site requirements. The supplier, however, typically needs to provide materials such as technical specifications and other documents relevant to system validation. This includes evidence on processes that are critical to quality, including the life-cycle methodology. DQ focuses on specifications, design documentation, requirements traceability from design to test, corrective action procedures, impact analyses, test plans, and test evidence. DQ responds to a requirement originally defined in GLP (21 CFR Part 58.61) that mandates that appropriate design and adequate capacity for consistent functioning as intended are assured for equipment used in activities subject to this regulation.

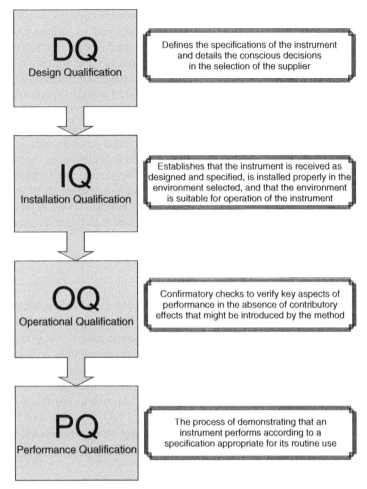

FIGURE 1   The four stages of instrument qualification and definition of terms according to the Valid Analytical Measurement Instrument Working Group. (From [5].)

Installation Qualification   IQ uses procedures that demonstrate, to a high degree of assurance, that an instrument or system has been installed according to accepted standards. IQ provides written evidence that the system has been installed according to the specifications defined by the manufacturer (supplier) and, if applicable, the user's organization. IQ checks the correctness of the installation and documents the intactness of the system, typically through system inventory lists, part numbers, firmware revisions, system drawings, and wiring and plumbing diagrams. Several organizations have provided specific guidance about the scope of an IQ and elaborated on the potential division of responsibilities between the system supplier and the user's organization. One important conclusion is that assembly checks performed at the supplier's factory

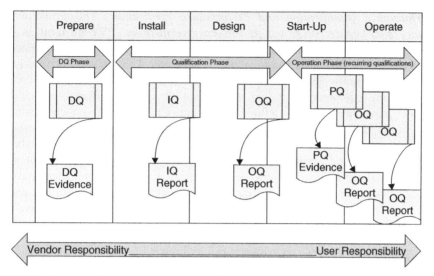

FIGURE 2   Activities, phases, and key deliverables during a system's validation life cycle. For the sake of simplification, system retirement (decommissioning) is not shown.

cannot be substituted for an IQ performed at the user's site [6]. The supplier's documented test results (e.g., factory acceptance tests), however, can be used to reduce the extent of validation activities performed during an IQ. The key is that IQ demonstrates and documents that the system has been received and installed in the user's domain according to the relevant specifications.

The IQ is usually provided by the vendor at a cost. The typical deliverables include the following information:

- System location
- Equipment model/serial numbers
- Documentation of basic function and safety features
- Documentation about compliance with site requirements

Operational Qualification   In contrast to an IQ, which challenges the installation process, operational qualification focuses on the functionality of the system. An OQ challenges key operational parameters and, if required, security functions by running a well-defined suite of functional tests. The OQ uses procedures that demonstrate, to a high degree of assurance, that an instrument or system is operating according to accepted standards. In most cases, the OQ is delivered as a paid service from the provider. It typically includes a suite of component and system tests that are designed to challenge the functional aspects of the system. The OQ deliverable needs to provide documented and auditable evidence of control. The frequency of the OQ is determined by the user's organization. In most laboratories, the typical frequency is once or twice a year after the initial OQ.

Performance Qualification and Performance Verification   The terms *performance qualification* (PQ) and *performance verification* (PV) are used as synonyms. PQ verifies system performance under normal operating conditions across the anticipated operating range of the equipment. This makes PQ mostly an application-specific test with application-specific acceptance limits. For chromatography equipment, ongoing verification of system performance includes *system suitability tests*, as defined in General Chapter ⟨621⟩ on chromatography of the USP [7], which outlines the apparatus tests as well as the calculation formulas to be used for quantification and the evaluation of system suitability. The European and Japanese pharmacopeias use a similar approach, but there are regional differences in how certain system suitability parameters have to be calculated. In the following chapters we focus on the holistic and modular tests required for operational qualification but do not elaborate in detail on application-specific performance qualification.

Requalification After Repair (RQ)   In essence, RQ is similar to OQ. RQ's goal is to verify the correctness and success of a repair procedure performed on a system, and to put the system back into the original qualified state by running a series of appropriate tests. RQ typically is a subset of an OQ, but for complex repairs to components that are critical to the overall performance of the system, it may be necessary to perform the complete suite of OQ tests.

3.2   Analytical Instrument Qualification: USP ⟨1058⟩

USP General Chapter ⟨1058⟩ is a step forward for the validation community [8]. It establishes the well-proven *4Q model* as the standard for instrument qualification and provides useful definitions of roles, responsibilities, and terminology to steer the qualification-related activities of regulated firms and their suppliers. The 4 Qs in the model refer to DQ, IQ, OQ, and PQ (see Figure 3).

The 4Q model helps answer the following critical questions:

- How can an analytical laboratory prove that a given analysis result is based on trustworthy and reliable instrument data?
- How can the analytical laboratory ascertain the validity of the analysis result and show appropriate evidence that the analytical instrument was really doing what the analyst thought it would do and that the instrument was within the specifications required for the analysis?

The AIQ chapter of the USP categorizes the rigor and extent of the qualification activities by instrument class. As an example, gas chromatographs are categorized as class C (complex instruments with highly method-specific conformance requirements). The acceptance limits (conformity bounds) are determined by the application. The deployment (installation and qualification) of such an instrument is complicated and typically requires assistance from specialists. In any case, USP ⟨1058⟩ class C instruments are required to undergo a full

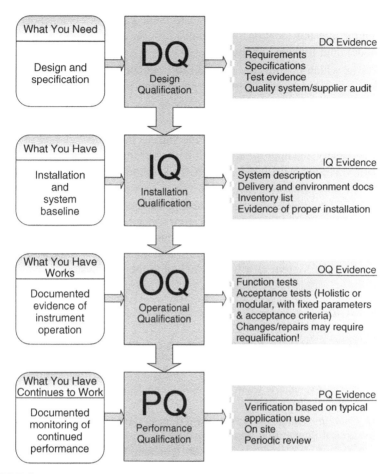

FIGURE 3 The 4Q model, consisting of DQ, IQ, OQ, and PQ, along with the key questions answered by each phase and its key deliverables.

qualification process, which requires structured and extensive documentation about the system and the approach used for qualification.

3.3 Recommendations for Analytical Instrument Qualification

1. Develop an SOP for AIQ according to the 4Q qualification model.
2. If you already have an SOP for AIQ, determine how it can be mapped to the 4Q model.
3. If your SOP proposes a different methodology than that of 4Q, you need to come up with a scientifically sound rationale. Document your rationale and explain how your methodology ensures trustworthy, reliable, and consistent instrument data.

4. Use a single procedure for an instrument category, independent of the vendor and the location. Acceptance criteria may have to vary by make, model, and intended application.

5. Assess which instruments are used for regulated activities and whether the data generated by the instrument are subject to a predicate rule.

6. Assess the risk of instrument failure or nonconformance, using scientific knowledge.

7. Define qualification protocols for the various instrument classes in your lab. If necessary and appropriate, work with your instrument suppliers or partner with someone who has a proven track record in the field of instrument qualification services.

8. The USP guidance is general regarding the use and impact of data systems. Therefore, plan additional qualification and acceptance tests to obtain a high degree of assurance that control, communication, and data are accurate and reliable. Your integrated validation and qualification approach needs to consider the system as a whole, including the data system.

REFERENCES

1. 21 *Code of Federal Regulations*, Parts 210 and 211. Part 210: *Current Good Manufacturing Practice In Manufacturing, Processing, Packing, or Holding of Drugs*; Part 211: *Current Good Manufacturing Practice for Finished Pharmaceuticals*. FDA, Washington, DC, 1996. Available at www.fda.gov/cder/dmpq/cgmpregs.htm#211.110. Accessed Aug. 17, 2007.

2. *GAMP Guide for Validation of Automated Systems*, 4th ed. International Society for Pharmaceutical Engineering, Tampa, FL, 2001.

3. S. K. Bansal, T. Layloff, E. D. Bush, M. Hamilton, E. A. Hankinson, J. S. Landy, S. Lowes, M. M. Nasr, P. A. St. Jean, and V. P. Shah. Qualification of analytical instruments for use in the pharmaceutical industry: a scientific approach. *AAPS PharmSciTech*, 5(1):article 22, 2004.

4. *U.S. Pharmacopeia*, General Chapter ⟨1058⟩, Analytical Instrument Qualification. USP, Rockville, MD.

5. P. Bedson. *Guidance on Equipment Qualification of Analytical Instruments: High Performance Liquid Chromatography (HPLC)*. Published by LGC in collaboration with the Valid Analytical Measurement Instrumentation Working Group, June 1998.

6. *Guidance Notes on Installation and Operational Qualification*. GUIDE-MQA-006-005. Health Sciences Authority, Manufacturing and Quality Audit Division, Centre for Drug Administration, Singapore, Sept. 2004.

7. *U.S. Pharmacopeia*, General Chapter ⟨621⟩, Chromatography. USP, Rockville, MD.

8. W. Winter. Analytical instrument qualification: standardization on the 4Q model. *BioProcess Int.*, 4(9): 46–50, 2006.

2

PHASE APPROPRIATE METHOD VALIDATION

CHUNG CHOW CHAN
CCC Consulting

PRAMOD SARASWAT
Azopharma Product Development Group

1 INTRODUCTION

The 2002 initiative "Pharmaceutical Quality for the 21st Century: A Risk-Based Approach" formally known as the "Pharmaceutical cGMP Initiative for the 21st Century," was intended to modernize the U.S. Food and Drug Administration's (FDA's) regulation of pharmaceutical quality for veterinary and human drugs, select human biological products such as vaccines, and capture the larger issue of product quality, with current good manufacturing practices (cGMPs) being an important tool toward improving overall product quality. This regulation acknowledges the need, and provides for, assessment of the risk and benefits of drug development and forms a foundation to provide guidance for reasonable, minimally acceptable method validation practices.

1.1 Cycle of Analytical Methods

The analytical method validation activity is a dynamic process, as summarized in the life cycle of an analytical procedure shown in Figure 1. An analytical

Practical Approaches to Method Validation and Essential Instrument Qualification,
Edited by Chung Chow Chan, Herman Lam, and Xue Ming Zhang
Copyright © 2010 John Wiley & Sons, Inc.

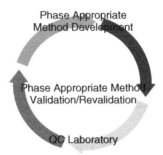

FIGURE 1   Life cycle of the analytical method.

method will be developed and validated for use in analyzing samples during the early development of a drug substance or drug product. The extent and level of analytical method development and analytical method validation will change as the analytical method progresses from phase 1 to commercialization.

In the United States, the FDA recognized that application of the cGMP regulations, as described in 21 *Code of Federal Regulations* (CFR) 211, is not always relevant for the manufacture of clinical investigational drug products. The FDA recognized the need to develop specific GMPs for investigational products and elected to address the progressive phase-appropriate nature of cGMPs in drug development for a wide variety of manufacturing situations and product types for compliance with cGMPs starting with phase 1 studies and progressing through phase 3 and beyond, referred to in this chapter as *phase appropriate method development and method validation*. The final method will be validated for its intended use, whether for a market image drug product or for clinical trial release.

1.2   Challenges of New Technologies

Regulatory agencies understand and encourage companies to apply new technologies to provide information on the physical, chemical (micro), and biological characteristics of materials to improve process understanding and to measure, control, and/or predict the quality and performance of products. New technologies [e.g., liquid chromatography–mass spectrometry (LCMS)] are being applied increasingly to support new products and new processes. Immunogenicity assay may be required for some biotechnology-derived products.

1.3   To Validate or Not to Validate?

Sometimes the question is asked: Should analytical method be validated as early as when going from preclinical to phase 1 studies? This question arose perhaps from a mis-interpretation of the guidance document "Content and Format of Investigational New Drug Applications (INDs) for Phase 1 Studies of

Drugs, Including Well-Characterized, Therapeutic, Biotechnology-Derived Products," which stated that validation data and established specifications ordinarily need not be submitted at the initial stage of drug development. The answer is, of course, that the analytical method should be validated. The validation data need not be submitted, but the validation must be completed so that the analytical methods used will assure the strength, identity, purity, safety, and quality (SISPQ) of the drug substance and the drug product.

In the cGMP guidance for phase 1 investigational drugs [1], it is stated explicitly that laboratory tests used in the manufacture (e.g., testing of materials, in-process material, packaging, drug product) of phase 1 investigational drugs should be scientifically sound (e.g., specific, sensitive, and accurate), suitable, and reliable for the specified purpose. The tests should be performed under controlled conditions and follow written procedures describing the testing methodology. Records of all test results, procedures, and changes in procedures should be maintained. Laboratory testing of a phase 1 investigational drug should evaluate quality attributes that define the SISPQ.

1.4 Quality by Design: A Risk-Based Approach

The focus of the concept of quality by design is to ensure that quality is built into a product, with a thorough understanding of the product and process by which it is developed and manufactured, along with a knowledge of the risks involved in manufacturing the product and how best to mitigate those risks. Regulatory bodies recognize that knowledge of a drug product and its analytical methods will evolve through the course of development. This is stated explicitly in ICH (International Conference on Harmonization) Q7A. Changes are expected during development, and every change in product, specifications, or test procedures should be recorded adequately. It is therefore reasonable to expect that changes in testing, processing, packaging, and so on, will occur as more is learned about the molecule. However, even with the changes, the need to ensure the safety of subjects in clinical testing should not be compromised.

The purpose in the early phase of drug development is to deliver a known dose that is bioavailable for clinical studies. As product development continues, increasing emphasis is placed on identifying a stable, robust formulation from which multiple bioequivalent lots can be manufactured and ultimately scaled up, transferred, and controlled for commercial manufacture. The method validation requirements of methods need to be adjusted through the life cycle of a method.

The development and validation of analytical methods should follow a similar progression. The purpose of analytical methods in early stages of development is to ensure potency, to understand the impurity and degradation product profile, and to help understand key drug characteristics. As development continues, the method should indicate stability and be capable of measuring the effect of key manufacturing parameters to ensure consistency of the drug substance and drug product.

Analytical methods used to determine purity and potency of an experimental drug substance that is very early in development will need a less rigorous method validation exercise than would be required for a quality control laboratory method at the manufacturing site. An early-phase project may have only a limited number of lots to be tested, and the testing may be performed in only one laboratory by a limited number of analysts. The ability of the laboratory to "control" the method and its use is relatively high, particularly if laboratory leadership is clear in its expectations for performance of the work.

The environment in which a method is used changes significantly when the method is transferred to a quality control laboratory at the manufacturing site. The method may be replicated in several laboratories, multiple analysts may use it, and the method may be one of many methods used daily in the laboratory. Late development and quality control methods need to be run accurately and consistently in a less controlled environment (e.g., in several laboratories with different brands of equipment). The developing laboratory must therefore be aware of the needs of the receiving laboratories (e.g., a quality control laboratory) and the regulatory expectations for successful validation of a method to be used in support of a commercial product.

Each company's phase appropriate method validation procedures, and processes will vary, but the overall philosophy is the same. The extent of and expectations from early-phase method validation are lower than the requirements in the later stages of development. The validation exercise becomes larger and more detailed, and it collects a larger body of data to ensure that the method is robust and appropriate for use at the commercial site.

2   PARAMETERS FOR QUALIFICATION AND VALIDATION

Typical analytical performance characteristics that should be considered in the validation of the types of procedures described here are listed in Table 1.

TABLE 1   Validation Parameters

Analytical Procedure Characteristic	Identification	Testing for Impurities Quantitation	Limit	Assay: Dissolution (Measurement Only), Content/Potency
Accuracy	$-^a$	$+^b$	$-$	$+$
Precision	$-$			
Repeatability	$-$	$+$	$-$	$+$
Intermediate precision	$-$	$+^c$	$-$	$+$
Specificity^d	$+$	$+$	$+$	$+$
Detection limit	$-$	$-^e$	$+$	$-$
Quantitation limit	$-$	$+$	$-$	$-$
Linearity	$-$	$+$	$-$	$+$

$^a-$, Characteristic is not normally evaluated.
$^b+$, Characteristic is normally evaluated.
^{c}In cases where reproducibility has been performed, intermediate precision is not needed.
^{d}Lack of specificity of one analytical procedure could be compensated by other, supporting analytical procedure(s).
^{e}May be needed in some cases.

3 QUALIFICATION AND VALIDATION PRACTICES

3.1 Phase Appropriate Method Validation

Regulatory agencies recognize that some controls, and the extent of controls needed to achieve appropriate product quality, differ not only between investigational and commercial manufacture, but also among the various phases of clinical studies [2]. It is therefore expected that a company will implement controls that reflect product and production considerations, evolving process and product knowledge, and manufacturing experience. The term *qualification* is sometimes used loosely to represent method validation in the early stage of method development.

3.2 Preclinical Method Validation

As described in more detail below, there is even less guidance on the requirements for method development for preclinical method validation. The scientist should qualify and not validate the method to the extent that the data generated to make decisions and provide information should be scientifically sound. A minimum study of linearity, repeatability, detection limit (for quantitation of impurities), and specificity will be required.

3.3 Phase 1 to Phase 2 to Phase 3: Drug Substance and Drug Product Method Validation

The typical process that is followed in an analytical method validation is listed chronologically below, irrespective of the phases of method validation. However, the depth and detail of treatment for each of the following activities will vary with the phase [e.g., an abbreviated validation protocol (Table 2) versus detail validation protocol (Table 3)].

1. Planning and deciding on the method validation experiments
2. Writing and approval of method validation protocol
3. Execution of the method validation protocol
4. Analysis of the method validation data
5. Reporting on the analytical method validation
6. Finalizing the analytical method procedure

Method validation experiments should be well planned and laid out to ensure efficient use of time and resources during execution of the method validation. The best way to ensure a well-planned validation study is to write a method validation protocol that will be reviewed and signed by the appropriate person (e.g., laboratory management, quality assurance). However, there are differences in the terms of agreement for writing a validation protocol and in the details of the protocol in the early stages of method development.

TABLE 2   Abbreviated Validation Protocol

Method	Experimental	Proposed Acceptance Criteria
Dissolution by high-performance liquid chromatography (HPLC)	Linearity for x mg capsules according to method (0.05–0.15 mg/mL) Accuracy and precision of six capsules of known release profile Standard and sample stability (e.g., at 24 and 48 h) Specificity	1. Meet system suitability. 2. Meet method requirement for linearity (R, slope, y-intercept, etc.). 3. $Q = 65\%$ at 60 min. 4. Report the % relative standard deviation (RSD) for precision. 5. Stability within 2% of initial. 6. There is no coeluting peak from the dissolution medium.
Assay by HPLC	Linearity for x mg capsules according to method (0.05–0.15 mg/mL) Accuracy and precision of six capsules Standard and sample stability (e.g. at 24 and 48 h) Specificity of degraded sample	1. Meet system suitability. 2. Meet method requirement for linearity (R, slope, y-intercept, etc.). 3. Accuracy recovery of 97–103%. 4. Report the % RSD for precision. 5. Stability within 2% of initial. 6. No coeluting peaks were detected. The peak of the drug substance is pure.

A normal validation protocol should contain the following elements at a minimum:

1. The objective of the protocol
2. Validation parameters that will be evaluated
3. Acceptance criteria for all the validation parameters evaluated
4. Information regarding the experiments to be performed
5. A draft analytical procedure

Phase Appropriate Validation of a Drug Substance in Early Development

Linearity   For most early-phase methods, the assay and impurity methods are combined. One possibility is to determine two linearity curves: one for impurity from LOQ level to 2.0% and one for assay from approximately 50 to 150%. There are different strategies using three to five concentrations for each set of linearity. Sample linearity plots are shown in Figures 2 and 3.

Accuracy   Accuracy can be inferred from the precision, linearity, and specificity. The overall mass balance of the drug substance peak and known impurities should be used to verify the accuracy of a method. An example of accuracy data for recovery of a drug substance and impurities are given in Table 4.

TABLE 3 Detailed Validation Protocol

A. Introduction

This protocol outlines qualification activities to be performed for the HPLC identity, assay, and impurity tests in the compound X capsule.

B. Specificity

Procedure: The following solutions are prepared and analyzed according to the identity and assay methods.
  Mobile phase A
  Mobile phase B
  Known impurity xxx, etc.

Acceptance Criteria:

  1. No peaks in the dissolving solvent, and mobile-phase injections should coelute or otherwise interfere with compound X or its known impurities.
  2. Known impurities should be resolved from compound X.

C. Linearity

Assay

Procedure: From a stock solution of compound X in diluent (e.g., 2.00 mg/mL), dilutions are performed to make five solutions at 50 to 150% of the assay concentration (0.50 mg/mL). For linearity of the method, plot the response vs. analyte concentration at each concentration level. Perform a linear regression analysis and determine the correlation coefficient.

Acceptance criteria:

  1. The correlation coefficient (r) should be NLT 0.99.
  2. Record the slope, y-intercept, residuals, and % y-intercept of response vs. nominal concentration.

Related Substances

Procedure: From a stock solution of compound X in diluent (e.g., 2.00 mg/mL), dilutions are performed to make five solutions from a limit of quantitation (LOQ) level of about 0.05 to 5% of the target compound X nominal sample concentration (0.50 mg/mL). For linearity of the method, plot the response vs. analyte concentration at each concentration level. Perform a linear regression analysis and determine the correlation coefficient.

Acceptance criteria:

  1. The correlation coefficient (r) should be NLT 0.98.
  2. Record the slope, y-intercept, residuals and % y-intercept of response vs. nominal concentration.

D. Accuracy

Assay, Impurity, and Identity

(Continued overleaf)

TABLE 3 (*Continued*)

Procedure:

1. Prepare triplicate sample preparations of compound X (total of nine sample preparations) according to the draft method procedure in x mL volumetric flasks that will yield 50%, 100%, and 150% nominal related substance sample solution.
2. Calculate the recovery of compound X, area% of compound X, individual impurities present at 0.05% (or greater) with respect to compound X peak, and total impurities at each concentration level.

Acceptance criteria:

1. The percent recovery values at each level should be between 98 and 102%.
2. The retention times of compound X should be greater than Y min.
3. Report the impurities $\geq$0.05% at each level.

E. Precision

Assay, Identity, and Related Substances

Procedure:

1. Three additional replicate sample solutions are prepared at nominal concentration and analyzed according to the draft analytical method.
2. Combine the results from the triplicate samples of the 100% accuracy level. Calculate the relative standard deviation (% RSD) of the following: compound X recovery, individual impurity present at 0.05% (or greater) with respect to compound X, and total impurities in the six samples.
3. Record the retention times and % RSD of compound X from all the results.

Acceptance criteria:

1. The % RSD of the six assay values should not be greater than 2.0%.
2. The % RSD of the six % area values for compound X should not be greater than 2.0%.
3. Report results of the individual impurity (>0.05% or greater) and total impurity.
4. The % RSD of the retention times of compound X should not be greater than 2.0%.

F. Stability of Sample and Standard Solutions

Use one of each standard solution and precision sample solution, and analyze these samples at different time points (e.g., 1, 2, and 4 days). Prepare fresh standard at each time point.

Acceptance criteria:

1. *Assay.* To be considered stable, the recovery of compound X should be 98 to 102% compared to the initial time point.
2. *Purity.* The profile of the sample at each time point should be similar to that at the initial time point. The individual impurity >0.05% and total % impurity should be NMT 0.1% (absolute) compared to the initial time point.

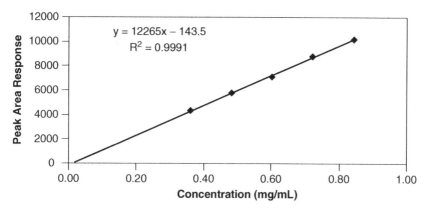

FIGURE 2   Linearity plot of peak area response vs. concentration (assay).

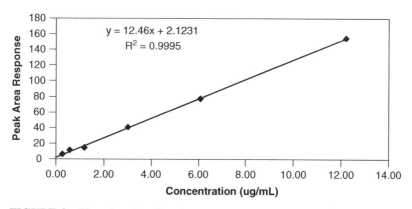

FIGURE 3   Linearity plot of peak area response vs. concentration (impurity).

System Precision and Method Precision

SYSTEM PRECISION   System precision will be derived from the system suitability requirements, which are tested at the time of method validation.

METHOD PRECISION
- *Method precision repeatability.* The precision repeatability of assay and impurity methods may be assessed by testing at least three sample concentrations or six replicate sample preparations at the nominal concentration. The latter strategy is often used. These results give analytical scientists a high level of confidence regarding the precision of the assay and impurity methods. Example sets of repeatability data for the assay and impurities are given in Tables 5 and 6.
- *Method intermediate precision and reproducibility.* At early stages of drug substance development, the methods are typically carried out in one

TABLE 4   Accuracy Recovery of Assay Validation

Sample Name	Recovery (%)	Avg. (%)
Accuracy 50%		
sample 1	101.1	
sample 2	102.3	102
sample 3	103.0	
Accuracy 100%		
sample 1	101.4	
sample 2	100.6	101
sample 3	102.2	
Accuracy 150%		
sample 1	100.6	
sample 2	101.4	101
sample 3	100.8	

TABLE 5   Repeatability Data of Assay Validation

Sample Name	Assay (%)
Precision 1	101.2
Precision 2	102.1
Precision 3	100.3
Precision 4	100.7
Precision 5	101.7
Precision 6	100.7
Average (%)	101.1
% RSD	0.7

laboratory by a few analysts. It is therefore usually not necessary to determine the intermediate precision and reproducibility of an assay or impurity method at the early stage of development.

Stability of Sample and Standard Solutions   At the early phase of development, validation should demonstrate that the standard and sample solutions are adequately stable for the duration of their use in the laboratory. Sample sets of data for stability of sample and standard solutions are given in Tables 7 to 9. During method development and early stages of drug development, the analyst should develop an experience base. This information is useful for later-stage development of specific robustness experiments and helps establish appropriate system suitability requirements for later methods.

Specificity   Assay and impurity method specificity should be evaluated during the early development stages. The specificity should be reviewed and reevaluated

TABLE 6   Repeatability Data of Assay and Impurities Validation

Peak Name	Area (%)	RRT	Peak Name	Area (%)	RRT
	Assay 1			*Assay 2*	
Compound X	97.09	1.00	Compound X	97.03	1.00
Unknown 1	0.13	1.11	Unknown 1	0.10	1.11
Unknown 2	2.53	1.20	Unknown 2	2.46	1.20
Unknown 3	0.17	1.31	Unknown 3	0.27	1.31
Unknown 4	0.08	1.40	Unknown 4	0.13	1.40
Total area	100.00		Total area	100.00	
	Assay 3			*Assay 4*	
Compound X	97.08	1.00	Compound X	96.97	1.00
Unknown 1	0.12	1.11	Unknown 1	0.10	1.11
Unknown 2	2.53	1.20	Unknown 2	2.57	1.20
Unknown 3	0.20	1.31	Unknown 3	0.26	1.31
Unknown 4	0.07	1.40	Unknown 4	0.10	1.40
Total area	100.00		Total area	100.00	
	Assay 5			*Assay 6*	
Compound X	97.09	1.00	Compound X	97.11	1.00
Unknown 1	0.15	1.11	Unknown 1	0.10	1.11
Unknown 2	2.52	1.20	Unknown 2	2.53	1.20
Unknown 3	0.16	1.31	Unknown 3	0.15	1.31
Unknown 4	0.08	1.40	Unknown 4	0.12	1.40
Total area	100.00		Total area	100.00	

Average of Six Assays		
Peak Name	RRT	Average (% RSD)
Compound X	—	—
Unknown 1	1.11	0.12 (19.3%)
Unknown 2	1.20	2.52 (1.4%)
Unknown 3	1.31	0.20 (25.0%)
Unknown 4	1.40	0.10 (27.6)

RRT, Relative retention time.

TABLE 7   Stability of Standard Solution (Assay)

Sample Name	Time	% Recovery	Change % Recovery
Std. 1–initial	Initial	100.6	NA
Std. 1–24 h	1 day	101.6	1.0
Std. 1–48 h	2 day	100.9	0.3

regularly as changes are made to the drug substance synthetic process. It is important that the method can demonstrate separation of the main component and impurities from the raw materials and intermediates. As the synthetic process continues to change with the progress of the development project, the scientist should constantly evaluate the potential for generating new impurities of side products and demonstrate the capability of the assay and impurity methods to separate new intermediates, side products, and raw materials. As appropriate, samples stored under relevant stress conditions (see drug product section) should

TABLE 8 Stability of Sample Solution (Assay)

Sample Name	Time	% Recovery	Change % Recovery
100%–sample 1	Initial	101.0	NA
100%–sample 2	1 day	100.2	0.8
100%–sample 3	2 day	100.8	0.2

TABLE 9 Stability of Impurity in Sample Solution

		24 h		48 h	
RRT	Initial Area (%)	Area (%)	Change (%)	Area (%)	Change (%)
1.00	97.09	97.13	—	97.11	—
1.11	0.14	0.10	0.04	0.08	0.06
1.20	2.51	2.49	0.02	2.51	0.00
1.31	0.18	0.18	0.00	0.17	0.01
1.40	0.10	0.10	0.00	0.13	0.03

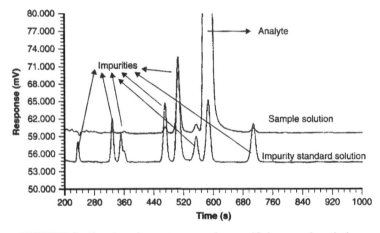

FIGURE 4 Overlay chromatogram of a specificity sample solution.

be used to demonstrate specificity. A chromatogram of the specificity is shown in Figure 4.

Phase Appropriate Validation of Drug Product At the beginning of drug development, it is anticipated that the formulation of the new drug will be the drug substance in a bottle, a simple immediate-release capsule or tablet formulation, and will be the assumption in the discussion below. The four principal methods of analysis of drug products are described in this section: assay, impurities, dissolution, and content uniformity. The general principles can be applied to other tests (e.g., the Karl Fischer test).

Linearity Linearity is a very simple experiment and should be performed when possible. For most early-phase methods, assay and impurity are combined. This test is performed as described earlier. There are different strategies that use three to five concentrations for each set of linearity.

Accuracy At an early stage of development, a minimum number of recovery studies are recommended. For the assay, recovery of the drug substance at 100% level at each strength is minimally required. ICH recommendations should be followed, as they require only preparing an additional two sets of spike solution (e.g., at 50 and 150%). For multiple strengths, a bracket at the lowest strength (e.g., 50% lowest strength) coupled with that at the highest strength (e.g., 150% highest strength) will cover the full range. For dissolution, recovery of the drug substance in the presence of excipients may cover 50, 100, and 120%. However, in certain cases a lower range may be required (e.g., 10% if a dissolution profile is required). For content uniformity, recovery of the drug substance from 70 to 130% is required, as this is the content uniformity range expected. For a multiple-strength product, bracketing may be used as for an assay. Degradation products may be scarce or not available at an early stage of development. Therefore, the accuracy of the impurity method is replaced by recovery of the drug substance in the presence of excipients from the reporting limit to specification or 150% of the specification level of the impurities.

System Precision and Method Precision

SYSTEM PRECISION   System precision will be derived from the system suitability requirements that are tested at the time of method validation.

METHOD PRECISION
- *Repeatability.* The repeatability of assay, impurity, content uniformity, and dissolution methods are generally carried out using triplicate sample preparations, and in combination with the triplicate preparations from the accuracy experiments, to give six replicates. For the impurity method, additional repeatability using the specification level of the impurity for the drug substance may be performed. For multiple strengths of similar formulation, the bracketing principle in the accuracy section may be applied.
- *Intermediate precision and reproducibility.* As for the drug substance, at the early stage of drug product development, the methods are typically carried out in one laboratory by a few analysts. It is therefore usually not necessary to determine the intermediate precision and reproducibility of an assay or impurity method.

Stability of Standard and Sample Solution The standard and sample solutions should be adequately stable for the duration of their use in the laboratory.

Robustness During method development and the early stages of drug development, the analyst should develop an experience base. This information is useful in later stages and helps establish appropriate system suitability requirements for later method revision.

Specificity   As for the drug substance, assay and impurity methods specificity should be evaluated during the early development stages. It is important that the assay method can demonstrate separation of the drug substance and impurities from the excipients and intermediates. For the impurity assay method, the drug product should be appropriately degraded to demonstrate that degradation peaks are resolved from the drug substance and synthetic impurities. Common degradation studies utilize 0.1 N HCl, 0.1 N NaOH, 3% hydrogen peroxide, or ultraviolet/visible light to give about 10 to 30% degradants. For the content uniformity and dissolution methods, the absence of interference from the extracting solvent, dissolution media, and excipients will need to be demonstrated.

LOD and LOQ   It is a regulatory requirement that the quantitation limit (LOQ) for the drug product impurity method be no greater than its reporting limit. The limit of detection (LOD) is not required at this stage (except for a limit test), but most companies determine it at the time of generating the LOQ by using an additional set of solution that is further diluted from the LOQ solution. For the assay, content uniformity, and dissolution methods, LOD and LOQ are not required.

3.4   Documentation for Phase Appropriate Method Validation

The fundamental concepts of cGMPs must be applied regardless of the details of the phase appropriate method validation strategy used [3]. Examples of these include:

1. Proper documentation
2. Change control
3. Deviations
4. Equipment and utilities qualification
5. Proper training

The raw laboratory data generated for the validation of analytical methods must be documented properly in a notebook or other GMP-compliant data storage device. Upon completion of all the experiments, all the data should be compiled into a detail validation report that will conclude the success or failure of the validation exercise. Depending on the company's strategy, only a summary of the validation data may be generated. Successful execution of the validation process will lead to a final analytical procedure that can be used by the laboratory to support future analytical work for a drug substance or drug product.

Information Required in an Analytical Procedure   To ensure compliance with traceability and GMP, the minimum information that should be included in a final analytical procedure is summarized below.

1. The rationale of the analytical procedure and description of the capability of the method. Revisions of the analytical procedure should include the advantages offered by the new revision.

2. The analytical procedure. This section should contain a complete description of the analytical procedure in sufficient detail to enable another analytical scientist to replicate it. The write-up should include all important operational parameters and specific instructions (e.g., preparation of reagents, system suitability tests, precautions, and explicit formulas for calculation of the test results).

3. The validation data. Either a detailed set or a summary set of validation data is included.

4. The revision history.

5. The signatures of authors, reviewers, management, and quality assurance.

4 COMMON PROBLEMS AND SOLUTIONS

4.1 Three-Point vs. One-Point Calibration for Quantitation

It is common and advantageous to use one-point calibration for a method, as it will save analyst time and create less documentation. However, the behavior of the drug substance may not allow the method to do that. Under the latter circumstance, it will be necessary to prepare three standard solutions at the time of each assay experiment.

4.2 Sink Condition for Dissolution

For a dissolution test, it is important to ensure that dissolution conditions are able to satisfy sink conditions. *Sink condition* refers to the ratio (usually 3:1) of the recommended volume of solvent or medium for dissolution testing to the volume required to dissolve the drug in the unit (tablet or capsule) to saturate the solution. The reason for this requirement is to assure that a sufficient volume of dissolution medium is available to dissolve the drug from the product. The solution should not reach saturation or it will affect the dissolution rate negatively. Without sink conditions, the dissolution results may lead to variable results during routine dissolution analysis.

REFERENCES

1. *Guidance for Industry: cGMP for Phase 1 Investigational Drugs*. FDA, Washington, DC, July 2008.

2. *Guidance for Industry: Content and Format of Investigational New Drug Applications (INDs) for Phase 1 Studies of Drugs, Including Well-Characterized, Therapeutic, Biotechnology-Derived Products*. FDA, Washington, DC, Nov. 1995.

3. *Guidance for Industry: Quality Systems Approach to Pharmaceutical cGMP Regulations*. FDA, Washington, DC, Sept. 2006.

3

ANALYTICAL METHOD VERIFICATION, METHOD REVALIDATION, AND METHOD TRANSFER

CHUNG CHOW CHAN
CCC Consulting

PRAMOD SARASWAT
Azopharma Product Development Group

1 INTRODUCTION

The applicable *U.S. Pharmacopeia* (USP) or *National Formulary* (NF) standard applies to any article marketed in the United States that (1) is recognized in the compendium and (2) is intended or labeled for use as a drug or as an ingredient in a drug. This applicable standard applies to such articles whether or not the added designation "USP" or "NF" is used. These standards of identity, strength, quality, and purity of the article are determined by official tests, procedures [1], and acceptance criteria, whether incorporated in the monograph itself, in the general notices, or in the applicable general chapters in the USP. *Method verification* refers to the experiments required to verify the suitability of the compendial procedure under actual conditions of use in the testing laboratory.

When manufacturers make changes to the manufacturing process of a drug substance (e.g., route of synthesis) or drug product (e.g., formulation), the changes

Practical Approaches to Method Validation and Essential Instrument Qualification,
Edited by Chung Chow Chan, Herman Lam, and Xue Ming Zhang
Copyright © 2010 John Wiley & Sons, Inc.

will necessitate revalidation of the analytical procedures [2]. Revalidation should be carried out to ensure that the analytical procedure maintains its analytical characteristics (e.g., specificity) and to demonstrate that the analytical procedure continues to ensure the identity, strength, quality, purity, and potency of the drug substance and drug product. The degree of revalidation is dependent on the nature of the change. When a different regulatory analytical procedure is substituted for the current method [e.g., high-performance liquid chromatography (HPLC) is used to replace titration] the new analytical procedure should be validated.

Analytical method transfer is the transfer of analytical procedure from an originator laboratory to a receiving laboratory. The analytical parameters that need to be considered for method transfer are similar to those for analytical method revalidation and verification.

2   CYCLE OF ANALYTICAL METHODS

Analytical method validation is not a one-time study. This was illustrated and summarized for the life cycle of an analytical procedure in Figure 1 of Chapter 2. An analytical method will be developed and validated for its intended use to analyze samples during the early development of an active pharmaceutical ingredient (API) or drug product. As drug development progresses from phase 1 to its commercialization, the analytical method will follow a similar progression. The final method will be validated for its intended use for the market image drug product and transferred to the quality control laboratory for the launch of the drug product. However, if there are any changes in the manufacturing process that have the potential to change the analytical profile of the drug substance and drug product, this validated method may need to be revalidated to ensure that it is still suitable for analyzing the drug substance or drug product for its intended purpose.

3   METHOD VERIFICATION PRACTICES

3.1   Method Verification Versus Method Validation

U.S. Food and Drug Administration (FDA) regulation 21 CFR 211.194(a)(2) states specifically that users of analytical methods in the USP and NF are not required to validate the accuracy and reliability of these methods [2] but merely to verify their suitability under actual conditions of use. USP has issued a guidance for verification in General Chapter ⟨1226⟩ [3]. This general guidance provides general information to laboratories on the verification of compendial procedures that are being performed for the first time to yield acceptable results utilizing the laboratories' personnel, equipment, and reagents.

Verification consists of assessing selected analytical performance characteristics, such as those that are described in Chapter 2, to generate appropriate, relevant data rather than repeating the validation process. Although complete revalidation of a compendial method is not required to verify the suitability of the method

TABLE 1 **Validation and Verification Requirements or HPLC Assay of Final Dosage Forms**

Analytical Validation Parameter	Drug Product Validation	Drug Product Verification
System suitability	Yes	Yes
Method precision	Yes	Yes
Accuracy	Yes	Yes
Limit of detection (LOD)	No	No
Limit of quantification (LOQ)	No	No
Specificity	Yes	Yes
Range	Yes	No
Linearity	Yes	No
Ruggedness	a	No

[a] May be required.

under actual conditions of use, some of the analytical performance characteristics listed in USP General Chapter ⟨1225⟩ [4] or ICH (International Conference on Harmonization) Q2 (R1) [5] may be used for the verification process. Only those characteristics that are considered to be appropriate for the verification of the particular method need to be evaluated. The degree and extent of the verification process may depend on the level of training and experience of the user, on the type of procedure and its associated equipment or instrumentation, on the specific procedural steps, and on which article(s) are being tested.

Table 1 compares the validation requirements with the verification requirements of an example HPLC assay of a finished dosage form. ICH requires validation of the analytical properties of accuracy, precision, specificity, linearity, and range. However, verification will require only a minimum of precision and specificity validation. The accuracy requirements will depend on the specific situation of the final dosage form.

An alternative validation method may be used to demonstrate compliance with a monograph. The alternative method may have advantages in accuracy, sensitivity, precision, selectivity or adaptability to automation or computerized data reduction, or in other special circumstances. It is important to note that such alternative or automated procedures must be validated. However, since pharmacopeial standards and procedures are interrelated; where a difference appears or in the event of a dispute, only the results obtained by the procedure given in USP is conclusive.

3.2 Verification Process of the Compendial Procedure

Users should have the appropriate experience, knowledge, and training to understand and be able to perform the compendial procedures as written. Verification should be conducted by the user such that the results will provide confidence that the compendial procedure will perform suitably as intended. If the verification of the compendial procedure is not successful, and assistance from the USP staff has not resolved the problem, it may then be necessary to develop and validate an alternative procedure as allowed in the general notices of the USP. Modification of a chromatographic procedure can be made as long as the system suitability requirements for the particular monograph test are met. The alternative procedure may be submitted to USP along with the appropriate validation data, to support a proposal for inclusion or replacement of the current compendial procedure.

3.3   Preparation of a Solution for Verification

Some adjustments may be necessary and are allowed in the process of verification of compendial methods. Proportionately larger or smaller quantities than the specified weights and volumes of assay or test substances and reference standards may be used [6]. However, the measurement should be made with at least equivalent accuracy, and any subsequent steps (e.g., dilutions) are adjusted accordingly to yield concentrations equivalent to those specified in the monograph.

3.4   Equivalent HPLC Columns

The L nomenclature to designate HPLC column type was introduced in the Fourth Supplement to USP XIX in 1978. The L1 designation is for columns with octadecylsilane as the bonded phase [7]. When USP XX was published in 1980, only seven columns were classified and given a brief description. Since then the list has grown to 56 descriptions, some of them very broad or with imprecise wording. For years, this classification system has generated an increasing number of inquiries to USP regarding which column brand is appropriate for a particular compendial procedure. Today, column packings are developed for specific applications, resulting in columns with distinct characteristics even though they belong to the same original USP classification. For example, more than 220 columns currently available in the worldwide market can be classified as L1, but not all of them have the same applications. This situation makes the process of selecting a column for a particular application very difficult. The problem is partially controlled by the system suitability test in most USP chromatographic procedures, but in many cases these tests are not conclusive, to ensure column interchangeability. USP had presented two approaches to help users find equivalent columns that they may use to verify or validate compendial procedures [8,9]: the USP approach and the PQRI approach. Both of these approaches have merit, and it is too soon to favor one over the other. The *USP approach* provides column performance characterization (e.g., theoretical plate count, good peak symmetry) and produces five data points to describe the column. The *PQRI approach* provides selectivity characterization (relative retention times), and the parameters are included in a searchable database that produces a list of suitable columns ordered by the column comparison function. The USP approach could also be provided in database form to permit ordering of columns based on a single factor derived from the measured parameters (analogous to the PQRI approach).

USP Approach to Equivalent Columns This procedure uses a mixture of five organic compounds (uracil, toluene, ethylbenzene, quinizarin, and amitriptyline) in methanol to characterize column performance. This test mixture is intended primarily for the characterization of C18 columns used in reversed-phase liquid chromatography. Selection of the components in the National Institute of Standards and Technology's (NIST's) SRM 870 was based on published testing protocols and commercial column literature to provide a broad characterization of column performance in a single, simple test. On the basis of the results obtained using SRM 870, four parameters were identified to be used in characterization of the columns: hydrophobicity (capacity factor of ethylbenzene), chelation (tailing factor of quinizarin), activity toward bases (silanol activity: capacity factor and tailing factor of amitriptyline), and shape selectivity (bonding density). Table 2 covers the characterization of 52 columns.

TABLE 2 Characterization of C18 Columns Using SRM 870

Column Number	Hydrophobicity Capacity Factor (k'), Ethylbenzene	Chelating Tailing Factor, Quinizarin	Silanol Activity		Shape Selectivity Bonding Density ($\mu mol/m^2$)
			Capacity Factor (k'), Amitriptyline	Tailing Factor, Amitriptyline	
1	2.8	No peak	No peak	No peak	3.4
2	2.1	1.4	8.2	6.7	3.5
3	2.0	1.1	7.3	2.3	2.0
4	2.4	1.0	6.1	1.8	4.0
5	2.4	1.1	5.9	3.4	3.8
6	1.0	6.0	7.5	4.0	1.1
7	1.5	7.5	4.6	3.0	2.7
8	2.2	1.7	5.1	1.7	3.2
9	1.6	1.5	3.1	1.2	3.3
10	0.7	No peak	23	3.0	1.7
11	2.0	No peak	11.5	7.0	2.6
12	1.0	1.2	1.7	1.1	2.3
13	1.5	1.1	3.3	1.3	2.2
14	2.0	No peak	35	8.0	2.7
15	1.7	1.1	5.1	2.4	1.6
16	2.0	2.0	6.3	1.9	3.5
17	1.5	1.9	23	2.8	2.2
18	1.6	6.6	4.1	2.7	3.2
19	4.2	1.6	11	3.9	3.6
20	3.2	1.6	7.6	2.0	3.6
21	0.9	1.3	2.2	2.1	4.2
22	0.4	2.5	1.0	4.9	Not available
23	1.5	1.5	3.5	2.0	3.1
24	1.5	3.4	4.3	3.6	3.2
25	1.5	2.0	5.6	4.1	2.4
26	1.2	2.2	12	2.6	4.6

(Continued overleaf)

TABLE 2   (*Continued*)

| Column Number | Hydrophobicity Capacity Factor (k'), Ethylbenzene | Chelating Tailing Factor, Quinizarin | Silanol Activity | | Shape Selectivity Bonding Density (μmol/m^2) |
			Capacity Factor (k'), Amitriptyline	Tailing Factor, Amitriptyline	
27	1.3	1.4	3.5	2.1	3.3
28	2.2	1.2	5.3	1.1	3.4
29	0.7	No peak	2.1	1.4	2.3
30	2.6	1.2	—	3.3	4.0
31	2.2	1.0	—	3.6	Not available
32	2.5	1.6	—	1.2	3.3
33	2.0	1.2	—	1.0	5.5
34	1.0	1.4	3.0	2.6	3.0
35	1.3	No peak	3.8	3.9	3.1
36	1.3	2.0	4.5	13	3.1
37	1.8	1.5	13.6	2.8	2.6
38	1.9	1.5	5.0	2.4	2.6
39	1.9	1.5	5.1	2.4	2.7
40	1.9	1.5	6.0	2.9	2.2
41	3.3	1.3	8.8	2.9	3.2
42	1.6	1.4	5.0	2.7	1.4
43	0.9	1.4	3.0	2.8	0.9
44	1.9	1.3	5.0	1.5	2.5
45	1.5	1.3	4.4	1.9	1.9
46	3.3	1.2	7.5	1.3	3.0
47	2.0	1.0	6.7	2.6	2.1
48	1.0	2.2	3.1	2.4	2.1
49	2.2	1.4	14.2	3.5	3.2
50	2.2	1.8	10.2	2.2	3.0
51	3.9	1.7	12.5	4.0	2.9
52	2.3	1.0	6.1	1.8	2.9

PQRI Approach to Equivalent Columns   Based on the retention data for a series of standard mixtures and the same separation conditions (50% acetonitrile/buffer; pH 2.8 and 7.0; 35°C), every reversed-phase column can be characterized by six column-selectivity parameters: relative retention (kEB), hydrophobicity (H), steric interaction (S^*), hydrogen-bond acidity (A) and basicity (B), and relative silanol ionization or cation-exchange capacity (C).

The ability to characterize column selectivity is of potential value for two different situations. First, routine HPLC procedures require replacement of the column from time to time due to deterioration of the column during use. Also, when an HPLC method is transferred, it is necessary to obtain a suitable column for that procedure. In either situation, there exists the possibility that an equivalent column from the original supplier may no longer be available. For this reason, two or more equivalent columns with different part numbers can be specified as part of method development. "Equivalent" columns will have similar (ideally, "identical") values of the six column-selectivity parameters discussed above.

A second use of the six column-selectivity parameters outlined above is for the selection of columns of very different selectivity. Columns of different selectivity are often required during HPLC method development (for a deliberate change in selectivity) or for the development of orthogonal procedures that can be used to ensure that no new sample impurity is present in some samples. The procedure was originally developed for application to type B C18 columns. It has since been extended to type B alkyl–silica columns with C1 to C30 ligands; type A C8 and C18 columns; columns with polar groups such as urea, carbamate, or amide that are either embedded in the ligand or used to end-cap the column; cyano columns; and phenyl and fluoro columns. Columns with identical values of H, S^*, A, B, and C are expected to give essentially identical selectivity (spacing of bands) for a given HPLC procedure (same mobile phase, temperature, and flow rate). Small differences in kEB values can be corrected by changes in flow rate. Although it is rare to find two reversed-phase columns that have identical values of H, S^*, and so on, small differences in these column parameters are still acceptable for any sample, and larger differences are allowable for some samples. A column comparison function (F_S) can be defined for two columns, 1 and 2, as follows:

$$F_S = \{[12.5(H_2 - H_1)]^2 + [100(S_2^* - S_1^*)]^2 + [30(A_2 - A_1)]^2$$
$$+ [143(B_2 - B_1)]^2 + [83(C_2 - C_1)]^2\}^{1/2} \qquad (1)$$

Here, H_1 and H_2 refer to values of H for columns 1 and 2, S_1^* and S_2^* are values of S^* for columns 1 and 2, and so on, for the remaining column parameters A, B, and C. If $F_S < 3$ for any two columns 1 and 2, the two columns should provide equivalent selectivity and band spacing for any sample or set of conditions. Equivalent separation may still be achieved for $F_S > 3$, but this is less certain. However, if it is known that the sample does not contain ionized compounds [e.g., no acids or (especially) bases], the term $C_2 - C_1$ in Eq. (1) can be ignored, which usually means a much smaller value of F_S for two columns. Similarly, if carboxylic acids (ionized or not) are absent from the sample, the term $B_2 - B_1$ can also be ignored, again reducing the value of F_S. In the event that columns of very different selectivity are desired, two columns with a very large value of F_S would be preferred. Figure 1 provides an example of the use of values of F_S to select columns of either similar (a–c) or different (d) selectivity.

3.5 Verification of Related Compounds and Limit Tests

Table 3 summarizes the analytical parameter requirements for verification of related compounds and limit tests on the drug substance and drug product. Assessment of specificity is a key parameter in verifying that a compendial procedure is suitable for use in the related compound and limit test of drug substances and drug products. Acceptable specificity for a chromatographic method may

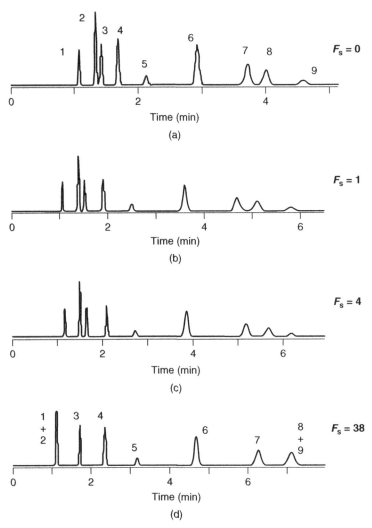

FIGURE 1   Use of F_S values to select columns of either similar (a–c) or different (d) selectivity. (Reprinted with permission. The USP Convention. Copyright © 2009. All rights reserved.)

be verified by conformance with system suitability resolution requirements (if specified in the method). However, drug substances from different suppliers may have different impurity profiles that are not addressed by the compendial test procedure.

Similarly, the excipients in a drug product can vary widely among manufacturers and may have the potential to interfere with the procedure directly or cause the formation of impurities that are not addressed by the compendial procedure. In addition, drug products containing various excipients, antioxidants, buffers, or

TABLE 3 Analytical Parameter Requirements for Verification of Related Compounds and Limit Tests

Analytical Validation Parameter	Drug Substance		Drug Product	
	Related Compound (Quantitation)	Limit Test	Related Compound (Quantitation)	Limit Test
System suitability	Yes	Yes	Yes	Yes
Method precision	Yes	No	Yes	No
Accuracy	No	No	a	No
LOD	No	Yes	No	Yes
LOQ	Yes	No	Yes	No
Specificity	Yes	Yes	Yes	Yes
Range	No	No	No	No
Linearity	a	No	a	No
Ruggedness	a	No	a	No

[a] May be required.

container extractables may potentially interfere with the compendial procedure. In these cases, a more thorough assessment of specificity will be required to demonstrate the suitability of the method for the particular drug substance or drug product.

3.6 Verification of Assay

Table 4 summarizes the analytical parameter requirements for verification of assay in drug substance and drug product. As for the related compound and limit test, specificity is the key parameter in verifying that a compendial procedure is suitable for use in the assay of drug substance and drug product. However, method precision may not be required for some drug substance if there is confidence and justification that method precision is not required.

TABLE 4 Analytical Parameter Requirements for Verification of Assay

Analytical Validation Parameter	Drug Substance	Drug Product
System suitability	Yes	Yes
Method precision	Yes	Yes
Accuracy	No	Yes
LOD	No	No
LOQ	No	No
Specificity	Yes	Yes
Range	No	No
Linearity	No	No
Ruggedness	a	No

[a] May be required, depending on the presence of new impurities.

3.7   Exception for Verification

Verification is not required for basic compendial test procedures that are performed routinely. Examples of basic compendial procedures include, but are not limited to, loss on drying (water), residue on ignition, various wet chemical procedures (heavy metals) (e.g., acid value), and simple instrumental methods (e.g., pH measurements). However, for the application of already established routine procedures to compendial articles tested for the first time, it is recommended that consideration be given to any new or different sample handling or solution preparation requirements.

4   METHOD REVALIDATION

4.1   When to Revalidate for a Drug Substance or Drug Product

If an analytical procedure meets the established system suitability requirements only after repeated adjustments to the operating conditions stated in the analytical procedure, the analytical procedure should be reevaluated, amended, and revalidated as appropriate.

In addition to the nonrobust procedure described above, there are other situations for which procedures need to be revalidated. During the development of drug substances, condition change will bring about a new source of the drug substance. The following are potential situations when the source of drug substance will change:

- The supplier is no longer in business.
- Intermediates are not available.
- The route of synthesis changes.

The same is true for the drug product as it goes through the various stages of drug development: from phase 1 to phase 2 to phase 3 to commercial production.

Validated assays and tests should be revalidated when significant changes are made to the equipment or conditions of analysis, or when the drug product or material being analyzed has changed.

4.2   Revalidation with a New Source of Drug Substances: Changed Route of Synthesis

This requirement for revalidation is very common in the development of a drug substance where the strategy of synthesis changes from phase 1 to phase 2 to phase 3 to commercial. This change is necessary, as the requirements of the drug substance increase from phase to phase. With an increased demand for the drug substance, a more efficient and higher-yielding synthetic route will be required in late-stage development. In phase 1 development work, it is important to get only

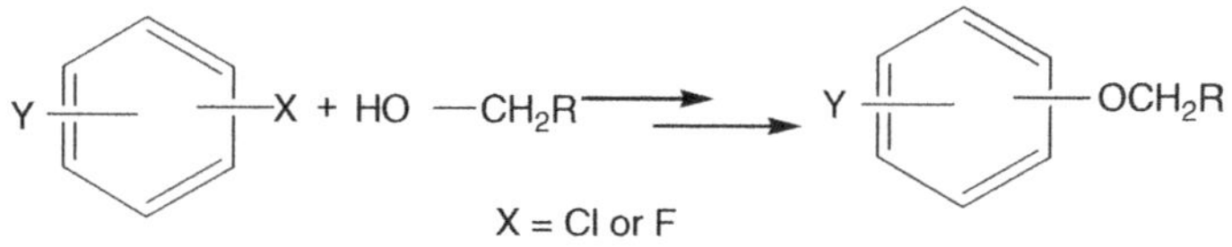

$$X = Cl \text{ or } F$$

FIGURE 2 Synthetic route of a drug substance.

enough cGMP (current good manufacturing practice) drug substance to conduct the first human clinical trial.

Figure 2 shows a change in synthetic route with a different starting material that increases the efficiency of the chemical reaction yielding the drug substance. In this case, a switch from a chloride starting material to a fluoride starting material was found to be beneficial. Table 5 summarizes the analytical parameters required for revalidation with a new drug substance source.

New system suitability will be required if new impurities or degradation products are present. Both method precision and accuracy need to revalidated to confirm that the method can still quantify the drug substances in the presence of the new impurities and/or degradation products. The LOD and LOQ do not need to be revalidated for the drug substance assay. However, the LOD should be reevaluated for the quantitation of impurities and the limit test. With the change in starting material, a new profile of impurities and degradation products will be expected. Therefore, the method specificity must be revalidated to ensure that it is still specific for its intended purpose. However, linearity needed to be established for quantitation of the new impurity. Ruggedness needs to be reestablished with the new impurity. The drug substance assay may need to be redeveloped if the new impurity elutes close to it.

TABLE 5 Analytical Parameter Requirements for Revalidation of Chromatographic Methods Involving Changes in Raw Material

Analytical Validation Parameter	New Source or Synthesis of Drug Substance (New Impurity Present)		
	Drug Substance Assay	Quantitative Impurities	Impurities Limit Tests
System suitability	Yes	Yes	Yes
Method precision	Yes	Yes	a
Accuracy	Yes	Yes	a
LOD	No	Yes	Yes
LOQ	No	Yes	No
Specificity	Yes	Yes	Yes
Range	No	Yes	No
Linearity	No	Yes	No
Ruggedness	a	Yes	Yes

[a]May be required.

4.3 Revalidation with Changes in a Drug Product and the Manufacturing Process

Drug product changes can come in the form of added strengths, new excipients, and changing levels of existing excipients. Adding strength to a drug product may be a result of new indications for the drug product, or the clinical trial experience of the drug product may necessitate a change in strength. New excipients may be introduced and replace an earlier-phase drug product. Formulation changes with new excipients or changes in the level of existing excipients will lead to product changes and warrant revalidation of existing methods. Changes in the manufacturing process (e.g., from direct blend to wet granulation) will require revalidation of some analytical parameters to ensure that the method is still fit for its intended purpose. Table 6 summarizes the analytical parameters required for revalidation with changes to drug product and the manufacturing process.

Revalidation with a Lower-Strength Drug Product   When a lower strength of drug product is required, a change in the manufacturing process may be required (e.g., from direct blending to wet granulation). With a change in the manufacturing process or a lower-strength drug product, new degradation products are commonly observed. In this case, the method must be revalidated with a new system suitability to ensure that it is still specific for its intended purpose. Both method precision and accuracy need to revalidated to confirm that the method can still quantify the drug substances in the presence of new impurities and/or degradation product. The LOD and LOQ of the new impurities need to be established. However, if there is a large increase in the excipient peak (due to a much higher level of excipients) that elute close to the drug substance or other peaks

TABLE 6 Analytical Parameter Requirements for Revalidation of Chromatographic Methods Involving Changes in the Drug Product and Manufacturing Process

Analytical Validation Parameter	Change in Drug Product			Change in Manufacturing Process
	Strength	New Excipients	Level of Existing Excipients	
System suitability	Yes	Yes	Yes	Yes
Method precision	Yes[a]	Yes	Yes	[b]
Accuracy	Yes[a]	Yes	Yes	[b]
LOD/LOQ	No	No	No	No
Specificity	No	Yes	No[a]	[b]
Range	Yes[c]	No	No	No
Linearity	Yes[c]	No	No	No
Ruggedness	No	No	No[a]	No

[a]Required in chromatographic methods when there are peaks that elute near peaks of interest, due to large increases in levels of excipient.
[b]May be required, depending on the presence of new impurities and the nature of the specific test.
[c]Unless adequately demonstrated in validation of the original method.

of interest, the method may need to be redeveloped and revalidated. The linearity of the drug substance needs to be revalidated for any change in strength unless it was demonstrated in the original method validation.

4.4 Revalidation with Modified Chromatographic Methods

Methodology changes in an analytical procedure may be required as the drug goes through the various stages of development and commercialization. It usually involves a change in one of three areas: sample preparation, chromatographic system components (e.g., detector changes), or chromatographic conditions (e.g., column). Sometimes changing sample preparations such as different extraction solvents and conditions may affect the specificity and sensitivity of the test method. Therefore, LOD/LOQ, specificity, and linearity may required to be revalidated. Table 7 summarizes the analytical parameters required for revalidation with modified chromatographic methods.

Revalidation with Changes in the Chromatographic Columns Changes in the chromatographic column of an analytical procedure may be the result of any of the following or other causes:

- The column manufacturer had discontinued the column.
- The packing material of the column has been discontinued.

TABLE 7 Analytical Parameter Requirements for Revalidation of Modified Chromatographic Methods

Analytical Validation Parameter	Changes in Methodology		
	Sample Preparation[a]	Chromatographic System Components[b] (Other Than Column)	Chromatographic Conditions[c]
System suitability	Yes	Yes	Yes
Method precision	Yes	Yes	[d]
Accuracy	Yes	Yes	Yes
LOD/LOQ	[d]	[d]	[d]
Specificity	[d]	[d]	Yes
Range	[d]	[d]	Yes
Linearity	[d]	[d]	Yes
Ruggedness	Yes	[d]	Yes

[a]Changes in the extraction solvents and conditions, derivatization, centrifugation, mixing, pH, filtration, or sample preparation.
[b]Changes in the detector, autoinjector, or data system.
[c]Changes in the column (supplier, packing material), flow rate, injection volume, column temperature, or mobile phase.
[d]May be required.

- New impurities detected from an ongoing stability study necessitate a change in column chemistry.

With changes in the chromatographic column and chromatographic conditions, these changes will require revalidation to confirm system suitability is still within the acceptance criteria. Method precision and accuracy both need to revalidated to confirm that the method can still quantify the drug substances in the presence of the new impurities and/or degradation products. However, changes in the chromatographic column had led to changes in the sensitivity of detection and require revalidation of this parameter. Changes in the column chemistry require revalidation of the linearity and range of analytical properties, and ruggedness will also need to be revalidated.

5   METHOD TRANSFER

5.1   Types of Method Transfer

Method transfer is a process that qualifies a laboratory to use an analytical procedure. There are four common ways to perform method transfer [10]: (1) comparative testing, (2) covalidation, (3) method validation and revalidation, and (4) transfer waiver.

Comparative Study   This is the most common method transfer practice in the pharmaceutical industry. Two or more laboratories will execute a preapproved method transfer protocol. The protocol details the experiments to be performed and the criteria by which the receiving laboratory will be qualified to use the procedure(s) being transferred. Results from the execution of experiments in the method transfer protocol are compared against a set of predetermined acceptance criteria. Generally, comparative testing is often used in late-stage development (transferring development procedures to the quality control function) and by contract research organizations.

Covalidation Between Two Laboratories   With this approach, the receiving laboratory will participate with the validation study from the beginning and is a part of the matrix of experimental design for the study. The results from the receiving laboratory will be part of the final validation report. As a result, the validation report will stand as proof of transfer of the analytical test procedure in the receiving laboratory.

Method Validation and Revalidation   This option requires the receiving laboratory to repeat some or all of the validation experiments. Upon successful method validation, the receiving laboratory is also considered qualified to perform the analytical procedure.

Transfer Waiver Under some special circumstances, the receiving laboratory might not need a formal method transfer:

- The receiving laboratory is already testing the product and is thoroughly familiar with the procedures.
- There is a comparable component or concentration of the drug substance in an existing drug product.
- The analytical procedures are the same or very similar in the receiving laboratory.
- The changes in the new method do not alter use of the method.
- The personnel who developed the method moved to the receiving laboratory.

5.2 Elements of Method Transfer

Various interrelated components are required for a successful method transfer.

Preapproved Test Plan Protocol An approved document that describes the general method transfer and specific acceptance criteria necessary for method transfer should be written by the scientists involved and approved by the quality unit and management. This document should clearly define the responsibilities of the originating and receiving laboratories. The selection of materials and samples used in the method transfer should be described. Generally, GMP-released materials will not be used for transfer activities because the results of not meeting the predetermined acceptance criteria could trigger an out-of-specification investigation. The identity and lot numbers of samples, certificate of analysis of all samples, and instrumentation parameters should be described in the document. Instrumentation should be held constant if possible (e.g., the same type of HPLC system).

Description of Test Procedures and Requirements It is important to share all records of the procedures, validation data, and any idiosyncracies of the procedures with the receiving laboratory. A detailed step-by-step procedure should be provided as well as clear equations and calculations.

Description and Rationale for Test Requirements Details of sample information (e.g., number of lots, replicates, injections) should be stated clearly to prevent misinterpretation of the protocol experiments. The rationale for the parameters chosen and system suitability parameters should also be included.

Acceptance Criteria The method transfer protocol should have acceptance criteria for the tests to be performed and evaluated. Both statistical and nonstatistical approaches to data evaluation are acceptable. Simple statistics, such as the mean of the various replicates, standard deviation, and t-tests, are commonly used for evaluation.

Documentation of Results   The results of method transfer should be documented in a method transfer report. The report will certify that the acceptance criteria were met and that the receiving laboratory is qualified to run the procedure(s). All observations during the method transfer exercise should be documented as part of the report. If the receiving laboratory fails to meet the established criteria, there should be a procedure that describes how to handle the failure. An investigation should be initiated for results that fail to meet the predetermined acceptance criteria. Corrective actions taken should be documented and justified.

6   COMMON PROBLEMS AND SOLUTIONS

6.1   Mobile-Phase Preparation

Issues arising from mobile-phase preparation constitute one of the most common problems encountered during method transfer. Details of mobile-phase preparation for the procedure and method transfer experiments should be detailed and clear. As an example, the description of mobile-phase 30% acetonitrile in 0.01 M monosodium phosphate buffer pH 4.0 could have at least three interpretations from different chemists.

- *Interpretation 1*: 300 mL of acetonitrile added to an aqueous solution of 0.01 M monosodium phosphate buffer pH 4.0 and made up to volume with the buffer solution
- *Interpretation 2*: 300 mL of acetonitrile combined with 700 mL of aqueous 0.01 M monosodium phosphate buffer pH 4.0
- *Interpretation 3*: either interpretation 1 or 2, with the final mixture pH adjusted to 4.0

These three solutions will potentially lead to different analyte separation in the HPLC column. Note that this 30% acetonitrile can also be achieved by mixing using HPLC pumps. There are also small differences in mixing between high- and low-pressure mixing pumps.

6.2   Dwell Volume Differences

Method transfer in the gradient method usually involves complicated differences in dwell volume between systems. *Dwell volume* is the system volume from the point of mixing to the column inlet. This results in an *isocratic hold* at the beginning of each gradient run, As a result, retention times are shifted by the difference in dwell time between the two systems. Typical dwell volumes for today's LC systems are 1 to 3 mL for high-pressure mixing systems and 2 to 4 mL for low-pressure systems, but can be 5 to 8 mL or even higher for some older HPLC equipment. A simple way to overcome the dwell volume issue is to add the difference in delay time as an isocratic hold at the beginning of each run.

6.3 Method Transfer Problems

Robustness studies performed in the later stages of method development and method validation are important in helping to identify critical elements in the procedure. It is important to highlight these method idiosyncrasies in the form of precautionary statements that can be shared by all users of the methods.

6.4 Column Temperature

Column temperature is another potential source of variability for method verification and method transfer. HPLC conditions using room temperature should be avoided. Methods should ideally be developed by thermostatting at least a few degress above the highest room-temperature to compensate for room temperature variability.

REFERENCES

1. *Draft FDA Guidance: Analytical Procedures and Methods Validation*. FDA, Washington, DC, Aug. 2000.
2. *Acceptable Methods*. Therapeutic Products Directorate, Health Canada, Ottawa, Ontario, Canada, July 1994.
3. *U.S. Pharmacopeia, Chapter ⟨1226⟩, Verification of Compendial Procedures.* USP, Rockville, MD.
4. *U.S. Pharmacopeia, Chapter ⟨1225⟩, Validation of Compendial Procedures.* USP, Rockville, MD.
5. *Validation of Analytical Procedures: Text and Methodology. Q2 (R1).* ICH, Geneva, Switzerland.
6. USP Pharm. Forum, 35(3), May–June 2009.
7. System suitability. USP Pharm. Forum, 27(5): 3073, 2001.
8. www.usp.org/USPNF/column.html.
9. USP Pharm. Forum, 31(2), Mar–Apr. 2005.
10. *PhRMA acceptable analytical practice for analytical method transfer. Pharm Tech*, Mar. 2002.

4

VALIDATION OF PROCESS ANALYTICAL TECHNOLOGY APPLICATIONS

ALISON C. E. HARRINGTON
ABB Ltd.

1   INTRODUCTION

The implementation of process analytical technology (PAT) in the pharmaceutical industry has grown over the past 18 years from individual analyses performed on laboratory-scale spectrometers to fully PAT-enabled manufacturing. The recent growth in PAT investment in the industry has been prompted by the U.S. Food and Drug Administration (FDA) Guidance for Industry PAT [1]. The FDA observed that manufacturing and associated regulatory practices did not adequately support or facilitate innovation and continuous improvement. They realized that an innovative regulatory process was necessary to transform pharmaceutical manufacturing to meet the current and future needs of the U.S. public [2]. The guidance that the FDA produced outlines a scientific, risk-based framework combining a set of scientific principles and tools supporting innovation and a strategy for regulatory implementation that aim to accommodate innovation while maintaining or improving the current level of product quality assurance. The PAT initiative is consistent with the current FDA belief that quality cannot be tested into products but should be built-in: quality by design (QbD) [3,4].

Practical Approaches to Method Validation and Essential Instrument Qualification,
Edited by Chung Chow Chan, Herman Lam, and Xue Ming Zhang
Copyright © 2010 John Wiley & Sons, Inc.

According to the FDA guidance, the desired state of pharmaceutical manufacturing is:

* Product quality and performance are ensured through the design of effective and efficient manufacturing processes.
* Product and process specifications are based on a mechanistic understanding of how formulation and process factors affect product performance.
* Quality assurance is continuous and in real time.
* Relevant regulatory policies and procedures are tailored to accommodate the most current level of scientific knowledge.
* Risk-based regulatory approaches recognize both the level of scientific understanding and the capability of process control related to product quality and performance.

The primary goal of PAT is to provide processes that consistently generate products of predetermined quality. By investing in PAT, industry can expect to attain improved usage of assets and equipment, and to achieve improved overall manufacturing efficiency. PAT allows the development of greater process and product understanding through the appropriate use of timely process and product measurement, the combination of many complementary disciplines, and the application of multivariate statistical analysis within an overall knowledge management framework deploying mechanisms for either feedforward or feedback process control. Thus, PAT has grown from the individual spectrometer to include the following PAT tools:

* Multivariate tools for design, data acquisition, and analysis
* Modern process analyzers
* Process control tools
* Continuous improvement and knowledge management tools [5]

There is also an industry trend to improve the use of assets and equipment and to generate improvements in overall manufacturing efficiencies. Manufacturing has become a cross-plant, cross-country, cross-enterprise process involving the entire supply chain. To remain competitive, manufacturers need to increase simultaneously understanding of the process to

* Reduce time to market
* Increase process visibility and production flexibility
* Optimize forecasting and scheduling
* Reduce rejects, stocks, and downtime
* Ensure optimal quality and production efficiency across global facilities

Meeting these goals requires an integrated information technology (IT) infrastructure that helps coordinate production on a global scale and, if necessary, in real time. This optimized coordination must cover such things as specifications, equipments, facilities, process and procedures, quality tests, and personnel resources. This trend leads to the removal of "islands of automation" in favor of integrated information. As enterprise resource planning (ERP) systems have stabilized, there is more interest in integrating the information from the automation system level to the business system level. A new set of international standards has evolved, the ISA-95 Manufacturing Enterprise System Standards [6–8], which establishes a clear definition of manufacturing enterprise system functionality and has developed models that describe and standardize manufacturing enterprise systems. Figure 1 illustrates the traditional five-layer computer-integrated manufacturing (CIM) model, which has evolved into the three-layer model in use in manufacturing today.

Further evolution will follow as ERP companies push downward with their products, e.g., SAP advanced planning and optimization modules, and automation companies extend their application suite to include traditional collaborative production management (CPM)/manufacturing execution system (MES) functionality, seamlessly integrated with all other plant-level information, e.g., the ABB 800xA system.

Does PAT-enabled manufacturing lead to a more complex validation package? Figure 1 shows that the number of system interfaces to consider is reduced and the FDA regulatory framework is a collaborative effort designed to lead to a successful filing, but compiling the validation package will involve much wider communication with people of different backgrounds and personalities, and the size of the financial investment required will make strong planning and management essential. FDA presentations indicate their anticipation that PAT

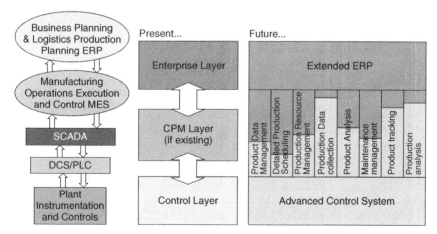

FIGURE 1 Evolution of the five-layer computer-integrated manufacturing model. ERP, enterprise resource planning; MES, manufacturing execution system; SCADA, supervisory control and data acquisition; DCS, distributed control system; PLC, programmable logic controller; CPM, collaborative production management.

implementation will eventually change the regulatory process. Documentation of quality by design (QdD) during the pre-IND (investigational new drug) meeting, the end of the phase 2 meeting, and in regulatory submissions will allow early review and analysis of the CMC (chemistry, manufacturing, and control) section of a new drug application (NDA) by the FDA. Addressing issues of concern and further QbD can result in classification of the drug substance and drug manufacturing process as low-risk. In some cases, this approach is expected to result in a less comprehensive or eliminated preapproval inspection. Although these procedural changes will not happen overnight, they present a possibility for more rapid regulatory approval and reduced time to market.

2   PARAMETERS FOR QUALIFICATION AND VALIDATION

2.1   Critical Process Parameters and Critical Quality Attributes

A PAT application is a system for designing, analyzing, and controlling manufacture through timely measurements (i.e., during processing) of critical quality and performance attributes of raw and in-process materials and processes with the goal of ensuring final product quality. Drug substance, excipients, container closure systems, and manufacturing processes that are critical to product quality should be determined and control strategies justified. Figure 2 illustrates the sources of product quality variation and identification of critical quality attributes (CQAs).

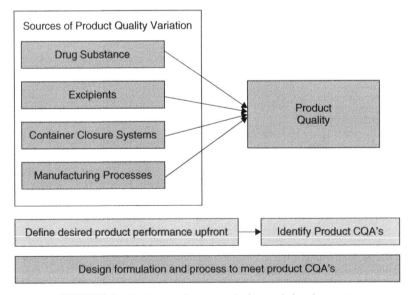

FIGURE 2   Product and process design and development.

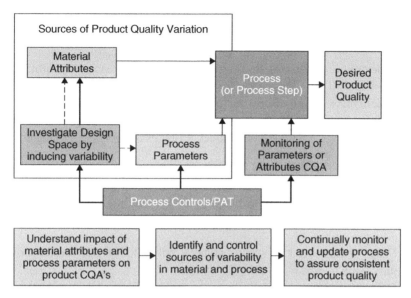

FIGURE 3   Risk assessment and risk control.

Critical formulation attributes and critical process parameters (CPPs) can generally be identified through an assessment of the extent to which their variation affects the quality of a drug product. Figure 3 illustrates how process controls and PAT can be used to control and measure CQAs and CPPs. Figure 3 also shows the design space, a multidimensional combination and interaction of input variables (e.g., material attributes), and process parameters that have been demonstrated to provide assurance of quality. Demonstrating scientific understanding of this expanded design space facilitates opportunities for more flexible regulatory approaches, for example, to facilitate:

- Risk-based regulatory decisions (reviews and inspection)
- Manufacturing process improvements, within the approved design space described in the dossier, without further regulatory review
- Reduction of postapproval submissions
- Real-time quality control, leading to a reduction in end-product release testing

This flexibility is realized by demonstrating an enhanced knowledge of product performance over a range of material attributes, manufacturing process options, and process parameters. Process understanding and product knowledge can be gained by the use of formal experimental designs, PAT, and/or prior knowledge. Appropriate use of risk management principles [9,10] is useful in prioritizing the additional pharmaceutical development studies to collect such knowledge.

Generating process understanding and product knowledge are the principles of implementing QbD. *QbD* is defined as:

- A scientific, risk-based holistic and proactive approach to pharmaceutical development
- A deliberate design effort from product conception through commercialization
- A full understanding of how product attributes and process parameters relate to product performance

CQAs are related to end-product properties such as identity, strength, potency, purity, and moisture. The following material attributes could be considered for an orally inhaled or nasal drug product:

- Drug substance (e.g., moisture content, polymorph form, surface morphology, particle size distribution)
- Delivery platform (if appropriate) [e.g., MDI (metered dose inhaler), DPI (dry powder inhaler), nasal spray, inhalation spray.]
- Formulation or device subtype (e.g., suspension vs. solution MDI or device-metered vs. pre-metered DPI)
- Excipients
  - Propellant(s) and ethanol (e.g., water content, impurities)
  - Surfactants (e.g., compositional profile, surface-active properties)
  - Lactose [e.g., hydrate form, amorphous content, surface morphology, water content, particle size distribution (PSD)]
  - Magnesium stearate (e.g., compositional profile, PSD)
  - Leucine, dipalmitoylphosphatidylcholine, water, buffers, salts, preservatives, and so on
- Container closure systems (CCSs) or device components are part of the drug delivery system, which is an integral part of the drug product. CCSs should have the following properties:
  - Be reliable and provide accurate dose delivery
  - Be materially compatible
  - Be stable and dimensionally consistent
  - Be mechanically robust
  - Offer protection to the formulation
  - Be readily manufacturable
  - Demonstrate user-friendly characteristics (ruggedness to variability in patient use)

The material choice for CCS components is driven by the CPPs and CQAs desired, including the types of metals, plastics, and elastomers; the fabrication methodology for each component; the additives in plastics and elastomers; and the processing aids used in forming, cleaning, and assembly.

Each unit operation should include an understanding of how CPPs affect CQAs and conduct risk analysis and assessment to generate manufacturing process understanding to:

- Identify critical process parameters and material attributes
- Understand direct and indirect interdependencies
- Develop risk reduction strategies
- Establish an appropriate control strategy to minimize the effects of variability on CQAs
- Evaluate risk in terms of severity, likelihood, and detectability

Table 1 provides a comparison of traditional formulation development vs. QbD/PAT-enabled unit operation. Implementing a PAT application with a QbD approach does not necessarily result in applying online analyzers to all measurements but to take a risk-based approach to consider those upstream parameters that are responsible for variation in the end product, enabling corrective action before out-of-specification problems occur. The PAT method will often require advanced analytical sensor techniques for direct CQA/CPP measurement. These PAT sensors must be specified and robust for the manufacturing environment: for

TABLE 1 Comparison of Approaches on the Unit Operation of Drug Substance Micronization: Traditional vs. QbD/PAT-Enabled

Traditional	QbD/PAT-Enabled
Time, temperature, and humidity are set at predefined ranges.	Combination and interaction effects of time, temperature, and humidity on design space CQAs studied and understood and design space established.
Fixed process, almost any change requiring agency approval.	Process adjustable within the design space without regulatory oversight.
Approach controlled but not robust. Tight controls over incoming nonmicronized drug substance usually necessary. Problematic with planned site, equipment, and scale changes. Sensitive to variability without being responsive to it. Data-laden but knowledge poor.	Control of the design space to the desired endpoints (PSD, polymorph limits, surface morphology, etc.) and is more robust.
Post-manufacture, off-line, lab-based testing for product quality.	Process-based, real-time monitor/control for product quality.

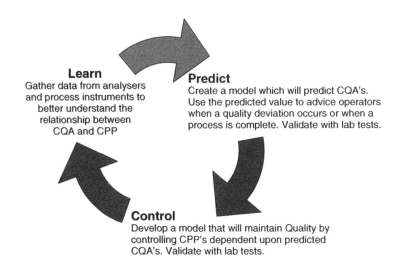

Learn
Gather data from analysers
and process instruments to
better understand the
relationship between
CQA and CPP

Predict
Create a model which will predict CQA's.
Use the predicted value to advice operators
when a quality deviation occurs or when a
process is complete. Validate with lab tests.

Control
Develop a model that will maintain Quality by
controlling CPP's dependent upon predicted
CQA's. Validate with lab tests.

FIGURE 4   The three stages of PAT application method development.

example, Fourier transform–infrared (FT-IR), near infrared (NIR) and Raman
spectroscopy, acoustic spectroscopy, fast chromatography, chemical and thermal
imaging, laser diffraction particle size monitoring, nuclear magnetic resonance,
and mass spectroscopy. These may be combined with more basic measurements
of rheological or other physical properties. The CQA/CPP measurement may be
qualitative or quantitative and usually requires a multivariate calibration for converting the PAT sensor measurement data into a physical property in the design
space. Figure 4 illustrates the three stages of PAT application method development, building product knowledge from process understanding in an environment
of continuous improvement.

Figure 5 illustrates a typical fluidized-bed dryer unit manufacturing process,
without PAT. Drying is controlled by using a fixed time followed by lab sampling
to confirm that the batch endpoint has been reached. There is no optimization
of the process to manage CQAs such as particle size or moisture. Product is
typically overdried to ensure consistent batch completion. Particles may "clump"
together, giving inconsistent particle size.

2.2   PAT Applications Modeling the Design Space

Looking at the same manufacturing process using a PAT application with a QbD
approach would identify the following CQAs of fluidized-bed drying:

- Moisture content
- Particle size

An experimental design should be used to understand the impact of the following material attributes and CPPs:

- In-process materials

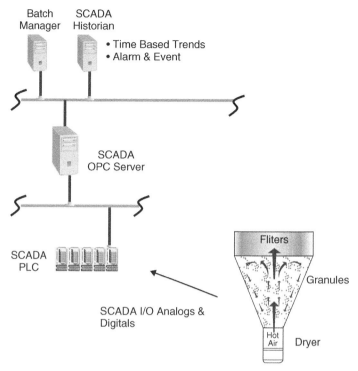

FIGURE 5 Typical fluidized-bed dryer process without PAT.

- Airflow
- Temperature of inlet air
- Moisture content of inlet air
- Drying time
- Load

This allows the manufacturer to identify and control the sources of variability in the material and the process. Figure 6 shows the same drying process with the addition of an NIR analyzer taking moisture readings of the fluidized-bed dryer in real time. The figure is an example of drying controlled using an NIR moisture measurement. Unit operation is stopped when the desired moisture level is reached. This offers a direct mechanism for improving batch-to-batch quality, consistency, and reduced cycle times. In the figure an ABB Bomem FT-NIR analyzer is used with FTSW100 software that provides PAT method configuration, univariate and multivariate data storage, and univariate data integration with open process control (OPC) and data acquisition (DA). However, if both CQAs' moisture and particle size are measured, a greater understanding of the design space can be achieved, as shown in Figure 7. Real-time measurement of both CQAs allows the design space to be investigated. It is now possible to use a

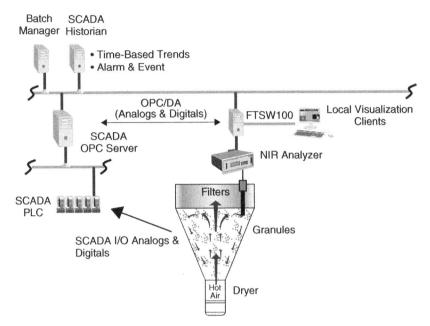

FIGURE 6 Typical fluidized-bed dryer process with real-time moisture measurement.

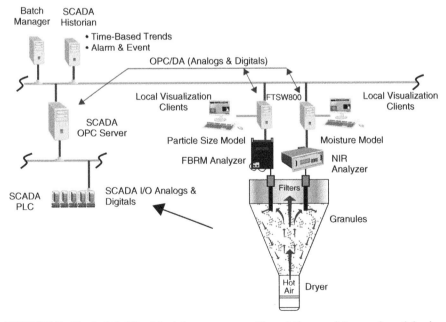

FIGURE 7 Typical fluidized-bed dryer process with real-time moisture and particle size measurement.

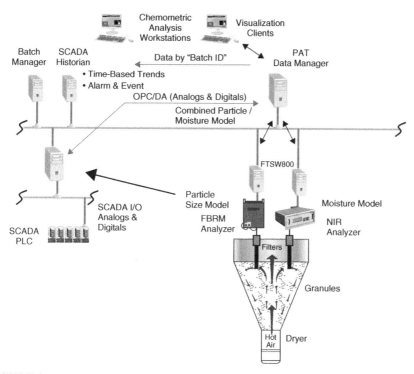

FIGURE 8 Typical fluidized-bed dryer process with real-time CQA measurement, CPP control, and PAT data manager.

multivariate solution to optimize dryer performance and reduce variability. In the figure an FBRM (focused-beam reflectance method) particle size analyzer is used with ABB FTSW800 software.

Figure 8 illustrates how a PAT data manager can be added to a PAT application to centralize data generated and maximize process and product knowledge harvested. In the figure the ABB PAT data manager is used in combination with a Bomem FT-NIR analyzer and an FBRM analyzer; this provides:

- Data stored by batch ID or other unique identifier
- Remote client support
- Data accessible by open interfaces (OPC, ODBC, etc.)
- Multianalyzer model support
- Time-based, alarm and event, spectral and method data stored and archived
- Centralized analyzer control

In this example the PAT data manager is used to manage, store, and serve up the data derived from, and the configurations for, all the instrument platforms and to link the data with those captured by the SCADA control system. Chemometric models that link a number of analytical sources together can be developed and the data used to advise or control.

Figures 6 to 8 include components of the manufacturing control system [e.g., batch manager, supervisory control and data acquisition (SCADA) historian, SCADA OPC server, SCADA PLC] to demonstrate the importance of integration with these systems. However, for the purpose of PAT application validation, they should be considered as being out of scope. The plant network, intranet, and Internet validation should also be considered but may be outside the validation scope of the PAT application. The control of CPPs and real-time measurement of both CQAs allows the design space to be understood, making it possible to reduce product variability and reduce overall regulatory approval time.

The many benefits to implementing PAT in a QbD approach include:

- CQAs for materials and products and CPPs for process parameters are better understood.
- Controls are rationally designed to fit end-use performance criteria.
- The entire manufacturing system is more flexible, accounting for and responding to variability in materials, environment, and process within a known design space.
- The more flexible regulatory framework relies on the demonstration and use of knowledge.
- The overall approval time (time to approval and launch) may be reduced.
- Product failures after approval associated with variability in ingredients and processes that would not otherwise have been considered may be reduced.

The system shown in Figure 8 identifies the level of complexity required in modern PAT application in a unit manufacturing operation, and Figure 9 shows a PAT solution suite for a full manufacturing process. The example shown in Figure 9 demonstrates the level of system integration required to achieve a full PAT application suite. Other available products are given in Table 2.

3   QUALIFICATION, VALIDATION, AND VERIFICATION PRACTICES

3.1   Defining the Regulatory Requirements

The level of validation, qualification, or verification required by a PAT application will depend on the end use. For this purpose the following definitions apply:

Validation Establishing documented evidence which provides a high degree of assurance that a specific process will consistently produce a product meeting its predetermined specifications and quality attributes [11].

Qualification Demonstrating the ability to fulfill specified requirements [12].

Verification Applying good engineering practices to the design, build, commissioning, and startup of manufacturing systems [13].

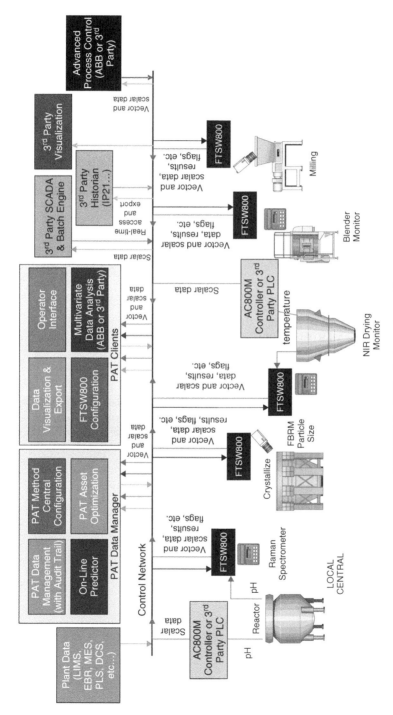

FIGURE 9 PAT solution suite architecture.

TABLE 2   PAT Application Suite Components

Description and Requirements	Example Products
PAT data manager Provides central storage of analytical and process data Provides central management of PAT methods Online prediction Asset optimization	ABB data manager consisting of: 800xA core 800xA information manager 800xA batch (produces IT) FTSW800 configuration editor and configuration loader
PAT client Operator interface Provides online status Main data analysis tools Provides data mining and information management for all types of data, including batch structure, scalar, alarm and event, and audit trail Visualization tool allowing data selection and data export Multivariate data analysis, including multivariate model building supporting MLR, PLS1, PLS2, PCA, and PCR models Direct access to SOPs (standard operating procedures)	ABB 800xA operator workplace with ABB aspect object technology The MathWorks MATLAB test and measurement suite
PAT analyzer control Application server Plug-in drivers for analyzer support Local console for use as a maintenance interface for local status and control, also allowing XP remote access Defines analyzer parameters for data collection and data processing Loads multivariate analysis models with plug-inarchitecture for chemometric prediction engines IO mapping for sending information to PAT data manager Plug-in architecture for communication protocols, including OPC, modbus, physical IO (4–20 mA) Displays control and status of all streams and all local alarms, events, and local history database	ABB FTSW800 analyzer control National Instruments' LabVIEW Intuitive Graphical Programming Language

TABLE 2   (*Continued*)

Description and Requirements	Example Products
PAT analyzer	ABB FT-IR and FT-NIR
	Axsun NIR
	Bruker FT-NIR
	Metler–Toledo FBRM particle size monitor
	Granumet acoustics
Chemometric prediction engines	GRAMS PLSplus/IQ
	Umetrics SimcaP
	CAMO unscrambler

There are no regulatory requirements for analyzers or data used for research use. This applies to analyzers used for research at any point in the product life cycle (in research, development, or production facilities). This also applies to data on a production process collected by an experimental process analyzer and is not inspected except under exceptional circumstances. In these circumstances the PAT application should be managed under the company quality management system and good scientific practice.

cGMP regulations do apply to PAT application used for either of the following situations:

- PAT analyzers used as part of a quality system for licensed product or investigational drug batches produced for clinical trial supply (phases 1 to 3)
- Where results generated by an analyzer may be used in support of a regulatory submission

Where cGMP regulations apply, the validation requirements are as follows:

- The analyzer must be appropriately validated (and calibrated).
- The computer systems must be validated.
- 21 CFR Part 11 applies to electronic predicate rule records and signatures.
- A system suitability test should be carried out before use.

3.2 Defining the Scope of the PAT Application

Define the scope of the PAT application as shown in Figure 10. Determine the scope of validation by specifying clearly what is in and out of the PAT system, and draw a boundary in the system landscape. If SCADA, batch management, control, and automations systems are already present and validated, it is only necessary to validate the interface with the PAT application. Once the system boundaries have been determined, a validation master plan can be written. This

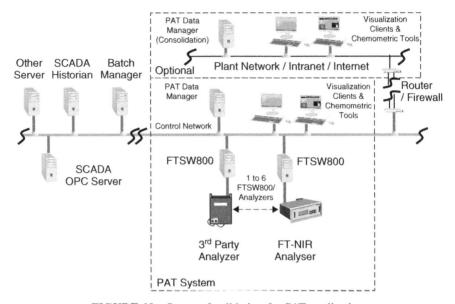

FIGURE 10   Scope of validation for PAT application.

should state clearly what regulations apply to the system and should also split the PAT environment into discrete units or components. Approaching the validation of each component separately will make validation planning more effective and will facilitate the validation of system changes in the future. Figure 11 shows a component approach to validation planning. System changes can be simple hardware or configuration changes, method changes, or changes of learning within the chemometric predictive models or their reference calibration data sets. The PAT application is a highly integrated system; defining the validation strategy during the planning stage to maintain the validated state of all the component parts is essential to sustaining overall regulatory compliance.

3.3   Discriminating Between the PAT Environment and the PAT Method

There are two parts to the PAT application: the PAT environment and the PAT method, and both parts may require validation. The *PAT environment* defines the infrastructure, control systems, analyzers, sensors, chemometric models, and data analysis tools used to perform, control, and monitor the manufacturing operation. The *PAT method* is a set of instructions used to dictate the operation, development, and validation of analyzer-based models used to monitor, predict, and/or control the CQA. The PAT environment can be validated for use with any PAT method. The PAT method will be specific to the material being manufactured and its critical quality attributes, which are identified by understanding the design space of the material being manufactured.

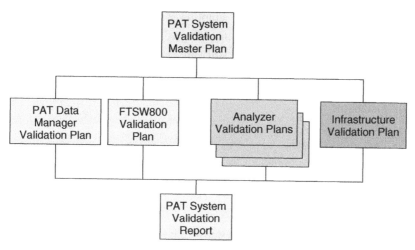

FIGURE 11 PAT application validation planning.

3.4 Inputs to the Validation Master Plan

Generate a validation master plan (VMP) to develop and capture the validation strategy for all the PAT system components. Initiate the VMP at the very beginning of the PAT application implementation project and then develop it as more information becomes available. Treat the VMP as a living document and review and update it at periodic intervals. Develop the VMP as suppliers are evaluated and the degree of confidence in supplier activities is assessed. Maintain a change history during subsequent reviews and ensure that the reasoning behind decisions is captured.

Compliance Requirements Validation of the PAT application may involve technical experts from several departments or even divisions. The various departments may have different internal quality guidelines or industry guidelines for validation, which must be taken into account. Ensure that QA representatives from each participating department or division approve the VMP to signify their acceptance of the quality guidelines adopted and ensure that any conflicts between guidelines are resolved. Other compliance inputs for the VMP will come from determining which external regulations (e.g., FDA, EMEA) the PAT application will need to comply with. Use the VMP to define a framework of roles and responsibilities for validation and quality assurance both during the implementation project and in operational use.

System Inputs Determine the system inputs to the VMP:

- List all hardware components of the PAT application and their interrelationships (systems and subsystems), including:

- PAT analyzers, sensors, and sampling systems
- List all software components, including:
  - Configurable and bespoke software applications (including sampling regimes)
  - Microsoft tools
  - Databases
  - Middleware
  - Operating systems
  - Embedded code items
  - Multivariate mathematical modeling and prediction software and associated software utilities
- List all interfaces between subsystems.
- List all interfaces to external systems (e.g., nonPAT instruments and sensors and the associated SCADA or PC-based computer systems) which are used to gather reference data that may be used to update chemometric models and in experimental design to model the design space.
- Include a categorization of all hardware and software components according to GAMP or other categorization scheme.

The GAMP system and subsystem categorization will form the basis of system risk assessment. External non-PAT analytical instruments which are used for the generation of reference analytical data should also undergo system suitability testing to demonstrate ongoing fitness for purpose.

GXP Assessment   Include a GXP Assessment of the PAT application against relevant regulations (e.g., 21 CFR Parts 210 and 211 cGMPs in the VMP):

- Identify GXP critical activities being performed by the system and subsystems.
- Determine GXP critical records generated and maintained by the system and subsystems.
- Compile a list of GXP critical records identified by the GXP assessment.
- Identify the systems and subsystems that will need to be assessed for 21 CFR Part 11 compliance.

It is important to identify where and when GXP records are initiated or created, manipulated, used, stored, and archived. Decisions must also be made or confirmed regarding what raw data are (data generated by analyzers or sensors prior to processing), naming conventions for data, method, and configuration files. Establishing clarity will facilitate compliance management and use of data helping to ensure that data, method, and configuration files are unique, have version management, and are traceable.

Security Inputs   Determine user roles and authorizations and assess where those user roles will interact with GXP critical operations or records; for example:

- *Regulatory affairs*. Secure CD export production.
- *IT infrastructure*. Monitor platform technology and architecture.
- *Quality assurance*. Investigate product variation in the period.
- *Operations manager*. Review the performance of the line.
- *Compliance manager*. Check compliance with 21 CFR Part 11, Usage and Privileges.
- *Chemometric analyst*. Investigate the performance of the process or product.
- *Business analyst*. Secure access to data for use in off-line study.

Validation should include the qualification of user roles and authorizations and ensure that training records and procedures for use are in place.

Risk Assessment   Execute high-level risk assessments to identify the scope and extent of verification, qualification, and validation activities. Apply a science- and risk-based approach to validation planning: for example, using the quality risk management process described in ICH (International Conference on Harmonization) Q9:

- Risk assessment
- Risk control
- Risk communication
- Risk review

Risk assessment can be performed using a number of techniques; for example:

- Qualitative assessment (risk impact, risk likelihood, probability of detection of defects, risk priority)
- Quantitative assessments (risk scores, percent risk)
- Assessment tools (e.g., word models, checklists, FMEA, HAZOP, CHA-ZOP)

Plan the number, type, and timing of risk assessments during project execution and use this as an input to the VMP. Include risk assessments for all subsystems and subfunctions for both: technical risk assessment (e.g., process chemistry, materials science, equipment, computer systems) and GXP/business impact risk assessment. Consider risks to product quality, to GXP data and records, and to the business posed by both the PAT application and the implementation project. Follow a formal protocol to identify sources and to analyze and evaluate risk. Determine what action is required to eliminate or reduce all risks to an acceptable level. Nominate a team leader to execute the risk assessment with a multidisciplinary team covering all the capabilities required to make an assessment across the entire PAT system. Any risks that cannot be reduced to an acceptable level should be controlled by further action; for example:

- Eliminate risk.
  - Change the mode of operation.
  - Use an alternative system.
  - Find, implement, and verify a technical or procedural solution.
- Contain risk.
  - Find a technical solution or manual process to reduce the likelihood of occurrence and/or impact to an acceptable level.
- Manage risk.
  - Document the risk identified and reasons for the decision made.
  - Investigate to ensure that the risk is fully understood.
  - Eliminate risk or apply available containment strategies.
  - Introduce a technical system or additional monitoring to increase the probability of detection to an acceptable level.
  - Put contingency plans in place, if required.

Develop the validation strategy based on the risk and criticality assessments. Plan the qualification and validation effort for maximum effect and risk reduction. Use a verification-only approach (e.g., good engineering practice) where appropriate. Use a risk based approach to determine the validation activities required; for example, a GXP risk assessment of the user requirement specification performed at both the system and component levels will determine if validation is required and will also determine and justify the validation approach. Risk assessment based on the results of supplier audits will also determine and justify the validation approach. Risk assessment of functional and technical design specifications will determine the test coverage required. Risk assessment of technical and procedural controls around the PAT system will determine the amount of validation effort to apply. Risk assessments of change control will determine the impact on the validation and test approach. Ensure that risk assessments are used as tools throughout the validation life cycle.

Efficiency and Effectiveness in Validation   Use the VMP to plan for efficiency and effectiveness in validation. Ensure that suppliers' contracts include validation deliverables. Leverage the available design and test documentation to avoid duplication. Ensure that supplier commissioning activities meet GEP (good engineering practice) and are documented appropriately. Participate as an observer in factory acceptance tests (FATs) and site acceptance tests (SATs) to verify execution according to preagreed procedures. FAT and SAT results can then be used to reduce duplicate activity at installation qualification (IQ) and operational qualification (OQ). For example, the PAT data manager from ABB has an associated streamlined validation documentation suite covering design, engineering, testing, quality control, and validation. The documentation is prepopulated with system-specific content and has designated repositories for the application-specific information defined by the project. Taking advantage of this can significantly accelerate validation delivery by starting a project with a complete document framework in place in relation to the control application scope. Validation efficiency and effectiveness can also be planned in the VMP by applying GEP to noncritical subsystems and activities, and by varying the depth and breadth of testing according to risk level and the criticality of subsystems and activities.

Cost-Effectiveness in Validation   The VMP can also be used to plan for cost-effectiveness in validation. Ensure the correct mix of competencies and capabilities in the validation team. Create an integrated validation and project team. Ensure that there are parallels between project system development life cycle (SDLC) stage gates and validation life cycle qualification activities. Ensure that all parties in the project and validation teams understand the quality management system (QMS) and applicable procedures. Invest in thorough validation planning to avoid costly and untimely failures. Investigate methods to ensure efficiency and effectiveness in the validation scope of work: for example, remote and robotic test management tools, document management systems with electronic signatures, and approval workflows. Plan to eliminate endless document review cycles and reduce the number of signatories per document, ensuring that document approvers understand the exact meaning of their signature. Use risk assessments effectively.

Phased Validation Approach   Use the VMP to map out the validation life cycle and define a phased validation approach, as shown in Figures 12 and 13. During the proof-of-concept phase and into early process R&D it is essential that experimental equipment is functioning, is installed correctly, and is fit for its purpose. Mistakes occurring during these early stages are business-critical and could cause unnecessary delays and future elevated costs. Therefore, basic system qualification and GEP should be applied. The validation strategy and detail plan should be developed during the middle phase as critical process parameters and the quality model are identified. Risk assessments at this stage will determine if any GXP critical data are produced that will require PAT system validation. Full PAT application validation should be performed in the third and final validation phase. This phased validation approach manages project risk and controls the cost of qualification in line with the increase in probability of project success as the project progresses.

PAT Method Validation   Apply the FDA/EMEA guidance on analytical methods to the validation of the PAT method. Work with scientists to define and specify accuracy, precision, specificity, detection limit, quantification limit, linearity, range, and robustness. There are additional challenges to overcome with PAT method validation because the analytical challenges are often new, use multiple data transformations, and are based on modeling techniques. PAT methods need

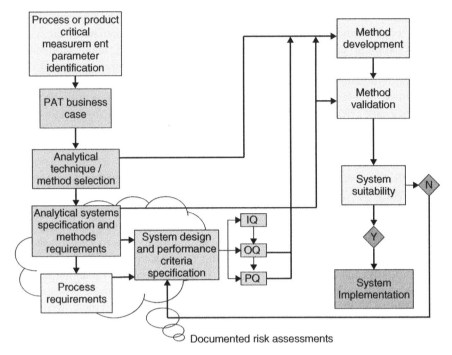

FIGURE 12   High-level PAT validation process flow.

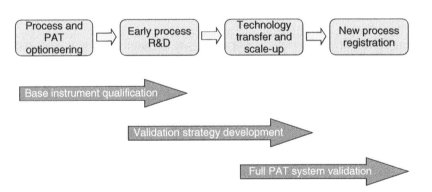

FIGURE 13   Phased validation approach.

to be documented and managed in line with regulatory requirements. Analytical methods should be validated during method development, either comparing their suitability against standard or published methods or their ability to match predicted results based on predictions calculated from first principles. A variety of techniques are used to calibrate analytical methods, including statistical analysis, standards and reference materials, blanks, spikes and fortifications, existing and incurred samples, and sample simulation technologies. The validation strategy for the analytical method must ensure that it meets the analytical method criteria above. Key method performance parameters may differ across methods and analytical techniques. Different method performance windows will be important in different analysis situations. Analytical methods should be revalidated for changes to their application, working environment, or following a long period of nonuse.

PAT methods require periodic performance evaluation to confirm that the systems and hardware continue to operate as intended and that they are fit for the purpose. Apply a science- and risk-based approach to make maximum use of normal, ongoing calibrations and system suitability tests and also to make the best use of instrument self-diagnostics and error detection and reporting systems. If unacceptable performance is identified, root-cause analysis and corrective action will be necessary. The root-cause analysis may entail experimentation or statistical investigations into the cause of the discrepancy and could lead to revision of the PAT method, maintenance of the calibration model, or revalidation of the method.

Process Control System and PAT Validation   Validation of process control systems based around process analytical technology raises several fundamental and interesting aspects. Referring to Figure 3, the control regime is likely to have a real-time process control loop and a process control optimization loop; the PAT application has an impact on both loops. The process control algorithm would optimize process parameters over time, based on the CQA measurements achieved. Therefore, for validation purposes the process control system is made up of a process model and a quality assurance (QA) model. The QA model provides the CQA acceptance criteria for the PAT method. Provided that the process and QA models are stable and synchronised, without process control optimization, the validation can be carried out largely by applying the principles of GAMP methodology. The QA model should be closely linked with the process model. When a process control optimization loop is reintroduced, running in "near time" or batch mode, the modeling becomes dynamic and more complex.

Change and Configuration Management   The change and configuration control procedure should be defined in the VMP, including at what stage it is implemented and the risk review process for assessing changes and hence revalidation requirements. Factors that may give rise to full or partial revalidation include:

- Changes in the manufacturing process
- Changes in raw materials (source and composition)
- Changes in instrumentation (repairs, replacements, reconfigurations)
- Changes in analytical methods
- Addition of a new material to the PAT library or capability
- Addition of new material characterization aspects
- Changes in the process control aspects

If any of these changes are expected, reasonable variation in the foregoing factors could be included in the experimental design so that these are already factored into the design space. If models are transferred between instruments, the relationship between the two instruments and the impact of the change on the PAT model needs to be fully explored and understood. In all cases, model transfer must be controlled and governed by clear procedures which ensure that the model remains fit for its purpose and valid for its intended purpose on the replacement instrument. This is an important issue for process instruments, for which a high level of availability and redundancy may be key requirements. It is also a consideration for scale-up where models developed on lab analyzers may be transferred to process analyzers.

Sampling System Validation   The online or at-line measurements on PAT applications can induce unreliable or erroneous performance if there are problems with the associated sampling systems. A fit-for-purpose sampling system should have the following attributes:

- Acquires a representative sample
- Does not unintentionally alter the physical or chemical properties of the sample
- Does not destroy the bulk sample
- Does not introduce an unacceptable time delay
- Conditions the sample reproducibly to meet the sample requirements of the associated analyzer
- Has fit-for-purpose reliability, safety, and environmental attributes
- Does not compromise the cGMP compliance of the parent process

These attributes should be considered in the design and validation of sampling systems. Sampling could be online, at-line, invasive, noninvasive, or extractive, depending on the core process, the sample, and the analysis technique. The following approach can be followed for the design and validation of sampling interfaces:

- Define the user requirements for sampling interface.
  - Establish the analyzer sample requirements.
  - Establish the control strategy sample requirements.
  - Establish the sample containment and handling requirements.
  - Establish the availability and engineering materials compatibility requirements.
  - Establish the sample system cGMP requirements.
- Specify or design the sampling system interface.
- Review the design and specifications of the sampling system interface.
- Conduct documented qualification of the sampling system interface (IQ and OQ) and confirm fitness for purpose and intended use.

• Review periodically the maintainability, reliability, and ongoing fitness for purpose.

4 COMMON PROBLEMS AND SOLUTIONS

4.1 Delivery of Business Goals

Many of the challenges in validating a PAT application are similar to those for large ERP (enterprise resource planning) or LIMS (laboratory information management system) projects, in that there is significant investment from the business, leading to the extra pressure of a high-profile project. Focus also needs to be maintained on the delivery of business goals. For PAT these will include shorter time to market, reduction of waste, increased yield, and reduction of work-in-progress inventory. Implementation of a PAT application will often require that business processes be redesigned or newly created and then mirrored in the technical solution. New operating procedures will be required. A different skill profile will also be required by those using and interpreting data from the application. The typical quality control role of the laboratory analyst will be either enhanced or replaced. QA personnel will also need training in the new science- and risk-based approaches to quality management in both validation and use of the PAT application. These different competency requirements will require training and/or recruitment of personnel to ensure reliable and effective operation and cGMP compliance of the PAT application.

4.2 Innovation and Risk

Innovation is required in the approach to both PAT application design and validation. Typically, PAT environments are complex, consisting of many configurable components and associated interfaces. There are often new combinations of automation, data management, and analyzer technology. There will be new process control strategies and new opportunities to change process parameters within the design space without regulatory review. These challenges and innovations bring new risks, which need to be managed to ensure the success of the project.

A PAT implementation project is highly complex and effort should be expended upfront to subdivide tasks into logical packages of manageable size and then to use risk assessment to focus validation effort effectively. Ensure that implementation project goals are adopted by both the validation team and the quality assurance organization. Innovate with QA personnel by discussing with them options within the QMS. Ensure that resources are available at the right time and with appropriate competence for the subject area.

4.3 Active Supplier Management

Manage suppliers effectively to ensure joint success. Analyzers and sample introduction interfaces may be prototypes or the first commercially manufactured systems from a particular supplier. Use supplier audits and/or gap analysis to drive out risks. Suppliers may have no experience in responding to a user requirement specification, weak system design documentation, poor formal testing records, a weak quality management system, or poor knowledge of pharmaceutical regulatory requirements. Determine whether and how to provide support to the supplier, or if risk can be mitigated within the implementation project. If necessary, encourage an innovator to partner with a reputable commercial supplier. Ensure that suppliers commit to providing long-term support for both the implementation project and lifetime of the PAT application.

4.4 Integration of PAT into the Business

Work with and involve affected employees to ensure that they feel ownership of the PAT application and use it most effectively. Validation does not take place in a vacuum; it affects people and practices. Corporate-level commitment to a PAT implementation project should also ensure that PAT is an integral element of product and process R&D, technology transfer, scale-up, and new product and process registration, although a phased implementation plan may focus on one or more of these areas first.

4.5 Effective Planning

Validating a PAT application is a complex and challenging task. Vulnerabilities are typical and can arise from different areas in different projects. Typical vulnerabilities in validation cause project delays, higher costs than planned or budgeted, duplication and rework, and inability of suppliers to provide validation deliverables. These can be addressed by thorough validation planning and management and the use of risk assessments to drive efficiency and effectiveness and active supplier management.

4.6 PAT Regulatory Framework

An open dialogue should be maintained with regulatory authorities through the PAT application life cycle. The FDA published the PAT framework [1] to define a more flexible regulatory strategy accommodating innovation by a PAT team approach to review and inspection with a jointly trained and certified FDA staff. The FDA offers two PAT system implementation options working under the production facility's own quality system.

1. *Supplement submission* [e.g., PAS (prior approval supplement), CBE-0 ("changes being effected" supplement), CBE-30 ("changes being effected in 30 days" supplement), AR (annual report)] prior to implementation, followed by an FDA inspection prior to implementation
2. *Comparability protocol submission*, outlining PAT research, validation, implementation strategies, and time lines

Following the approval of the comparability protocol by the FDA, one or a combination of the regulatory pathways above can be adopted for implementation of the PAT application. Additionally, a preoperational review of a PAT manufacturing facility and process(es) by the PAT team may be requested by the manufacturer to facilitate adoption or approval of a PAT application.

PAT submission information should include:

- Process description
- Analytical properties
- Testing and rationale
- Risk assessment
- PAT system and sampling description
  - Type of measurement technology (e.g., NIR, spectral region), sampling system [e.g., fibers (if any), sample location, sample–product interface], sampling plan
  - Risk management, including identification of system failure and strategy for managing system failure
- Experimental design protocol, including a table of experiments with justification, and references to documents (experimental design and conclusions)
- Factors identified as critical, factors chosen as critical and chosen for control justification, references to experimental design
- Modeling strategy and criteria for management of outliers
- Change control strategy for model maintenance
- Performance verification and calibration
- Process monitoring and control strategy
- Acceptance criteria

The FDA expects an inverse relationship between the level of process understanding and the risk of producing a poor-quality product. A well-understood process leads to less restrictive regulatory approaches to manage change. Focus on process understanding and the facilities quality system can facilitate risk-managed regulatory decisions and innovation.

The comparability protocol (CP) should be a well-defined, detailed written plan for assessing the effect of specific postapproval chemistry, manufacturing, and control (CMC) changes in the identity, strength, quality, purity, and potency of a specific drug product. It can be a plan for future CMC changes, or it can be submitted in an original application or prior approval supplement (postapproval). The CP concept was first introduced for biotechnology products in 1997. A well-planned protocol provides sufficient information for the FDA to determine whether the potential for an adverse effect on the product can be adequately evaluated and can lower the risk for implementing the change with FDA's prior approval.

The FDA team approach to CPs augments the scale-up and postapproval changes (SUPAC) and changes to approved NDA or ANDA guidance documents and is consistent and complementary to FDA initiatives on pharmaceutical cGMP for the twenty-first century. It has been designed to promote continual process and product improvement and innovation by facilitating CMC changes. It could also allow an applicant to implement CMC changes and place a product in distribution sooner than without the use of a comparability protocol [14].

A CP used for changes in analytical procedures should indicate whether the protocol is being used to modify or change an analytical procedure. It will demonstrate that proposed changes improve or do not significantly change characteristics used in methods validation that are relevant to the type of analytical procedure (e.g., accuracy, precision, specificity, detection limit, quantitation limit, linearity range). It should include a plan for validation of the changed analytical procedure, including:

- Suitability of the analytical procedure
- Prespecified acceptance criteria for validation parameters
- Susceptibility to matrix effects by process buffers or media, product-related contaminants, or other components present in dosage form
- Statistical analyses to determine the comparability of two procedures

When a CP is used for release or process control, the use of a new or revised analytical procedure should not result in deletion of a test or relaxation of acceptance criteria that are described in the approved application.

The success factors for the FDA PAT framework are:

- Open, frank, and science-based dialogue, building trust and mutual understanding, throughout the project
- A commitment to share knowledge as learned
- A team approach

4.7   Core Competencies in the PAT Team

The team approach requires that an ongoing dialogue with the regulators be planned into the PAT application project life cycle. This means that personnel with this core competency should be included in the project team very early on. In conclusion, the PAT application project and validation team for a successful implementation should contain all of the following core competencies:

- Plant and process improvement
- Technical project management
- Feasibility studies and technology risk management
- Analytical chemistry and experimental design

- HAZOP/CHAZOP
- System build; integration, testing, and commissioning
- Organizational change management
- Training
- Process control strategies
- Chemometrics and multivariate statistical analysis competence
- Regulatory and validation
  - PAT validation strategy
  - Regulatory liaison
- Data and information/knowledge management
- Process analytical solutions
  - NIR/FT-IR
  - Technology and interfacing
  - Sampling (sterile/toxic/hazardous)
  - Application development
- System support and development
- System improvement and development

If a team is weak in any of these areas, it should supplement them by the best means available for that business. This may include setting up a collaborative project with a consultancy supplier with a complementary skill set.

In summary, the validation of a PAT-enabled manufacturing suite is complex and requires detailed planning, risk assessment, and communication. The implementation project can also be technologically challenging and require a large capital expenditure. Therefore, the business case should also be detailed and approved by all project sponsors, so that the business objectives and scope of the project are visible to all and can be measured.

Acknowledgments

We thank Jennifer Thompson, Principal Consultant, ABB Life Sciences; Keith Beresford, Principal Consultant, ABB Engineering Services; and Frédéric Despagne, Industry Manager Life Sciences, ABB Analytical, for their contributions.

REFERENCES

1. *Guidance for Industry PAT: A Framework for Innovative Pharmaceutical Development, Manufacturing and Quality Assurance*. FDA Pharmaceutical cGMP's. FDA, Washington, DC, Sept. 2004.
2. A. Hussain. FDA Science Board meeting, Nov. 5, 2004.

3. *ICH Harmonised Tripartite Guideline*. Specifications: Test Procedures and Acceptance Criteria for New Drug Substances and New Drug Products: Chemical Substances. Q6A. ICH, Jeneva, Switzerland.

4. *ICH Harmonised Tripartite Guideline*. Specifications: Pharmaceutical Development. Q8. ICH, Jeneva, Switzerland.

5. V. Shah. *FDA PAT Forum*. Royal Pharmaceutical Society, London, Dec. 14, 2004.

6. *Enterprise-Control System Integration*, Part 1, *Models and Terminology*. ANSI/ISA-95.00.01–2000. ISA, Research Triangle Park, NC.

7. *Enterprise-Control System Integration*, Part 2, *Object Model Attributes*. ANSI/ISA-95.00.02–2001. ISA, Research Triangle Park, NC.

8. *Enterprise-Control System Integration*, Part 3, *Models of Manufacturing Operations Management*. ANSI/ISA-95.00.03–2005. Also, ISA Technical Papers on manufacturing execution systems. ISA, Research Triangle Park, NC.

9. *ICH Harmonised Tripartite Guideline*. Specifications: Quality Risk Management. Q9. ICH, Jeneva, Switzerland.

10. *Draft Consensus Guideline*. Pharmaceutical Quality System. Q10. ICH, Jeneva, Switzerland.

11. *Notes for Guidance on Process Validation*. CPMP/848/96. European Agency for the Evaluation of Medicinal Products, London.

12. *A Risk-Based Approach to Compliant GXP Computerised Systems*. GAMP 5. International Society for Pharmaceutical Engineering, Tampa, FL.

13. *ASTM Standard for Specification, Design and Verification of Pharmaceutical and Biopharmaceutical Manufacturing Systems and Equipment*. ASTM, West Conshohocken, PA.

14. *FDA Draft Guidance for Industry: Comparability Protocols—Chemistry, Manufacturing and Controls Information*. FDA, Washington, DC, Feb. 2003.

5

VALIDATION OF NEAR-INFRARED SYSTEMS FOR RAW MATERIAL IDENTIFICATION

LENNY DASS

GlaxoSmithKline Canada Inc.

1 INTRODUCTION

Near infrared (NIR) is a spectroscopic technique employed in analytical chemistry. The spectral region is from about 1000 to 2500 nm or 10,000 to 4000 cm^{-1}. This region is in the NIR area and has historically been termed the *fingerprint region*. This is because the spectra in this area of a compound represent its fingerprint (i.e., are unique to that compound). As there are no two human beings (albeit identical twins) with the same fingerprint, there are no two compounds with identical spectral signatures. Even polymorphic compounds with the same chemical structure but with different crystal structures can be distinguished in this area.

Reference spectra are collected for raw material identification. In reflection mode, the signal received by the detector has been reflected off the target material. The spectra in the infrared region can be overlaid against reference spectra and a visual comparison made (Figure 1). This comparison can result in a positive identification—the sample and standard spectra are the same—therefore, the compound is identified as the standard. Or, it can result in a negative identification—the sample and standard spectra differ—therefore, the sample compound is different from the standard compound. This has been the historical

Practical Approaches to Method Validation and Essential Instrument Qualification,
Edited by Chung Chow Chan, Herman Lam, and Xue Ming Zhang
Copyright © 2010 John Wiley & Sons, Inc.

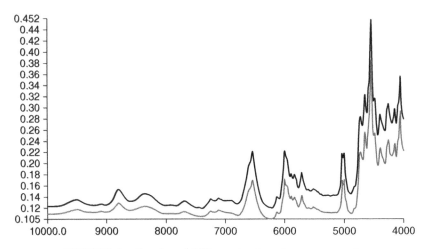

FIGURE 1   Overlay of NIR spectra from two compounds.

method for spectral identification. This technique has been very successful for pure compounds in this spectral region.

However, the NIR spectra are not as distinct. Spectral peaks are neither sharp nor distinct. Features are overtones and combination bands of organic functional groups and are weaker vibrational artifacts. Overlaps also create broad spectral features. A comparison against a reference standard and a visual determination as to a positive or negative identification would not be viable, as the features cannot be visually resolved. To overcome this limitation, multivariate analyses are used.

Due to the limitations of visual comparison, the power of the computer has been employed. Computers and software enable scientists to utilize this spectral region for qualitative and quantitative applications using statistical analysis or chemometrics [1–4]. *Wikipedia* defines *chemometrics* as the science of relating measurements made on a chemical system or process to the state of the system via application of mathematical or statistical methods. Modern NIR instruments are PC controlled and have some chemometric features. Soft independent modeling of class analogy (SIMCA) is a popular algorithm used for NIR [5].

It is beyond our scope here to go into the depths of chemometrics and multivariate analysis. However, the essential elements to know are that the spectra are rendered into its principal components in multidimensional space. A model or standard can be created by addition of many spectra of the same compound. The instrument can be taught a spectral pattern for each compound that may be encountered. The instrument software can then compare an unknown against a library to determine which spectral pattern it matches. This pattern matching is the basis of identification (i.e., the distance of the sample point from the library collection in *n*-dimensional space).

It is important to note that the instrument must be taught the pattern of a compound. There is no off-the-shelf NIR spectral library that can be purchased against which to compare your samples. It is necessary to build a specific library.

The scientists are required to obtain known examples to prepare the library. These known samples are typically the reference samples that are retained after quality control analysis. This means that these samples have already been identified using a primary method such as mid-IR or ultraviolet (UV). Thus, NIR is a secondary identification technique: that is, a method that is predicated or contingent on a primary source.

The most significant benefits of the system are from the speed and nondestructive nature of the analysis. Typical analysis time is in the order of a few seconds and can be done in situ (i.e., through a transparent bag containing the material). None of the material is consumed, so there is zero waste and no sample required for the analysis. It can be said that NIR is a green technology and environmentally friendly. This prodigious ability has seen NIR use increase significantly over the past few years in many industries [6–8]. It is an ideal system for rapid identification of materials. This technology is leveraged by many manufacturing companies and is not specific to pharmaceutical applications. NIR systems are typically deployed at the receiving dock to replace classical chemical or instrumental analysis that would normally take place in the laboratory. This is advantageous, as materials are available for use almost immediately on receipt.

One of the significant benefits of NIR is that it not only identifies the material but can differentiate different grades of the same material based on variations in their physical properties. As indicated previously, even crystal structures between identical structures can be differentiated. This is very beneficial. Material that is identified correctly but with some characteristics that different from those of material received previously may not behave as anticipated in your manufacturing process. The costs associated with the manufacture, packaging, and testing of a product are significant. Any opportunity for waste avoidance should be explored. This does not mean that the material is not usable, but there should be some investigation of why this batch is different from those received previously.

This variation in physical property can be seen in the following illustration. The principal components plots of sample spectra are plotted along three principal components axes in Figure 2. There is a tight cluster of samples in the right-hand quadrant, while there are seven outlier samples to the left. Although these samples may be identified as the tight cluster, there is some characteristic that separates them from the rest. To understand the scatter, it would be necessary to study the characteristics of the materials outside the cluster. The variation may be due to moisture content, particle size, polymorphic variations, or a combination of all.

2 VALIDATION OF AN NIR SYSTEM

Regulatory agencies require that laboratory instrumentation used in the analysis of material or products be validated. Validation has been defined by the U.S. Food and Drug Administration (FDA) as "establishing documented evidence, which provides a high degree of assurance that a specific process will consistently produce a product meeting its predetermined specifications and attributes" [9].

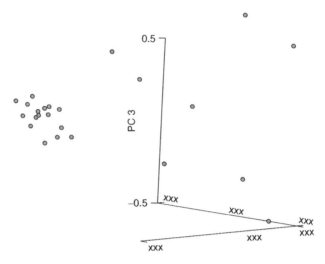

FIGURE 2   Principal components plot of a collection of spectra from samples of a single compound.

The main output of any validation exercise is documentation that proves that user requirements have been met. NIR validation is no different. The approach is typical of analytical instrument validation where there is installation, operational, and performance qualifications (GAMP 4) [10].

The benefit of validation and good documentation lies in the in-depth understanding of the system. Validation of systems promotes a deeper understanding of the way that systems work. Taking the time to prepare documentation, a test case, and execution leads to a level of intimacy with the system that would not be available to the user simply by rushing to use it. This is important not only for proper maintenance of the system but also to enable accurate communication during GMP (good manufacturing practice) audits. There are two tiers to this validation. The first is instrument acceptance testing and the second is the NIR spectral library, which would need validation to demonstrate accuracy, precision, and specificity in the identification process. The validation approach for an NIR system is shown in Figure 3.

The plan, requirements, installation, acceptance testing, and reporting are consistent with general instrumental validation and are discussed in detail throughout this book. It is in the spectral library validation that there are differences, which is the focus of this chapter.

3   VALIDATION PLAN

Every validation should begin with planning. The validation plan should contain the following elements:

1. An introduction, to describe the purpose of the project and plan

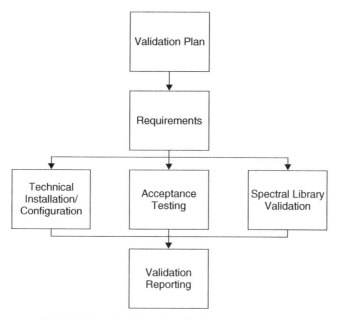

FIGURE 3   High-level validation approach flow.

2. The scope, to define the boundaries of the project
3. A compliance assessment, to describe why the system requires validation
4. An approach to describing a high-level validation (as in Figure 3)
5. A risk analysis
6. Key deliverables
7. Change control
8. Document management
9. Configuration management
10. Incident management
11. Criteria for project completion

The introduction and scope should be clear and concise. There should be unambiguous boundaries to the initiative described earlier. The compliance assessment should be indicative of the need for validation. If the compliance determination indicates that the system does not require validation, proceed no further. If the system does require validation, you would continue as described below. The validation strategy should include the approach described in Figure 3 and risk analysis should include the following considerations (GAMP 4, Appendix M3):

• Does the system require validation? (covered in the compliance assessment)
• How much validation is required?

- What aspects of the system or process are critical to product quality or patient safety for drug products?
- What aspects of the system are critical to business?

The amount of validation effort should be proportional to the risk level of the system. Risk evaluation should include the following questions:

- Did the material received include a certificate of analysis?
- Has the vendor been approved or certified by the quality assurance staff?
- Has the vendor been audited?
- Has any material been received that has failed identification testing?
- Is there adequate testing of the finished product that would detect if incorrect materials were used?
- How sever would be the repercussions of a misidentification?
- What is the cost to the business of an identification error?

The risk associated with the use of the system could be low if the materials received from a vendor have been audited, certified, were accompanied by a certificate of analysis, and never failed identification testing or specifications. Also, testing of the material is included in the final release testing of the product. The risk is low because the probability of receiving incorrect or substandard material is low and any problems can be detected prior to release of the product.

Conversely, the risk associated with the use of the system would be high if material received from a vendor had not been audited, did not have a certificate of analysis, was not the material ordered, or had failed to meet the release specification in the past. Also, detection of the material in the final product was not carried out or was not specific. This would be the case for excipients, where there is no test for, say, methylcellulose in the end product. The risk is high because the probability of receiving incorrect or substandard materials is greater, and problems may not be uncovered during final testing of the product.

3.1 Requirements

Requirements documents should include the following items:

1. Operational requirements for the spectrophotometer, including sampling accessories, wavelength range, wavelength accuracy, wavelength repeatability, photometric noise, and photometric linearity
2. Hardware specifications for computers
3. Software specifications
4. Regulated electronic record and signatures (RERS) requirement
5. Spectral acquisition and processing
6. Qualitative analysis for routine identification

7. Data requirements (e.g., report types, compatibility with other systems)
8. Environmental requirements
9. Training
10. Vendor validation package
11. Any other requirements
12. Constraints

Requirements should be written so that they can easily be transferred to an acceptance test case. This would facilitate the need for traceability between requirements and testing. Operational requirements can be lifted directly out of the compendia for NIR identification. *U.S. Pharmacopeia* (USP) 32 Monograph ⟨1119⟩, Near-Infrared Spectrophotometry, details the performance specifications recommended [11]:

- *Uncertainty*. USP near IR system suitability reference standard or National Institute of Standards and Technology (NIST) SRM 2036 for reflectance and SRM 2035 for transmittance: ± 1 nm, 700 to 2000 nm; ± 1.5 nm, 2000 to 2500 nm.
- *Photometric linearity and response stability*. Four reference standards, ranging from 10 to 90% reflection or transmission. Applications measuring absorbance greater than 1.0 may require standards with reflective properties in the range 2 to 5%. Slope $= 1.00 \pm 0.05$; intercept $= 0.00 \pm 0.05$.
- *Spectrophotometric noise*. Typically, an instrument diagnostic to determine signal-to-noise ratios.
- *High flux noise*. High-light flux is evaluated by measuring reflectance or transmission of the reference standard as both the sample and background (typically, using a 99% reflection standard).
- *Low flux noise*. Low-light flux is evaluated by measuring the reflectance or transmission of the reference standard as both the sample and background (typically, using a 10% reflection standard).

The regulated electronic record and signatures requirement (RERS) is described in 21 CFR Part 11, Electronic Records and Electronic Signatures.

Instrument control software is commercial off-the-shelf software (COTS). It is not practical to test every function within the software code after it is developed. It is not possible to test every combination of programmed routines, nor is it recommended or value added. A practical approach to software validation and the most significant action that can be performed to achieve a "high level of assurance" is to perform a vendor audit, which would disclose if the software was developed following established processes to ensure quality in the development of the software and hardware. A positive audit of the instrument vendor will also decrease your risk level, as you have examined how the software was developed firsthand.

Most equipment vendors have available validation packages that can be used to qualify their systems. These packages simplify the validation process. Their purchase should be negotiated at the time of acquisition of the instrument. These test protocols must be reviewed and approved by a representative of the purchaser prior to execution.

3.2 Technical Installation

As indicated above, instrument vendors generally have documentation packages for installation and operational qualification of their systems. The vendor or a qualified engineer at the user site may perform the installation. The location, date, serial number, and other details of the device must be documented.

3.3 Acceptance Testing

There are specific tests that should be conducted. These include the following test cases:

- User access and electronic records electronic signature verification
- Instrument setup and spectral acquisition
- Spectral library development
- Qualitative method development
- Qualitative analysis
- USP instrument checks
- Backup and restoration of spectral data

The structure of the test case should include the following:

- Introduction
- Prerequisites for the test
- Test objectives
- Traceability of that test case to requirements
- Test environment (instrument ID, software versions, etc.)
- Test steps with acceptance criteria
- Overall test disposition section to be completed after execution
- References
- Version history

Typically, test cases are prepared using document templates. These should be followed to ensure a consistent format for all your validation exercises. The test steps with acceptance criteria are the essence of the document (Table 1). The test case steps should be such that the requirement can be demonstrated through the actions. To test user access, a user with an analyst profile cannot approve results.

TABLE 1 Acceptance Test Steps

Test Step	Expected Result	Actual Result	Pass or Fail
1. Log in as analyst.	Login successful.		
2. Select menu option. Review/approve results.	Menu option is not available.		
3. Next steps.			
End of test			

Well-designed test cases may be able to capture a significant number of requirements in fewer than 10 tests. Developing a single test case to demonstrate all requirements is not recommended, because a single protocol would be long, and revision and maintenance would be more frequent. One person would have to execute the test, and another review it, so testing and review would take more time.

3.4 Configuration Management

An item typically overlooked for instrumentation is configuration management, the tracking of the components and their versions for the specific system. There should be a broad configuration management plan that describes the intention of the plan. This is to maintain and control system integrity by tracking components of the system and any changes. This plan should contain the following sections:

- Introduction
- Purpose
- Scope
- Configuration identification (items and their relationships)
- Configuration control (control of changes to configurable items)

An index that supports configuration management is the master configurable item list (MCIL). This is typically a spreadsheet that has the list of all the components of the system along with their software versions. This spreadsheet can contain as well the list of materials and spectral libraries that are validated. This spreadsheet is version controlled so that changes to the system over time can be recorded. A baseline should be taken when validation of the system is undertaken. A baseline is a snapshot of the configurable components of the system at the beginning. Another snapshot (new version) of the MCIL can be taken when validation is concluded and probably annually thereafter.

3.5 Spectral Library Validation

In validation of the spectral library, NIR departs from the norm of analytical instrument validation. However, since the technology is now mature, standard approaches to validation have been developed. Library validation is described by the flowchart in Figure 4. The sections of this process are usually documented in a library development plan.

The process would be described within the plan document. Sections would include:

- Introduction and purpose.
- Scope (section A in Figure 4).
- List of raw materials to be tested (section A).
- Entry criteria for development of the library. These would include completion of the installation, acceptance testing, and availability of samples.
- Samples for calibration. These can either be retained samples that have already been tested and meet release specification, or new samples just received (section B).
- Sample authentication (section C). A primary method must be used to identify the material. This method must be the current approved method for identification testing. All samples must be identified as positive and meet the remaining release specifications.
- Acquisition and display of spectra (section D). The parameters for the acquisition of spectra should be defined. These are the instrument parameter settings.
- Calibration and validation sample selection (section E). Only samples that pass quality control testing should be used to prepare the library. In addition, the samples for a particular material should ideally be coming from the same supplier.
- Data preprocessing (section F). This section lists the instrument preprocessing parameters. As a general rule, use a second derivative and multiplicative scatter correction (MSC). These treatments can reduce the effects of particle size variations in spectral features. Also, certain spectral areas can be omitted during this step. When samples are scanned directly through a plastic bag, the bag region of the spectra should be removed from the analysis. This region can be determined simply by scanning an empty bag. Samples scanned through glass vials do not need this treatment, as the glass does not contribute to the spectra in the NIR region.
- Library construction (section G). The material model should contain enough representative samples to describe the batch-to-batch variation that may be encountered. At a minimum, six distinct batches are required to characterize a material: four used for chemometric model construction, one held for validating the model, and one used as an unknown to validate the library, which contains all the materials to be identified as listed in the validation plan based on the chemometric model developed. The model will be more robust, with more samples.
- Algorithm selection and threshold determination (section H). SIMCA for PerkinElmer (PLS for other vendors) should be used as the algorithm. Some

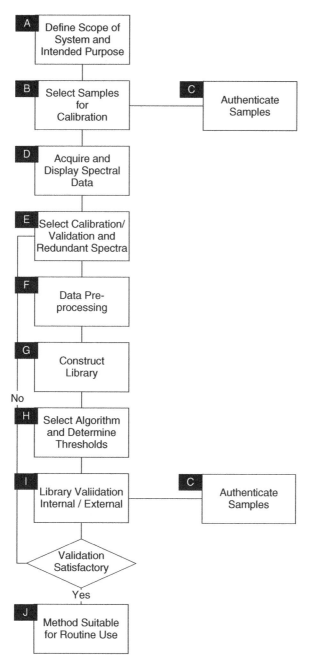

FIGURE 4 PASG (Pharmaceutical Analytical Sciences Group) guidelines for the development and validation of NIR spectroscopic methods. (From [12].)

applications allow selection of other algorithms. However, SIMCA should be sufficient. Thresholds should be set to 1.000. The threshold typically is the amount of variation allowed.

- Library validation, internal/external (section I). This is where the specificity, accuracy, and precision of the library are demonstrated. As indicated previously, six distinct lots are used to characterize a material model. For groupings of materials, called methods, one of the materials must have an additional three distinct batches for specificity challenges. The purpose of these additional batches is discussed later in the chapter.
- Method suitable for routine use (section J). This is the identification workflow.
- Acceptance criteria for the conclusion of library validation.

The library validation plan includes a list of test cases. These test cases will validate the acquisition, development, and testing of the spectral library. Test cases should include the following:

1. *Acquire NIR spectra.* This test case should describe the process for acquiring the spectra. It should also list the materials and batches to be scanned.
2. *Develop method.* This test case would describe the process for method development. Methods are collections of materials analyzed by the same set of parameters.
3. *Generate and validate models.* This test case would describe the process for adding material scans to develop models for each material.
4. *Validate methods.* This test case would demonstrate the specificity of the identification technique.
5. *Execute user procedures.* This test case would execute draft user procedures that are prepared for routine use of the instrument. These procedures would include routine analysis, administration, maintenance, and so on.

All of these test cases are pretty standard, with the exception of cases 3 and 4. The ideal model would incorporate the normal variation that would be encountered. Figure 5 shows a typical material model. The variation of 1.000 standard deviation is the exterior of the spherical region. This variation is measured from an average center. The theoretical center of the sphere is the mathematical average of the samples. The perimeter of the sphere is 1 standard deviation about this mean.

During this test case the internal validation is performed by the software. Samples used to develop the model are tested against it. One sample at a time is excluded from the chemometric model building and treated as unknown, to test the reliability of the model. The process will be repeated until all the samples used in construction of the model have been tested. As the samples in the model are already "in the model," this testing usually provides a good indication of the usefulness of the model but does not challenge the model sufficiently as to its

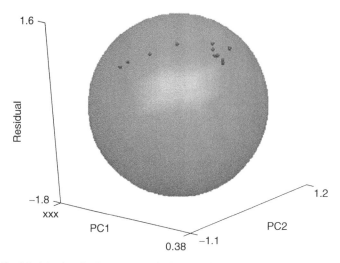

FIGURE 5 Model of a single compound plotted with three principal components. Individual samples are distributed within the sphere.

ability to identify materials in the library correctly in day-to-day operations. A sample that is not part of the model group should be used: external validation. This would confirm that the model is valid. Remember, all samples are lots previously tested and released that met quality specifications. All the batches should be within the predication power of the model and identified correctly.

Thus far, only the model has been tested. However, the spectral library method (a grouping of models) must show specificity in determining identification. A method with the following material models will not show specificity. Figure 6 shows a principal components plot of a library of materials. Different compounds appear as different spherical regions. Samples of the same compounds are distributed within each of their respective spheres. The first plot shows closely grouped spheres, indicating that the model is not yet specific enough. The second plot shows distinct spheres with clearer separation, indicating a higher level of specificity.

This view is dependent on which principal components are plotted. There is obvious region overlap. If a lot of material under test were to be located in the intersection of regions, a clear identification could not be made. Samples that are plotted within the individual spheres would be identified as that respective material. The validate method test case will demonstrate that a robust method will identify the compounds with great confidence. This is done by challenging the materials in the method with additional distinct batches outside of those used for the model: the external validation method.

An example method or library is depicted in Figure 7.

There is clear separation between the models of the compounds from A, B, C, D, E, F, and G. There are four scenarios to challenge the specificity of the method, as indicated in Figure 8.

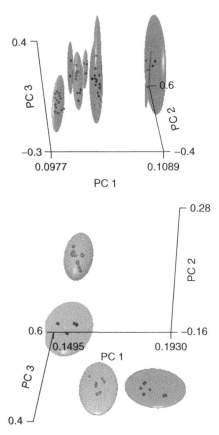

FIGURE 6   Principal components plot of a library of materials. Different compounds appear as different spherical regions. Samples of the same compounds are distributed within each of their respective spheres. The first plot shows closely grouped spheres, indicating that the model is not yet specific enough. The second plot shows distinct spheres with clearer separation, indicating a higher level of specificity.

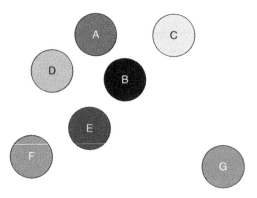

FIGURE 7   Library plotted in two dimensions: two principal components.

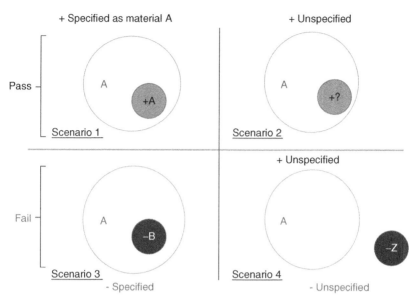

FIGURE 8 Model challenge scenarios for an identification method.

- *Scenario 1: positive specified.* Materials in the test method should have a sample analyzed as a specified sample. A sample belonging to material group A is presented to the system for testing and its identity as been specified as material A. The system should be able to confirm the identity of the sample as material A.
- *Scenario 2: positive unspecified.* Each material in the test method should have a sample presented for analysis as an unknown. A sample belonging to material group A is presented to the system for testing, without specifying the sample as material A. The system should be able to identify the sample as material group A.
- *Scenario 3: negative specified.* Materials in the test method should have a sample analyzed as a specified sample that is not what it really is. A sample belonging to material group B is presented to the system for identification, specifying the sample as material A. The system should reject the identity of the sample as material A and be able to identify the sample correctly as material group B.
- *Scenario 4: negative unspecified.* A material that is not part of the model should be analyzed as an unknown. The sample under test is compound Z and is not in the method. The identification should fail and the device should indicate that the compound is unknown.

Once the test cases are completed, a library validation report should be generated. This report should summarize the results of the test cases executed. Supporting information for the primary method of identification should not be omitted.

However, this can be appended to the test case for spectral acquisition. All output reports from the identifications should be included as evidence for the test case.

3.6   Validation Reporting

At the end of the validation exercise a report is generated. This report should be reviewed and approved by the same or equivalent signatories as the plan. Typically, the report confirms that the plan was executed without deviation. A table listing the deliverables (documents) generated for the validation should be included. A rehash of the acceptance test report or library development report is not required, nor is it recommended.

A specific section of the report should include limitations and restrictions on use. It is very likely that the validation did not include all the materials that your organization receives. The limitations and restrictions on use section should indicate that the identification method is valid only for materials that have been validated. The conclusion should indicate that the acceptance criteria have been met. A clear statement indicating that the system is validated should also appear. The subsequent system release should indicate the date on which first use will begin.

REFERENCES

1. R. Kramer. *Chemometric Techniques for Quantitative Analysis*. Marcel Dekker, New York, 1998.
2. J. Miller and J. Miller. *Statistics and Chemometrics for Analytical Chemistry*, 5th ed., Pearson Prentice Hall, Upper Saddle River, NJ, 2005.
3. G. Reich. Near infrared spectroscopy and imaging: basic principles and pharmaceutical applications. *Adv. Drug Deliv. Rev.*, 57:1109–1143, 2005.
4. E. L. Willighagen, R. Wehrens, and L. M. C. Buydens. Molecular chemometrics. *Crit. Rev. Anal. Chem.*, 36(3):189–198, 2006.
5. A. Candolfi, R. DeMaesschalck, D. L. Massart, P. A. Hailey, and A. C. E. Harrington. Identification of pharmaceutical recipients using NIR spectroscopy and SIMCA. *J. Pharm. Biomed. Anal.*, 19:923–935, 1999.
6. R. C. Lyon, E. H. Jefferson, C. D. Ellison, L. F. Buhse, J. A. Spencer, M. M. Nasr, and A. S. Hussain. Exploring pharmaceutical applications of near-infrared technology. *Am. Pharm. Res.*, 6(3):62–70, 2003.
7. P. Borman, P. Nethercote, M. Chatfield, D. Thompson, and K. Truman. The application of quality by design to analytical methods. *Pharm. Technol.*, 31:142, Oct. 2007.
8. K. Kramer and S. Ebel. Application of NIR reflectance spectroscopy for the identification of pharmaceutical recipients. *Anal. Chim. Acta*, 420:155–161, 2000.
9. *General Principals of Validation*. FDA CDER, Washington, DC, May 1987.

10. *GAMP 4 Guide: Validation of Automated Systems*. Updated NIR Method Development and Validation Guidelines. Pharmaceutical Analytical Sciences Group, International Society for Parmaceutical Engineering, Tampa, FL, 2001.

11. *U.S. Pharmacopeia*, 32, Chapter ⟨1119⟩, Near-Infrared Spectrophotometry. USP, Rockville, MD.

12. www.pasg.org.uk/NIR/NIR_Guidelines_Oct_01.pdf.

6

CLEANING VALIDATION

XUE MING ZHANG
Apotex Inc.

CHUNG CHOW CHAN
CCC Consulting

ANTHONY QU
Patheon Inc.

1 INTRODUCTION

Cleaning validation is a critical part of current good manufacturing practice (cGMP), ensuring that active pharmaceutical ingredients (APIs) and other production materials that come in contact with equipment surfaces are not contaminated or adulterated. Throughout all stages of pharmaceutical development and manufacturing, every piece of equipment surface must be verified as clean before it can be used, unless it is dedicated. In cleaning validation, equipment surfaces (e.g., stainless steel, glass, rubber, Teflon) are sampled after cleaning and the samples analyzed for trace residual contaminants, including APIs and cleaning agents. The analysis must be quantitative, and pass–fail levels are determined according to the regulatory guidelines and company policy.

In this chapter we provide guidance on issues and topics related to cleaning validation, with the objective of facilitating compliance with regulatory requirements. The information provided, which can be applied in limited and defined

Practical Approaches to Method Validation and Essential Instrument Qualification,
Edited by Chung Chow Chan, Herman Lam, and Xue Ming Zhang
Copyright © 2010 John Wiley & Sons, Inc.

circumstances, is for guidance purposes only. Appropriate scientific justification should be used for individual cases.

2 SCOPE OF THE CHAPTER

In this chapter we address special considerations and issues pertaining to validation of cleaning procedures for equipment used in the manufacture of pharmaceutical products, radiopharmaceuticals, and biological drugs. Principles incorporated in regulatory guidance [1–5] have been taken into account in the preparation of this chapter. We also cover validation of analytical methods of equipment cleaning for the removal of contaminants associated with previous products, residues of cleaning agents, and the control of potential microbial contaminants.

3 STRATEGIES AND VALIDATION PARAMETERS

3.1 Cleaning Validation Strategies

Cleaning validation is product and equipment specific and must be considered for each product for all the major pieces of manufacturing and packaging equipment. Verifications of cleaning for three consecutive lots are to be included in the study. A cleaning procedure is considered validated if all three cleaning verification results pass.

Prior to the introduction of any new product, an evaluation is to be made regarding the solubility and toxicity of its ingredients. If any of the APIs are deemed to be more potent or less soluble in water than the ingredients for which cleaning validation has already been completed, or are deemed to be particularly difficult to remove from equipment, a new validation study will be carried out. A matrix approach will be used to select the worst-case API for use in cleaning validation studies, based on the API deemed the most difficult to clean.

When a new piece of equipment is added to a process train for which cleaning validation has been performed, the need for cleaning verification of this piece has to be determined, based on the similarity of this equipment design and cleaning procedure to the other pieces in the train.

1. Products containing the same API can be grouped, and cleaning validation needs to be completed for the least water-soluble product in the group.
2. Equipment can be grouped based on similarity of design. Only the smallest and largest sizes from each equipment group are to be included in cleaning validation studies, provided that the same cleaning procedure is used for all sizes.
3. Used and recommissioned equipment for which product history is not known or analytical methods are not available will be cleaned three consecutive times and tested for detergent residue only prior to use. Used

equipment must be accompanied by a certificate stating that it has not been used for penicillin, cephalosporins, or cytotoxic products.

4. To commission newly purchased equipment, only passivation and verification of removal of passivating agents are required for cleaning verification.

5. Cleaning validation for utensils will be performed for products deemed to be the most difficult to clean, and should be covered in a separate protocol.

6. Nonproduct-contact surfaces are to be monitored periodically (e.g., once a month for the most stringent cases) for API (allergenics, potent steroids, and cytotoxics) residue. The APIs selected are to be used to monitor nonproduct-contact surfaces. This will be part of a separate report for containment.

3.2 Elements of Cleaning Validation

There are at least four elements in cleaning validation:

1. *Visual examination.* Visual examination of both product-contact and nonproduct-contact surfaces of equipment for the presence of drug product and detergent residues.

2. *Testing for API (chemical) residue.* Swab samples from critical and difficult-to-clean product-contact surfaces of equipment are to be taken and tested for residual API of the last drug product manufactured. Generally, high-performance liquid chromatography (HPLC) is the analytical method used to quantify active residues. Other methods may be used as applicable. If total organic carbon is used to quantify residual API, rinse water samples may be collected.

3. *Testing for cleaning agent residue.* The use of appropriate detergent should be indicated in a cleaning standard operation procedure (SOP). Verification of removal of detergents from processing and packaging equipment is required. The entire product-contact surface of equipment is to be sampled by rinsing with USP (*U.S. Pharmacopeia*) purified water, and tested for conductivity of the cleaning agents.

4. *Microbial load test.* Swab sampling of selected areas of the equipment train will be done to determine the number of colony-forming units (CFU) present. If any colonies are found to be present, appropriate tests will be conducted to identify the organisms.

3.3 Validation Practices

cGMP requires that "equipment and utensils shall be cleaned, maintained and sanitized at appropriate intervals to prevent malfunctions or contamination that would alter the safety, identity, strength, quality, or purity of the drug product beyond the official or other established requirement." The guidelines also require that "written procedures shall be established and followed for cleaning and maintenance of equipment, including utensils, used in the manufacture, processing, packaging, or holding of a drug product." Moreover, the methods used must be validated.

The objective of cleaning validation is to verify the effectiveness of the cleaning procedure for removal of product residues, degradation products, preservatives, excipients, and/or cleaning agents so that the analytical monitoring may be reduced to a minimum in the routine phase. In addition, one needs to ensure there is no risk associated with cross-contamination of APIs. Cleaning procedures must strictly follow carefully established and validated methods.

Appropriate cleaning procedures must be developed not only for the product-contact equipment used in the production process but consideration should also be given to noncontact parts into which product may migrate (e.g., seals, flanges, mixing shaft, fans of ovens, heating elements).

This is more applicable for biological drugs because of their inherent characteristics (proteins are "sticky" by nature), and for parenteral products since they have high-purity requirements and involve a broad spectrum of equipment and materials during the manufacturing process.

Cleaning procedures for products and processes that are very similar do not need to be validated individually. This could depend on what is common, the equipment and surface area, or an environment involving all product-contact equipment. It is considered acceptable to select a representative range of similar products and processes. The physical similarities of the products, the formulation, the manner and quantity of use by the consumer, the nature of other products manufactured previously, and the size of the batch in comparison with products manufactured previously are critical issues that justify a validation program.

A single validation study under consideration of the worst case can then be carried out which takes account of the relevant criteria.

For biological drugs, including vaccines, bracketing may be considered acceptable for similar products and/or equipment provided that appropriate justification, based on sound, scientific rationale, is given. Examples: cleaning of fermenters of the same design but with different vessel capacity used for the same type of recombinant proteins expressed in the same rodent cell line and cultivated in closely related growth media; a multiantigen vaccine used to represent the individual antigen or other combinations of them when validating the same or similar equipment that is used at stages of formulation (adsorption) and/or holding. Validation of cleaning of fermenters should be done on an individual pathogen basis. As a general concept, until the validation of the cleaning procedure has been completed, the product-contact equipment should be dedicated.

In a multiproduct facility, the effort to validate the cleaning of a specific piece of equipment that has been exposed to a product and the cost of dedicating the equipment permanently to a single product should be considered. Equipment cleaning may be validated concurrently with actual production steps during process development and clinical manufacturing. Validation programs should be continued through full-scale commercial production.

It is usually not considered acceptable to test until clean. This concept involves cleaning, sampling, and testing, with repetition of this sequence until an acceptable residue limit is attained. For a system or equipment with a validated cleaning procedure, the practice of resampling should not be utilized. Products that simulate the physicochemical properties of the substance to be removed may be considered for use instead of the substances themselves when such substances are either toxic or hazardous.

Raw materials sourced from different suppliers may have different physical properties and impurity profiles. These differences should be considered when designing cleaning procedures, as the materials may behave differently. All pertinent parameters should be checked to ensure that the process is validated as it will ultimately be run. Therefore, if critical temperatures are needed to effect cleaning, these should be verified. Any chemical agents added should be verified for type as well as quantity. Volumes of wash and rinse fluids, and velocity measurements for cleaning fluids, should be measured as appropriate.

If automated procedures are utilized [cleaning in place (CIP)], consideration should be given to monitoring the critical control points and the parameters with appropriate sensors and alarm points to ensure that the process is highly controlled. CIP is a system designed for automatic cleaning and disinfecting without major disassembly and assembly work. CIP covers a variety of areas but its main purpose is to remove solids and bacteria from tanks, vessels, and pipework in the food, dairy, beverage, nutraceutical, pharmaceutical, and biotechnology processing industries. Additionally, a well-designed CIP system [employing double-seat valve (block and bleed) technology and a bit of process integration] will make it possible to clean one part of the plant while other areas continue to produce products. The cleaning can be carried out with automated or manual systems and is a reliable and repeatable process that meets the stringent regulations demanded by the food, dairy, biotechnology, and pharmaceutical industries.

The validation of cleaning processes should be based on a worst-case scenario, including:

• Challenging the cleaning process to show that the challenge soil can be recovered in sufficient quantity or to demonstrate log removal to ensure that the cleaning process is indeed removing the soil to the level required
• The use of reduced cleaning parameters, such as overloading of contaminants, overdrying of equipment surfaces, minimal concentration of cleaning agents, and/or minimum contact time of detergents

At least three consecutive applications of the cleaning procedure should be performed and shown to be successful to prove that the method is validated.

3.4 Equipment and Personnel

All processing equipment should be specifically designed to facilitate cleanability and permit visual inspection, and whenever possible the equipment should be made of smooth surfaces of nonreactive materials. Critical areas (i.e., those hardest to clean) should be identified, particularly in large systems that employ semiautomatic or fully automatic CIP systems. Dedicated product-contact equipment should be used for products that are difficult to remove (e.g., tarry or gummy residues in bulk manufacturing), for equipment that is difficult to clean (e.g., bags for fluid-bed dryers), or for products with a high safety risk (e.g., biologicals or products of high potency which may be difficult to detect below an acceptable limit).

In a bulk process, particularly for very potent chemicals such as some steroids, the issue of by-products needs to be considered if equipment is not dedicated. It is difficult to validate a manual cleaning procedure (i.e., an inherently variable cleaning procedure). Therefore, operators carrying out manual cleaning procedures should be adequately trained and monitored and assessed periodically.

3.5 Microbiological Considerations

Whether or not CIP systems are used for cleaning of processing equipment, microbiological aspects of equipment cleaning should be considered. This consists largely of preventive measures rather than removal of contamination once it has occurred. There should be some documented evidence that routine cleaning and storage of equipment do not allow microbial proliferation. For example, equipment should be dried before storage, and under no circumstances should stagnant water be allowed to remain in equipment subsequent to cleaning operations. Time frames should be established for the storage of unclean equipment prior to commencement of cleaning, as well as time frames and conditions for the storage of cleaned equipment.

The control of the bioburden through adequate cleaning and storage of equipment is important to ensure that subsequent sterilization or sanitization procedures achieve the necessary assurance of sterility. This is particularly important from the standpoint of the control of pyrogens in sterile processing since equipment sterilization processes may not be adequate to achieve significant inactivation or removal of pyrogens.

3.6 Documentation

The most important documentation in a cleaning validation is the planning, writing, and approval of the cleaning validation protocol that is to be used. A cleaning validation protocol should describe the procedure used to validate the cleaning process. In addition to other information it should include: a description of the equipment used; the interval between the end of production and the beginning of the cleaning procedures; the cleaning procedures to be used for each product, each manufacturing system, or each piece of equipment; the sampling procedures to be used, with rationales; the analytical methods to be used, including limit of detection and limit of quantification; and the acceptance criteria to be used, with rationales and conditions for revalidation.

Depending on the complexity of the system and cleaning processes, the amount of documentation necessary to execute various cleaning steps or procedures may vary. When more complex cleaning procedures are required, it is important to document the critical cleaning steps. In this regard, specific documentation on the equipment itself, including information about who cleaned it, when the cleaning was carried out, and the product that was processed previously on the equipment being cleaned should be available. However, for relatively simple cleaning operations, mere documentation that the overall cleaning process was performed might be sufficient.

Other factors, such as the history of cleaning, residue levels found after cleaning, and variability of test results may also dictate the amount of documentation required. For example, when variable residue levels are detected following cleaning, particularly for a process that is believed to be acceptable, one must establish the effectiveness of the process and the operator performance. Appropriate evaluations must be made and when operator performance is deemed a problem, more extensive documentation (guidance) and training may be required.

3.7 Use of Cleaning Agents

When cleaning agents (e.g., detergents, surfactants, solvents) are used in the cleaning process, their composition should be known to the user and their removal should be demonstrated.

Detergents should be easily removable, being used to facilitate the cleaning during the cleaning process. Acceptable limits should be defined for detergent residues after cleaning. The possibility of detergent breakdown should also be considered when validating cleaning procedures.

In such cases, the following requirements must be addressed in the cleaning validation study:

- The validation protocol must include supporting data and the rationale for each cleaning agent used or a reference to support the choice of cleaning agent(s).
- The validation protocol must include testing for the residual cleaning agent. This test may be designed to detect the specific agent or it may be a non-specific test.
- The maximum cleaning limit for the cleaning agent is required (e.g., not more than 100 ppm).
- Toxciolology opinion may be required in some cases.

3.8 Justification of Swab Positions

The following assumptions should be made in the design of the cleaning validation program, selection of sampling locations, and establishing acceptance criteria:

1. A contaminant is distributed uniformly in the next product batch.

2. There is a potential for localized contamination to exist. This factor is addressed by the selection of "hardest to clean" sample sites, safety factors, and development of the cleaning procedure.

3. There is no inherent difference between automated and manual cleaning methods. Both methods are able to clean to the same predetermined limit.

4. Cleaning agents are not part of the manufacturing process and are water-soluble, thus are easily removable from the equipment.

5. Excipients used in manufacture are considered nontoxic, and as such, verification of their removal is by visual inspection only.

3.9 Types of Swabs

The major consideration in selecting types of swab is freedom from of contamination under test conditions. If there is contaminant with one type of swab, other types have to be tested, or pretreatment of the swab may be needed. Polyester swabs are those used commonly for swabbing equipment surfaces. Both woven (e.g., TX714A, TX761, TX758B) and nonwoven polyester (e.g., TX759B, TX662) can be selected. Comparatively speaking, woven polyester is the better choice. This type of swab exhibits the following advantages:

- Ultralow nonvolatile residue
- Low particle and fiber generation
- Good sorbency
- No contaminating adhesives
- Excellent chemical resistance

The Texwipe large Alpha Swabs 714A and 761 are widely used and are available in kits with clean sample containers. Cotton-tipped cleaning sticks, gauze sponges, and mop clothes are also used.

4 ANALYTICAL METHODS IN CLEANING VALIDATION

4.1 Analytical Method Selection

The analytical methods used to detect residuals or contaminants should be specific for the substance or class of substances to be assayed (e.g., product residue, detergent residue, and/or endotoxin), such as high-performance liquid chromatography (HPLC), ion-selective electrodes, ion mobility, flame photometry, derivative ultraviolet spectroscopy, enzymatic detection, and titration, or it can involve nonspecific methods that detect the presence of a blend of ingredients, such as total organic carbon, pH, and conductivity. The regulatory agencies prefer specific methods but will accept nonspecific methods with adequate rationales for their use. For investigations of failures, a specific method is usually preferable.

For biological drugs, the use of product-specific assay(s) such as immunoassay(s) to monitor the presence of biological carryover may not be adequate; a negative test may be the result of denaturation of protein epitope(s). Product-specific assay(s) can be used in addition to total organic carbon for the detection of protein residue.

4.2 Analytical Method Validation

Selectivity or specificity is a parameter that needs to be investigated first during method validation to determine if the method is suitable for the analysis or if the swab selected is suitable for use in the cleaning process. Usually, the diluting solvent, mobile-phase swab blank, placebo solution, surface blanks prepared by swabbing, or rinsing should be checked for interference. The sensitivity of the analytical methods [detection limit (DL) and quantitation limit (QL)] should be determined. If levels of contamination or residue are not detected, it does not mean that there is no residual contaminant present after cleaning. It only means that levels of contaminant greater than the sensitivity or detection limit of the analytical method are not present in the sample.

In the recovery study for direct surface sampling, a sampling and detection method on known spiked surfaces at representative levels, typically spiked at 50%, 100%, and 150% of the acceptable limit, should be used. For example, a specific product with its acceptable residue limit of 0.5 μg/4 in^2 (equivalent concentration of 0.1 μg/mL since the method requires 5 mL of diluent to extract the swab) is established based on the calculations methods described in Section 6. During recovery study, 50%, 100%, and 150% of the acceptable limit were spiked on four different surfaces (stainless steel, Teflon, Viton, and Tefsteel). Data in Table 1 illustrates percent recovery at 100% spiked. It clearly demonstrates that the method is accurate for stainless steel and Teflon surfaces. For Tefsteel surface, a correction factor of 1.23 is needed. The method is not suitable for a Viton surface. If it is used for Viton, a correction factor of 2.07 should be applied.

In the recovery study for the rinse samples, the rinseability profiles showing the complete rinsing of the active or individual detergent ingredients should be undertaken if the solubility of the active or any detergent ingredients or the rinseability after drying is in doubt. For a small rinsing surface area (e.g., filling needle), the volume of the rinsing diluent should be the same as the volume used for direct surface sampling or adjusted according to the surface area provided by validation personnel (see Section 5.2). The linearity study should cover the concentration from QL to 150%, with an acceptable regression coefficient (see Table 2). During method validation it is also the right time to establish wipe or rinse sample storage conditions and time limits to get the sample analyzed. In some cases bioburden and endotoxin levels may need to be validated. It is recommended that this process be done separate from the cleaning process so that the cleaning validation can be completed while the lengthier bioburden/endotoxin evaluation is done.

TABLE 1   Percent Recovery at 100% Spiked

Surface/Sample	Amount (μg/mL) Spiked	Recovered	% Recovery
Stainless steel sample 1	0.100	0.103	103.0
sample 2	0.100	0.098	98.0
sample 3	0.100	0.109	109.0
Mean			103.3
% RSD (relative standard deviation)			5.3
Teflon sample 1	0.100	0.098	98.0
sample 2	0.100	0.098	98.0
sample 3	0.100	0.097	97.0
Mean			97.7
% RSD			0.6
Viton sample 1	0.100	0.051	51.0
sample 2	0.100	0.051	51.0
sample 3	0.100	0.043	43.0
Mean			48.3
% RSD			9.6
Tefsteel sample 1	0.100	0.082	82.0
sample 2	0.100	0.083	83.0
sample 3	0.100	0.078	78.0
Mean			81.0
% RSD			3.3

TABLE 2   Linearity study

% Acceptable Limit	Actual Concentration (μg/mL)	Peak Area of API
QL (20)	0.0205	1,686
30	0.0308	2,601
50	0.0513	4,218
80	0.0821	6,840
100	0.1026	8,517
150	0.1540	12,818
Slope		83,261
y-intercept		-11.101
r^2		1.000

5   SAMPLING TECHNIQUES

Two general types of sampling are considered to be acceptable: direct surface sampling (swab method) and indirect sampling (use of rinse solutions). A combination of the two methods is generally the most desirable, particularly in circumstances where accessibility of equipment parts can mitigate against direct surface sampling.

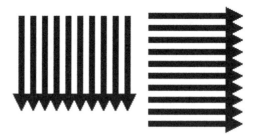

FIGURE 1 Swabbing techniques.

5.1 Direct Surface Sampling

Areas hardest to clean and which are reasonably accessible can be evaluated by direct sampling, leading to establishing a level of contamination or residue per given surface area. Additionally, residues that are "dried out" or are insoluble can be sampled by physical removal. The suitability of the material to be used for sampling and of the sampling medium should be determined. The ability to recover a sample accurately may be affected by the choice of sampling material. It is important to assure that the sampling medium and solvent (used for extraction from the medium) are satisfactory and can be used readily.

The swab method involves using a moistened wipe or swab that is typically wiped over a defined area in a systematic multipass way, going from clean to dirty areas to avoid recontamination. Equipment should be allowed to dry after the cleaning procedure before swabbing begins. Generally, on flat regular surfaces, swabbing is performed on a 4-in^2 area (e.g. 2 in. by 2 in.). On irregular surfaces (e.g., filling needle, propeller shafts) where a 2-in. by 2-in. square area may not be available, swabbing should be done in such a manner as to meet the required surface area. On flat surfaces (Figure 1), swabbing begins at the top left corner of the square and is continued by wiping horizontally from left to right 10 times. Then the swab is turned over and wiping continues at the top left corner vertically from top to bottom 10 times. After each swabbing is completed, the swab is replaced in diluent to dislodge residual material. A blank swab preparation is required. For TOC analysis, very clean low background swabs or wipes and sample vials should be used.

5.2 Rinse Samples

Rinse samples allow sampling of a large surface area and of inaccessible systems or systems that cannot be disassembled routinely. However, consideration should be given to the fact that the residue or contaminant may be insoluble or may be physically occluded in the equipment. The residue or contaminant in the relevant solvent should be measured directly when rinse samples are used to validate the cleaning process: for example, in cleaning validation of liquid production equipment, for an item such as a narrow gauge or a filling needle, where the product-contact surfaces are not accessible with a swab and the item can be readily disassembled. Validation personnel usually remove the equipment and forward to a test laboratory to collect a rinse sample. Validation personnel should advise the lab of the surface area of equipment that is subjected to the rinse sample procedure. Based on the surface area provided, the test laboratory will rinse a known volume of diluent through the equipment, collecting the rinse solution as prepared for a swab sample according to the test method. The number of times to pass the diluent through the equipment is determined based on the equipment size and amount of diluent. The rinse samples will be analyzed for the active material by a selective test method, normally an HPLC method. The values will be utilized in the calculation of residual active left on the equipment surface after cleaning.

Indirect testing such as conductivity testing is sometimes of value for routine monitoring once a cleaning process has been validated. This could be applicable to reactors or centrifuges and piping between large pieces of equipment that can be sampled only using a rinse solution. If the placebo method is used to validate the cleaning process, it should be used in conjunction with rinse and/or swab samples. It is difficult to provide assurance that the contaminate will be dispersed uniformly throughout the system or that it would be worn off the equipment surface uniformly. Additionally, if the contaminant or residue is of large enough particle size, it may not be dispersed uniformly in the placebo. Finally, the analytical power of the direct measurement of the active may be greatly reduced by dilution of the contaminant. Therefore, it is important to use visual inspection in addition to analytical methodology to ensure that the process is acceptable.

Water for injection should be used in the last rinse for product-contact equipment to be utilized in the fabrication of sterile products. Purified water is considered acceptable for product-contact equipment used in the fabrication of nonsterile products. Because of the presence of varying levels of organic and inorganic residues as well as of chlorine, tap water should never be used in the last rinse of any cleaning procedure for product-contact equipment.

6 ACCEPTANCE CRITERIA OF LIMITS

In the past, carryover of residues from one product to the next in a pharmaceutical manufacturing environment was virtually undetectable. In addition, drug products were much less potent, and as a result, whatever carryover did occur was of little practical consequence. In today's environment, however, the situation is very different. Detection of minute quantities of residual product on manufacturing equipment is possible even after intense cleaning procedures have been undertaken. At the same time, as many more potent products enter the marketplace, there is a need to determine acceptable levels of product carryover. The U.S. Food and Drug Administration (FDA) has stated that acceptance limits must be established and that there must be a scientific rationale for the choice of these limits. A number of factors must be considered in determining acceptable limits, such as batch sizes produced, dosing, toxicology, and equipment size.

6.1 Visual Examination

The acceptance limits for visual examination require that no visible residue be observed on an equipment surface. It does not seem appropriate that residues could be visible on GMP equipment that is labeled "Clean." As a result, the equipment must be cleaned until no residues are visible.

According to the observations by Jenkins and Vanderwielen [7], drugs, cleaning agent residue, and various buffer salts on a flat stainless steel surface can be observed visually down to 1.0 $\mu g/cm^2$ with the aid of a light source. Earlier studies by Fourman and Mullen [6] reported drug residues observed down to 4.0 $\mu g/cm^2$.

6.2 Residue of Active Drug and Cleaning Agents

The fabricator's rationale for selecting limits for product residues should be logical and based on the materials involved and their therapeutic dose. The limits should be practical, achievable, and verifiable. In establishing product residual limits, it may not be adequate to focus only on the main reactant since by-products and chemical variations (active decomposition material) may be more difficult to remove.

The approach for setting limits can be:

1. Product-specific cleaning validation for all products
2. Grouping into product families and choosing a worst-case product
3. Grouping into groups of risk (e.g., very soluble products, similar potency, highly toxic products, difficult to detect)
4. Setting limits on not allowing more then a certain fraction of carryover
5. Different safety factors for different dosage forms

Carryover of residues of active drug and cleaning agents should meet defined criteria: for example, the most stringent of the following criteria:

- No more than 0.1% (1/100) of the minimum dose of any product will be present in a maximum daily dose of subsequent product. This method assumes that carryover product is considered inactive at 10% of minimum dose. A further 1/100 is applied as a safety factor. Therefore, the 0.1% method is applicable only for residue of active drug and not for the residue of cleaning agents or chemicals.
- No more than 10 ppm of any product is to appear in another product.

- LD_{50} (oral rat, mg/kg) calculation. This refers to a maximum amount of product 1 in the maximum daily dose of subsequent product 2 based on the LD_{50} per kilogram of product 1 multiplied by 10 kg (the average body weight of a child) and divided by a safety factor of 2000.
- No residue is to be visible on equipment after a cleaning procedure has been performed. Spiking studies should determine the concentration at which most active ingredients are visible.
- For certain allergenic ingredients, penicillins, cephalosporins, or potent steroids and cytotoxics, the limits should be below the limit of detection by the best available analytical methods. In practice this may mean that dedicated plants are used for these products.

Some limits that have been mentioned in the literature or in presentations by industry representatives include analytical detection levels such as 10 ppm, biological activity levels such as 1/1000 of the normal therapeutic dose, and organoleptic levels such as no visible residue. The U.S. Environmental Protection Agency and toxicologists suggest that an acceptable level of a toxic material may be that which is no more than 1/1000 of a toxic dose, or 1/100 to 1/1000 of an amount not known to show any harmful biological effect in the most sensitive animal system known (e.g., no effect).

6.3　Calculation of the Acceptance Limits

Generally, the following information is needed for calculation of the acceptance limits for the carryover residues of active drugs and cleaning agents or chemicals:

1. The equipment surface area for product-contact surfaces
2. The smallest therapeutic dose of the product under consideration
3. LD_{50} for active drug and cleaning agents or chemicals
4. The smallest production-scale batch size of potential subsequent products
5. Production equipment for potential subsequent products
6. Surface area for product contact surfaces of equipment shared by both the product under consideration and potential subsequent products
7. Maximum daily dose of potential subsequent products
8. Minimum dose of subsequent product

Calculation of carryover residue based on the 10-ppm method:

$$10\,\text{ppm (µg/swab)} = \frac{10(\text{mg}) \times \text{SNB(kg)} \times S(\text{in}^2) \times 1000(\text{µg})}{\text{kg} \times A(\text{in}^2) \times \text{swab} \times \text{mg}} \tag{1}$$

Calculation of carryover residue based on the 0.1% method:

$$0.1\%\text{limit (µg/swab)} = \frac{0.1\%(\text{mg}) \times \text{SNB(units)} \times S(\text{in}^2) \times 1000(\text{µg})}{M(\text{units}) \times A(\text{in}^2) \times \text{swabs} \times (\text{mg})} \tag{2}$$

Calculation of carryover residue based on the LD_{50} method:

$$LD_{50} \text{ limit } (\mu g/Swab) = \frac{LD_{50}(mg/kg) \times C(kg) \times SNB(units) \times S(in^2) \times 1000(\mu g)}{M(units) \times F \times A(in^2) \times swabs \times (mg)}$$

(3)

where

$\quad$ SNB $=$ smallest next batch size of the subsequent product (kg)

$\qquad A =$ shared equipment surface area (in^2)

$\qquad S =$ swab area $(4in^2/swab)$

$\qquad M =$ maximum daily dosage (units) of subsequent product

$\quad LD_{50} =$ toxicity value of product for consideration (mg/kg)

$\qquad C =$ 10 kg (average body weight of a child)

$\qquad F =$ safety factor of 2000

$\quad 1000 =$ convert into micrograms

6.4 Microbial Burden

The possibility of microbial growth on clean equipment is limited by the use in the cleaning procedure of detergent and isopropyl alcohol, which have bacteriocidal properties. The equipment is considered clean if the total microbial count is less than 10 CFU per $inch^2$ of equipment surface, and there are no USP objectionable organisms present (e.g., *E. coli*, *Salmonella*, *Pseudomonas aeurginosa*, *Staphylococcus aureus*). If any growth is observed, appropriate tests to identify the organisms are to be conducted according to standard operating procedures. If the acceptance limit is exceeded or an objectionable microorganism is identified, an investigation must be performed to identify the source of microbial contamination. Contaminated equipment must be recleaned and reswabbed, and it must meet acceptance limits prior to release for further manufacturing.

7 CAMPAIGN CLEANING VALIDATION

Finally, when product demand does not allow for a complete cleanup after each batch, a campaign cleaning validation must be conducted. A back-to-back cleaning is performed between batches in a campaign, and a complete cleaning at the end of the campaign. The equipment is sampled after the last batch in a campaign, and all test results must meet predetermined acceptance criteria according to campaign cleaning validation protocol. Samples for a residual active may be taken after back-to-back cleanings where possible, for information only. Microsamples may be taken after back-to-back cleanings in a campaign to ensure that there is no excessive microbial growth on the equipment. If an extension of a campaign is necessary, data gathered previously will be summarized and a new protocol for extending the campaign length will be generated. Upon completion of the number of batches in the campaign, cleaning validation of product-contact surfaces is to be conducted. Three successful cleaning verifications after campaign completion will validate the cleaning procedures for the manufacturing campaign. A matrix approach may be used for campaign cleaning validation. A successful campaign cleaning validation has been completed for campaigns of at least two-dozen batches. The need for campaign cleaning validation for new products will be evaluated based on toxicity category and active ingredient solubility.

8 COMMON PROBLEMS AND SOLUTIONS

8.1 Residues Other Than APIs

When HPLC analysis of swab samples from stainless steel surfaces indicates the presence of residues other than APIs, an investigation may be conducted. This investigation may include, but is not limited to, comparison of a sample HPLC chromatogram to that of a placebo preparation, degraded API, and cleaning agents (gloves and swab material). If the investigation is deemed not necessary, or if the source of the extraneous peaks cannot be identified, the sum of all extraneous peak areas will be calculated and quantitated against the reference standard peak for the active compound. This quantity will be used to calculate an estimated carryover (ECO) value using the same calculations as those used for API peaks. The ECO value obtained for the extraneous peaks will be compared to the acceptance limit for extraneous peaks, which is NMT 10 ppm of unknown substances in the next product batch. If the ECO value for the extraneous peaks exceeds the acceptance limit of NMT 10 ppm, recleaning and reswabbing will be done at all sample locations. The equipment is considered clean if the ECO value for the API residue is less than 10 ppm, and the ECO value derived from the extraneous peaks (if detected) is NMT 10 ppm.

8.2 Nonuniform Contamination

Sometimes contamination from a prior lot is not distributed uniformly in the subsequent product lot. For example, residue in a filling line is distributed primarily to the first few vials to be filled in the next product lot to be manufactured. When risk analysis of equipment cleaning is performed, nonuniform contamination of processing equipment is identified as the greatest risk to the cleaning process. When considering equipment in the process train, equipment used after the final mixing or blending of the product is most likely to have some potential for nonuniform contamination. For example, consider product A and product B, both aqueous liquid solution products containing a different API and manufactured using the same equipment:

1. Product A is mixed in the mixing tank.
2. Product A is filled in bottles using liquid-filling equipment.
3. The mixing tank and filling equipment are cleaned after product A manufacturing is completed.

4. Product B is mixed in the mixing tank.
5. Product B is filled into bottles using liquid-filling equipment.

There are two possible sources of contamination: the mixing tank and the filling equipment (with associated transfer lines). When product B is mixed in the mixing tank, any product A residue in the tank is distributed uniformly in product B. The acceptable level of product A residue in product B is calculated according to the method of Fourman and Mullen. When product B is filled using the filling equipment, any product A residue in the filling equipment will be dissolved in product B solution when product B contacts the equipment. The first bottles of product B will be contaminated with product A residue. After some volume of product B passes through the filler, all product A residue will have been dissolved. Product B bottles will be nonuniformly contaminated with product A residue. This situation is possible for tablet compressing after blending, capsule filling after blending, tablet coating after compressing, or any unit operation that does not contact the entire subsequent lot.

The risk of nonuniform contamination can be eliminated by calculating the residue level in the high-risk equipment and then determining how many product units would be subject to nonuniform contamination. For example, if the filling equipment above, including connecting piping, had a total surface area of $10,000$ cm^2, and product A contamination in the equipment was determined to be 1 µg/cm^2, the total contaminant on the equipment $= 10,000$ µg. This residue could then be transferred to the first bottle of product B to be filled. If all product A residue was transferred to the first bottle of product B, this single bottle would contain $10,000$ µg of product A.

The number of product B bottles that should be discarded can be calculated. The acceptable contamination level of product A in product B should be known (used for the Mullen equation). The number of bottles potentially exceeding the maximum allowable contaminant level can then be calculated. For example, if the acceptable residue level of product A in product B is 100 µg in each bottle, the first 100 product B bottles could potentially have product A contamination exceeding 100 µg. The first 100 bottles (plus a safety factor) should then be discarded since these bottles may potentially contain unacceptable high levels of product A contamination. The bottles used in filling equipment setup are part of the bottles potentially contaminated since they may contain initial product A residue. Bottles to be discarded are often not a problem because the first bottles filled are discarded as part of equipment setup.

The same calculation method can be applied to initial tablets compressed or capsules filled following blending to calculate the number of dosage units to be discarded. If an excessive number of dosage units must be discarded because of potential contamination (as might be possible if product A were a highly potent drug), a study could be conducted to demonstrate that sufficient contamination was eliminated after a fewer number of dosage units. For example, testing every tenth bottle of the first 100 bottles of product B might indicate acceptable product A (below the target limit) residue after discarding the first 20 bottles of product B.

It is recommended that the number of units discarded in normal equipment setup procedures exceed the number of units calculated with high contamination and that this calculation be performed and documented to demonstrate that consideration for potential nonuniform contamination has been addressed [8].

REFERENCES

1. *Guide to Inspections of Validation of Cleaning Processes*. FDA Guidance Document. FDA, Washington, DC, 1993.
2. *Guide to Inspection of Bulk Pharmaceutical Chemicals*. FDA Guidance Document. FDA, Washington, DC, 1991.
3. *Biotechnology Inspection Guide*. FDA Guidance Document. FDA, Washington, DC, 1991.
4. *Cleaning Validation Guidelines*. Guidance Document. Health Products and Food Branch Inspectorate, Health Canada, Ottawa, Ontario, Canada, 2002.
5. *Recommendations on Validation Master Plan, Installation and Operational Qualification, Non-sterile Process Validation and Cleaning Validation*. Draft Document. Pharmaceutical Inspection Convention, 1998.
6. G. L. Fourman and M. V. Mullen. Determining cleaning validation acceptance limits for pharmaceutical manufacturing operations ? *Pharm. Technol.*, 17(4):54–60, 1993.
7. K. M. Jenkins and A. J. Vanderwielen. Cleaning validation: an overall perspective. *Pharm. Technol.*, 18(4):60–73, 1994.
8. www.ivtconferences.com/forum/forum_posts.asp?TID=148&PN=1.

7

RISK-BASED VALIDATION OF LABORATORY INFORMATION MANAGEMENT SYSTEMS

R. D. McDOWALL

McDowall Consulting

1 INTRODUCTION

The purpose of this chapter is to present the options available for the validation of a laboratory information management system (LIMS) within a regulated good manufacturing practice (GMP) environment. It is important to note that the content of this chapter describes good computing practices if applied outside the pharmaceutical industry. The only difference for the pharmaceutical industry is the need for quality assurance to approve some of the key documents written during a project.

First, we define a LIMS and introduce and discuss briefly the laboratory and organizational environment under which a system will operate. This is important, as the entity that will be validated is not a single LIMS application but will include the other applications to which it will interface, as well as the analytical instruments and systems connected to the LIMS. Second, we look at the life cycles and software that constitute a LIMS, followed by the roles and responsibilities of the personnel involved in a validation project. Next we discuss the processes that a LIMS could automate and how to use the LIMS to make them more efficient and effective. We then present the life-cycle stages and discuss the documented

Practical Approaches to Method Validation and Essential Instrument Qualification,
Edited by Chung Chow Chan, Herman Lam, and Xue Ming Zhang
Copyright © 2010 John Wiley & Sons, Inc.

evidence from concept to roll-out of the system, and finally, describe the measures necessary to maintain the validated status of an operational LIMS.

It is assumed that a commercial system will be implemented and validated. Depending on the specific LIMS being implemented, there will be additional configuration of the software and/or writing of custom software using either a recognized commercial language or an internal scripting language. It is no longer justifiable, on the basis of time, cost, and support effort, for an organization to write its own LIMS application.

1.1 4Q's Terminology for Computerized System Validation

The following 4Q's terminology and abbreviations are used in this chapter:

Design qualification (DQ): definition of the intended purpose of the system to be validated. The document written is a user requirements specification (URS) rather than a DQ.

Installation qualification (IQ): installation and integration of the system components (hardware, software, and instruments).

Operational qualification (OQ): testing that the software works as the supplier intended.

Performance qualification (PQ) *or user acceptance testing*: testing the system in the way it is intended to be used and against the requirements in the URS and verification that requirements in the URS have been fulfilled.

Note that this terminology is the same as that of *U.S. Pharmacopeia* (USP) General Chapter ⟨1058⟩ on analytical instrument qualification but has different meanings. The terminology used in this chapter relates solely to computerized system validation.

1.2 LIMS Do Not Have On Buttons

Mahaffey [1] stated in his book that LIMS do not have ON buttons. This is a vital concept, as true today as it was when it was written nearly 20 years ago. This concept must be understood by management and project team members involved in the implementation or operation of any LIMS. There is substantial software configuration and/or customization involved in getting the system to match the current or planned laboratory working practices as well as populating the database with specifications and migrating data from legacy systems if required. This takes time, which management is not likely to appreciate, but the work still needs to be done and to be included in the project plan for the system. The size of the data population may mean that the implementation is phased over time.

1.3 Project Time Scales and Phased Implementation

As a consequence of having no ON button, the project time scales for a LIMS implementation can be rather lengthy and may be implemented over a number of phases. It is important that management understand this and that there can be no quick roll-out unless quality is compromized or project resources increased accordingly. The first phase of the project should deliver a functioning system for at least part of the laboratory (e.g., raw materials, finished products) and also have the major instruments interfaced to the system.

Typically, the phases of the project will take the following times:

- Selection may take between six and nine months. This could be reduced to zero if the laboratory has to implement a system that has been selected by the organization.
- After key personnel training, there will be six to 12 months before the first phase of the system is implemented, validated, and rolled out.

So a typical project could run between 12 and 24 months for the first phase of work and therefore will need to be resourced adequately in terms of money, time, and personnel. To maintain project momentum it is better to have shorter rather than longer time scales. Further phases of the LIMS project will follow, extending the time line, depending on the complexity of the work to be performed.

1.4 Project Risk

Risk management is a key element in any LIMS project. However, it is important to understand that there are two types of risk to be managed. The first is regulatory risk associated with compliance with applicable external regulations and corporate policies, which are covered in this chapter. However, space does not permit a detailed discussion of the second, which is risk associated with the project and its business risk. As over 50% of LIMS implementations fail to meet initial expectations, the reader is referred to a paper by McDowall [2], which contains risk tables where various business risk scenarios for laboratory automation projects are presented and discussed. This will help project teams identify potential risks and develop plans to mitigate them before they occur.

1.5 References for LIMS Validation

There are a number of references that can be used for help when considering the validation of a LIMS. There is *US. Pharmacopeia* General Chapter ⟨1058⟩ on analytical instrument qualification [3], which references the U.S. Food and Drug Administration (FDA) guidance on the general principles of software validation [4]. Using the FDA guidance is a flawed approach for a LIMS, as it is written in the context of medical devices that are not configured or customized by the users. Therefore, omitting this phase of work from the validation opens the laboratory to unacceptable business and regulatory risks.

The GAMP Forum has published a good practice guide on the validation of laboratory computerized systems [5]. However, this suffers a number of drawbacks in terms of overly complex risk management and its simpler implementation life cycle for a LIMS [6]. In the author's view, it is better to adapt the various life-cycle models presented in GAMP 5 [7], which are discussed later in this chapter, or adapt the validation approach for chromatography data systems [8] for the additional work required for a LIMS.

2 LIMS AND THE LIMS ENVIRONMENT

Before we begin a discourse on the validation of a LIMS, it is important to understand two terms: *LIMS* and *LIMS environment*. The first refers to the LIMS application that is purchased from a commercial supplier, which is then implemented in the laboratory. The term *LIMS* implies that only the application is to be validated, but this will not be the case in most LIMS implementations.

2.1 LIMS Defined

A LIMS is a computer application designed for the analytical laboratory that is designed to administer samples, acquire and manipulate data, and report results via a database [9]. It automates the process of sampling, analysis, and reporting and in its simplest form is shown in Figure 1. Here the samples are generated outside the laboratory and submitted for analysis. The laboratory analyzes these samples, generates data that are interpreted, and produces information in the shape of a report to those who will use the information to make decisions. It is therefore important to realize that a LIMS should affect both the laboratory where it is implemented and the organization that the laboratory serves. To be effective a system should deliver benefits to both the laboratory and the organization. Thus, a LIMS has two targets: the laboratory (the information generator) and the organization (the sample provider and the information user).

Figure 1 shows a LIMS sited at the interface between a laboratory and an organization. Samples are generated in the organization and received in the LIMS followed by laboratory analysis. The data produced during analysis are reduced within the LIMS environment to information that is transmitted back into the organization. This represents the ideal placement of the LIMS: both the organization and the laboratory benefit via the system. The line dividing the organization and the laboratory shows that the LIMS is of equal benefit to both. There are other versions of this diagram that can be implemented that provide virtually little benefit to the laboratory and the organization that are discussed in more detail in the literature [10]. Therefore, to hit these two targets, integration of the LIMS with other applications both inside and outside the laboratory is the key to success. Hence, we need to consider the term the *LIMS environment* as encompassing the information technology (IT) environment inside and outside the laboratory.

2.2 LIMS Environment

In reality, a LIMS is more complex than just a single application, and hence the term *LIMS environment* is preferred to describe at least two of the following elements:

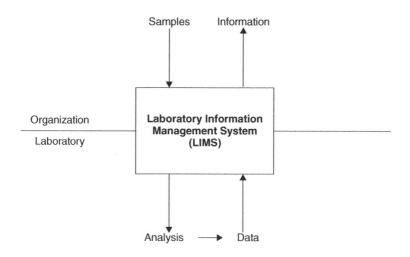

FIGURE 1   A LIMS has two targets.

- LIMS application
- Analytical instruments interfaced directly with the LIMS
- Laboratory data systems and computer systems interfaced with the LIMS (i.e., chromatography data systems, scientific data management systems, electronic laboratory notebooks, etc.)
- Applications outside the laboratory that are also interfaced with the LIMS (e.g., enterprise resource planning systems)

This is the full scope of the computerized system that could be validated within a LIMS project and is shown in Figure 2.

Designing the LIMS environment means that you need to consider all the other systems in the laboratory that must interface with the LIMS. This includes other applications, such as scientific data management systems, a chromatography data system (CDS), and electronic lab notebooks, as well as various data systems that may be attached to those or run independently. It also includes analytical instruments, chromatographs, and laboratory observations, shown in the lower half of Figure 2. Data can be transferred to the LIMS by a variety of means:

- Direct data capture from an instrument connected directly to the LIMS
- Data capture from an instrument with analysis and interpretation by the attached data system, with only the result transferred to the LIMS
- As above, but with the results or electronic records transferred to the LIMS via a scientific data management system
- Laboratory observations written into a notebook, then entered manually into the LIMS or captured electronically via an electronic laboratory notebook and transferred electronically to the LIMS

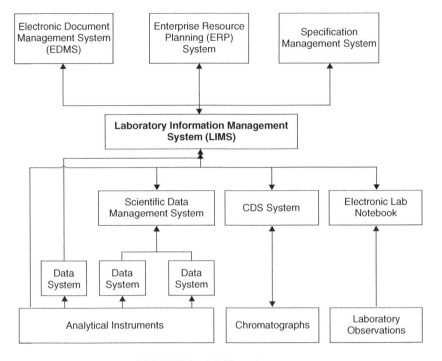

FIGURE 2   LIMS environment.

Once the laboratory side of the LIMS environment has been designed, the LIMS needs to be integrated into the organization. Some of the systems used to construct the LIMS environment and interface with the system are:

- E-mail for transmission of reports to customers or for keeping them aware of progress with their analysis
- LIMS Web servers for customers to view approved results
- Applications maintaining product specifications
- Enterprise resource planning systems for linking the laboratory with production planning and batch release
- Electronic document management systems
- Electronic batch record systems

These are just a few of the possible applications with which a LIMS could be interfaced; the list of potential candidates will be based on the nature of the laboratory and the organization it serves.

2.3   LIMS Application

We also need to consider the LIMS application itself in more detail and what functions the software can undertake on its own. Here the discussion is general, as individual commercial LIMS applications will differ in their scope and functional offerings. In addition, there may be functional overlap between the LIMS and applications that could be interfaced with it; either the LIMS or another application could automate a specific portion of a process. This overlap needs to be resolved with an overall strategic plan for the LIMS environment to determine if the LIMS or another application will undertake a specific function. In more detail, the functions that could typically be automated within a LIMS are:

- Specification management
- Scheduling analytical work
- Sample management, including sample labeling
- Analysis management: definition of methods and procedures
- Instrument interfacing and communication
- Results calculation, management, and reporting
- Stability studies management and reporting, including calculation of storage times
- Environmental sample planning, analysis, and reporting
- Instrument calibration and maintenance records
- Trending results vs. specifications
- Laboratory out-of-specification investigations

Figure 3 shows a high-level process flow of a typical LIMS that starts with the development and validation of an analytical method followed by sample analysis. Highlighted in dashed lines around the "develop and validate method" and "analyze samples" boxes is the implied need to interface with instrument data systems such as a CDS and analytical balances. However the, options implemented for a specific LIMS installation will depend on the functions of the laboratory being automated and the other applications, operational or planned, to be installed in the LIMS environment.

2.4 LIMS Matrix

In addition to process mapping, another tool to help plan the implementation of a system is the LIMS matrix [10], which is useful in documenting and visualizing the high-level needs of an LIMS environment. This is a three six matrix; nine of the cells represent the laboratory and the other nine the organization. This is a way of depicting the LIMS application together with the LIMS environment as each application, either planned or existing, is mapped onto the matrix. The matrix is most useful in getting senior management and the LIMS project team to agree on the overall scope of the system.

3 UNDERSTANDING AND SIMPLIFYING LABORATORY PROCESSES

A prerequisite before implementing a LIMS, or indeed any major computerized system, is to map and optimize the laboratory processes that the LIMS will

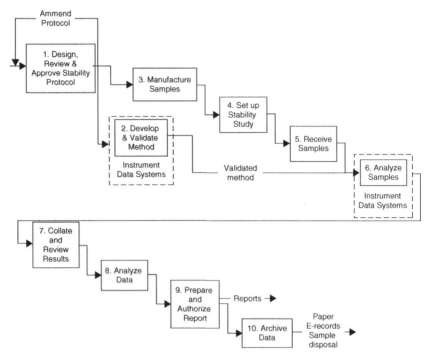

FIGURE 3 LIMS workflow for method development and sample analysis.

automate. The laboratory needs to understand the process and to identify any bottlenecks in the process and the reasons for the occurrence of each. This is especially important when moving from a paper base to an electronic environment. The reason for this is that most laboratory processes have evolved over time rather than having been designed; moreover, the processes are paper based rather than electronic. (A publication of McDowall's [10] is a good source for further information about the LIMS matrix.) The goal of any LIMS implementation is a simplified and streamlined electronic process rather than automation of an inefficient, paper-based status quo.

3.1 The Way We Are

The first stage in considering a paperless laboratory is to look at the basic processes and computerized systems: How do they operate currently, and how are they integrated? A laboratory may have many computerized systems, such as a chromatography data system and data systems associated with the main analytical techniques, such as MS, mass spectroscopy, ultraviolet, and near infrared. As such, the laboratory can appear on the surface to be very effective, but in practice these are islands of automation in an ocean of paper. The main way that data are transferred from system to system is via manual input using paper as the transport medium, a slow and inefficient process. Furthermore, the process will have evolved over time and may have additional tasks that do not add value to the laboratory output, and it becomes very slow and inefficient.

The diagnostic process is to map your current process and then redesign and optimize your laboratory process to use IT systems, including LIMS, effectively and efficiently to ensure that they deliver business benefit and regulatory compliance. Therefore, the process maps for the current working practices understand what you do and why you do it. In many instances it will be due to one or more of the following:

- Custom and practice (we have always worked this way)
- Evolution over time (we have had new projects or new tasks to do)
- Extensive quality control checks (the FDA did not like our previous way of working)

For example, Figure 4 shows a process map for sample receipt in a laboratory. There are two process flows for samples: the first for those generated internally and the second for those generated externally. Originally, the two process flows were the same, but when a shipment of samples was lost after delivery to the site, the manager of the laboratory instigated the second process flow to prevent this from happening again. However, also shown in Figure 4 is an undocumented process developed by one person to streamline the work. Although simple and easy to perform, it is, nevertheless, noncompliant, as no standard operating procedure has been written for it. The problem when implementing a LIMS is that to incorporate all three process flows into the system brings three times the effort in specifying, implementing, and validating. In all probability the resulting workflows will be less efficient than those of the paper system that replaces it: hence, the need to map and redesign, standardize, harmonize, or optimize the processes in the laboratory to ensure simpler implementation.

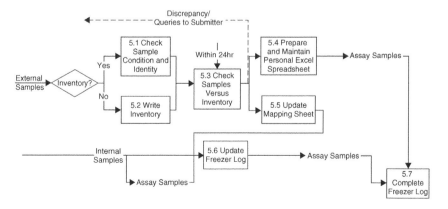

FIGURE 4   Multiple current processes for sample management.

3.2 Operating Principles of an Electronic Laboratory

There are three basic operating principles of an electronic laboratory that should be used to redesign or optimize the laboratory processes [4]:

1. *Capture data at the point of origin.* If you are going to work electronically, data must be electronic from first principles. However, there is a wide range of data types, including observational data (e.g., odor, color, size), instrument data (e.g., pH, liquid chromatography, ultra-violet), and computer data (e.g., manipulation or calculation of previous data). The principle of interfacing must be balanced with the business reality of cost-effective interfacing: What are the data volumes and numbers of samples coupled with the frequency of instrument use?

2. *Eliminate transcription error checks.* The principles for design are as follows: Never retype data, and design simple electronic workflows to transfer data and information seamlessly between systems. This requires automatic checks to ensure that data are transferred and manipulated correctly. Where appropriate, implement security and audit trails for data integrity and only have networked systems for effective data and information sharing.

3. *Know where the data will go.* Design data locations before implementing any part of the LIMS and the LIMS environment. The fundamental information required is: What volumes of data will be generated by the instrumentation, and where will the data be stored: in an archive system, with the individual data systems, or on a networked drive? The corollary is that security of the data and backup are of paramount importance in this electronic environment. In addition, file-naming conventions are essential to ensure that all data are numbered uniquely, either manually or automatically. If required, all archiving and restoring processes must be designed and tested so that they are reliable and robust.

The key message when designing electronic workflows is to ensure that once data are acquired, they are not printed out or transcribed again but are transferred electronically between systems using validated routines.

3.3 The Way We Want to Be

The main aim is to understand where there are bottlenecks and issues in the process. Analyze and find the basic causes of these bottlenecks, as they will help you to challenge and improve the process. When the current process is redesigned and optimized, the aim must be to have, as far as it practicable, electronic ways of working and effective and efficient hand-offs and transfers between applications and organizational units. This will enable a laboratory to get the process right. Figure 5 shows an improved sample management process that will be implemented by a LIMS. Just by visual comparison of the current and new processes, you can see that the new process is simpler and easier

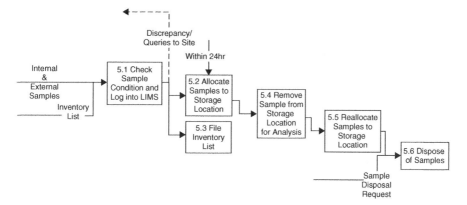

FIGURE 5 Optimized sample management process prior to LIMS implementation.

to understand. Here, the existing formal and informal process flows have been merged into a single, simplified process, and sample labels contain bar codes to enable better sample tracking and management. There is no differentiation between the source of the samples—all samples are treated the same. This means that user training and validation work is spent only on a single unified process, saving time and effort which more than pays for the cost of redesigning the process.

Look at your basic laboratory process and design electronic ways of working: See what changes could be made to your way of working to remove inefficient tasks and improve the speed. Knowledge and interpretation of the good laboratory practice (GLP) or good manufacturing practice (GMP) regulations with which the laboratory works is also very important: knowing which records need to be signed and when. However, trying to work electronically requires that any application used is technically compliant with 21 CFR Part 11.

Furthermore, it is essential that the LIMS be interfaced with the analytical systems in the laboratory to ensure that data are captured electronically. Unfortunately, this is not always the case, as one report has stated that less than 50% of systems are connected to instruments [12]. In such a situation, it is doubtful if any LIMS implementation will be cost-effective, as the resulting manually driven process around the LIMS will have few advantages over the paper system it replaces.

4 GAMP SOFTWARE CATEGORIES AND SYSTEM LIFE CYCLE FOR A LIMS

To define the risk and amount of work that we need to do when validating a LIMS, we need to understand the categories of software present in a LIMS. Once this is determined, the life cycle that is necessary to implement a LIMS can be defined.

4.1 GAMP Software Categories in a LIMS

The good automated manufacturing practice (GAMP) guidelines constitute an industry-written document for the validation of computerized systems used in the pharmaceutical industry (now in its fifth version [7]). In all versions there is a classification of software into one of five categories (further discussion and debate on the GAMP software categories as applied to laboratory computerized systems can be found in a paper by McDowall [13]):

- *Category 1:* infrastructure software
  - Established or commercially available layered software, including operating systems, databases, office applications, etc.
  - Infrastructure software tools, including antivirus tools, network management tools, etc.
- *Category 2:* firmware
  - Discontinued; firmware now treated as category 3, 4, or 5
  - Clash with USP ⟨1058⟩ over the approach to group B laboratory instruments: Validate or qualify?
- *Category 3:* nonconfigured products
  - Off-the-shelf products that cannot be changed to match business processes.
  - Can include products that are configurable, but only if the default configuration is used.
- *Category 4:* configured products
  - Configured products provide standard interfaces and functions that enable configuration of the application to meet user-specific business processes.
  - Configuration using a vendor-supplied scripting language should be handled as custom components (category 5).
- *Category 5:* custom applications
  - These applications are developed to meet the specific needs of the regulated company.
  - Implicitly includes internal application macros, LIMS scripting language customizations, and VBA spreadsheet macros.
  - High inherent risk with this type of software.

A LIMS could therefore contain the following categories of software:

- LIMS application software which is configured (category 4)
- Customization of the product using the internal scripting language (category 5)
- Writing custom code using a recognized computer language to connect the LIMS to another application or instrument (category 5)

At a minimum, a LIMS could consist only of category 4 software (option 1 above), but in a GMP environment it will also contain at least one type of category

5 software, with the scripting language option for customization. This mixed environment affects the life cycle and will lengthen the time for implementation of the system. Therefore, when at all possible, the laboratory should change the business process to match the LIMS to reduce implementation time and validation cost, which was discussed earlier.

4.2   LIMS Life Cycle Model

The combination of category 4 and category 5 softwares in a typical LIMS means that a more complex life cycle is required to accommodate these two categories. This means that the life cycle must control implementation of the configuration of the commercial LIMS software as well as control the writing of the custom elements. The GAMP guide version 5 [7] and the GAMP good practice guide on testing [14] have a number of life-cycle models, including a model of software category 4 integrated with category 5 extensions. It is this model that we will consider for a LIMS life-cycle model. A life-cycle model adapted for a LIMS is presented in Figure 6. The category 4 life cycle follows the stages outlined in bold, and the category 5 software modules are nested within the category 4 life cycle for the extensions to the configured system and are linked by thinner lines. This is the most common approach to a LIMS system life cycle.

Some explanation of the life cycle is required. The category 4 and 5 life cycles are highly integrated and are dependent on one another. The main part of the life cycle is for the category 4 portion of the application and requires the following phases connected by the heavy solid line:

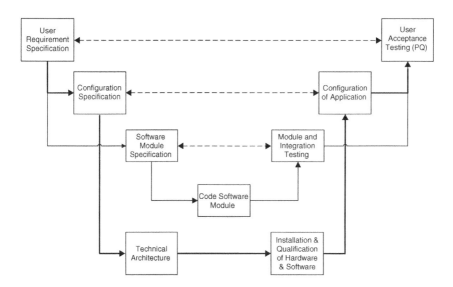

FIGURE 6   LIMS life-cycle model for category 4 and category 5 software. (Adapted from [7].)

- Writing a user requirements specification (URS) to define the overall system and business requirements of the LIMS.
- Configuration of the system is documented in a configuration specification (CS) that details how the application will be configured to meet the requirements.
- A technical specification will define the computing platform that will run the production system and any other environments.
- The hardware platform and infrastructure software, including the operating system, will be installed and qualified (IQ and OQ) in a data center, followed by the installation and qualification of the database and LIMS software (IQ and perhaps the OQ).
- After qualification of the LIMS software, it will be configured as specified in the configuration specification. This process needs to be documented either at this stage or during the PQ.

Category 5 custom elements will have the nested life cycle shown in Figure 6 by the light solid line:

- Customization of the system using the LIMS scripting language needs to be specified in a software module specification.
- Modules of custom code will be written using the scripting language and follow the coding conventions devised by the LIMS vendor.
- Testing of each module will be undertaken against the software module specification to show that it works and will include tests to show that the code is integrated with the main LIMS application.

To ensure that the configured application (category 4) and the custom modules (category 5) work together, the entire system undergoes user acceptance testing, or PQ is carried out by the users to show that the entire system (configured application and custom modules) works as specified in the URS. Note that not all the documents required for a LIMS validation are shown or indicated on the life-cycle model depicted in Figure 6—only the main phases of a project. A functional specification is not shown in the diagram, as this will typically be written by the LIMS supplier and can be used by the laboratory to reduce the amount of testing based on whether the LIMS function is standard, configured, or customized.

5 VALIDATION ROLES AND RESPONSIBILITIES FOR A LIMS PROJECT

Because a multidisciplinary approach is essential for the implementation of a LIMS, a number of people will be involved in a LIMS project, in the range of roles discussed below. Any of the following may be involved, depending on the company, its organizational structure, and the project size.

Senior management. This person or persons will be responsible for the budget and authorization of the project. Summary reports of the progress of the project against the project plan should be generated regularly to ensure continued support.

Laboratory management. A manager determines resource allocation between the project and the normal work. Typically, the laboratory manager will be the system owner and will be responsible for the LIMS and the overall approach taken in the validation.

Laboratory users. Users will constitute the bulk of the project team and typically will split their time between their normal work and the LIMS. Allowance needs to be made by management to allow time on the LIMS project; otherwise, the project will suffer and fail to meet milestones and deadlines.

LIMS project manager. This person is the single point of responsibility for the entire project and is responsible for planning and organizing the work with the resources available. Ideally, this role should be full time [2], to ensure dedication to the delivery of the project on time and on budget. The project team members will be tasked through the project manager. If there is a conflict between the normal role of the project team members and the LIMS, the project manager will have to negotiate with the laboratory manager. It is these conflicts that will result in delays to the LIMS project and possible budget escalation.

Sample providers and information users. These are the customers of the laboratory who generate samples and use the information provided by the laboratory. They must be included in the scope of the project, as otherwise the project will benefit the laboratory but not the organization.

Quality assurance. The role of QA personnel is to ensure compliance with applicable regulations and company procedures. To ensure efficient document production, not every document needs to be authorized by QA before release. At a minimum, the validation plan, summary report, and user requirements specification and any test plans should be authorized by QA, but not all the individual test scripts, as there is limited value of a QA review before execution as opposed to a postexecution review.

IT department. There will be input into the technical architecture and ensuring that corporate IT standards are met by the new system. IT personnel will be involved with the installation and qualification of the computer platform and the operation of the platform of the validated system. There may be a role in allocating user identities and access privileges, backing up the system, and maintaining the platform, including patching and database administration.

LIMS supplier. The supplier provides the LIMS application, supports company audits to see how the system was developed, and supports and provides technical expertise to help configure or customize the system to meet the business requirements. The vendor may offer consultants or contractors to carry out this work, who will work alongside the project team members and users to achieve a configured and customized system. Familiarity with the LIMS application, its configuration possibilities, and the scripting language are key factors in deciding whether to use the supplier in a larger role.

Internal validation group. If involved in the LIMS project, this team will be involved in providing validation advice and may also be writing many of the documents. In many cases the use of a validation group may be an unwritten statement to the users that validation is not their problem—but nothing could be further from the truth. The responsibility for validation, as stated above, belongs with the system owner. This group is a repository of validation expertise that can be used in the validation of any system, but they rely on input from users, supported by management, to achieve their aims and objectives.

Consultants. Consultants can be used either for overall direction and phasing or a LIMS project or for advice about specific topics, such as validation of the system. The laboratory should use consultants to add value to a project and advise the laboratory of better ways of undertaking the work. This group can also be used in place of an internal validation group, or work alongside them. Care should be taken to ensure that corporate standards are followed when using consultants unless there is a good business reason for not doing so.

6 SYSTEM LIFE-CYCLE DETAIL AND DOCUMENTED EVIDENCE

The life cycle and the documented evidence discussed in this section are based on the validation of a number of systems, but need to be understood in the context of an organization's computer validation policies and procedures. Each organization can have different approaches and terminology, so the terminology used here may differ from that used in some organizations. What matters is that you have performed the work described in each section rather than what a specific document is called. The key message is that you can demonstrate that the system was developed under control and has been validated. The main documents needed for validation of a LIMS are presented in Table 1.

6.1 System Risk Assessment

Before starting a LIMS project, a risk assessment should be undertaken at the level of the system itself to determine if any or all of its functions are under GMP regulations. One such methodology has been described [15,16] which is based on the records generated by the system and the GAMP category of software. This process determines if the system needs to be validated, and if so, the approach to validation. There are alternative approaches used by GAMP [7] and Scolofino and Bishop [17], among others.

TABLE 1 Typical Documentation for a LIMS Validation

Document Name	Outline Function in Validation
System risk assessment	• Documents the decision to validate the LIMS or not to validate, and the extent of validation work to be undertaken
Validation plan	• Documents the scope and boundaries of the validation effort • Defines the life-cycle tasks for the system • Defines the documentation for the validation package • Defines the roles and responsibilities of the parties involved
Project plan	• Outlines all tasks in the project • Allocates responsibilities for tasks to persons or functional units • Several versions as progress is updated
User requirements specification (URS)	• Defines the functions that the LIMS will undertake • Defines the scope, boundaries, and interfaces of the system • Defines the scope of tests for system evaluation and qualification
System selection report	• Outlines the systems evaluated on paper or in-house • Summarizes the experience of evaluation testing • Outlines the criteria for selecting the system chosen
Functional risk assessment and traceability matrix	• Prioritizes system requirements: mandatory and desirable • Classifies requirements as either critical or noncritical • Traces testable requirements to specific PQ test scripts
Vendor audit report	• Defines the quality of the software from a supplier's perspective (certificates) • Confirms that quality procedures match practice (audit report) • Confirms the overall quality of a system before purchase
Purchase order	• From a supplier quotation, selects software and peripherals to be ordered • Uses a delivery note to confirm actual delivery against a purchase order • Defines the initial configuration items of the LIMS
Configuration specification	• Defines the configuration of the system policies • Defines user types and access privileges • Defines default entries into the audit trail
Software module specifications	• Specifies a custom module and how it will integrate within the LIMS • Codes and documents the module to predefined standards • Informal developer testing and correction of the module code

(Continued overleaf)

TABLE 1 *(Continued)*

Document Name	Outline Function in Validation
Technical architecture (technical specification)	• Defines IT platform(s) (e.g., terminal servers, database server) together with resilience features • Features operating systems and service packs • Features operating environments: production, validation, etc.
Installation qualification (IQ)	• Installs the components of the system by the IT and the LIMS supplier after approval • Tests individual components • Documents the work carried out
Operational qualification (OQ)	• Tests the system installed • Uses an approved supplier's protocol or test scripts • Documents the work carried out
LIMS application configuration	• Configuration of the LIMS application according to the configuration specification
Data base population	• Controlled input of methods to the LIMS • Controlled input of raw material, intermediates, and in-process control sample and finished product specifications to the LIMS
Module testing and integration of custom software	• Formal testing of the module against the software design specification • Integration testing with the LIMS application
Data migration	• Identifies the data elements and fields to migrate from an old LIMS (e.g., specifications, results, ongoing stability studies) • Plans and executes the work • Confirms successful data migration
User acceptance test (e.g., PQ) test plan	• Defines user testing on the system against the URS functions • Highlights features to test and those not to test • Outlines assumptions, exclusions, and limitations of the approach
PQ test scripts	• Confirms software configuration • Covers key functions defined in the test plan • User scripts to collect evidence and observations as testing is carried out • Documents any changes in the test procedure and if the test passed or failed

TABLE 1 (*Continued*)

Document Name	Outline Function in Validation
User training, SOPs, and system documentation	• Procedures defined for users and system administrators, including definition and validation of custom calculations, input of specifications, account management, and logical security • Procedures written for IT-related functions • The practice must match the procedure
Service-level agreement (SLA)	• Agreement between the laboratory and IT for IT and infrastructure services for the LIMS
User training material	• Makes available initial material used to train superusers and all users • Documents refresher or advanced training • Updates training records accordingly
Validation summary report	• Summarizes the entire life cycle of the LIMS • Discusses any deviations from validation plan and quality issues found • Authorizes management to use the system • Releases the system for operational use (in some organizations this can be a separate release certificate)

6.2 Initial Definition of User, Business, and System Requirements

The key document in the entire LIMS validation is the user requirements specification (URS), as this defines the user acceptance tests and can also influence the validation strategy to be outlined in the validation plan. From the life-cycle model shown in Figure 6, the user and system requirements are linked to the tests carried out in the user acceptance tests or performance qualification. Therefore, it is important to define the requirements for the basic functions of the LIMS, the adequate size, 21 CFR Part 11 [18] requirements, and consistent intended performance in the URS. Remember that the URS provides a laboratory with the predefined specifications to validate the LIMS; without this document the system cannot be validated [4].

The main elements in an URS should include the following major areas; each requirement must be numbered and written individually so that it can be traced to where it is either tested or verified later in the life cycle.

- Overall system requirements, such as number of users, locations where the system will be used, instruments connected to the system, and whether terminal emulation will be used

- Compliance requirements from the predicate rule and 21 CFR Part 11, such as logical security, audit trail, user types and access privileges, requirements for data integrity, time and date stamp requirements, and electronic signature requirements
- LIMS functions defined using the workflow outlined in Figure 3 (ensureing that capacity requirements are defined), such as maximum number of samples to be run, custom calculations, and reports for the initial implementation and roll-out.
- IT support requirements such as database support, backup and recovery, and archiving and restoration.
- Interface requirements (e.g., will the LIMS be a stand-alone system or will it interface with instruments, data systems such as a CDS within the laboratory, or ERP system outside it?)

The completed document will be an initial step for the selection of a system, as it will be refined as the project progresses and after the final system has been purchased. Developing requirements is a continual process, and the final version of the document will reflect the specific LIMS and version number that will be implemented. A URS is a living document [4,19].

6.3 Vendor Audit

The majority of the system development life cycle for a commercial LIMS will be undertaken by a third party: the vendor. The European Union GMP Annex 11 on computerized systems [20] states in Section 5: "The software is a critical component of a computerised system. The user of such software should take all reasonable steps to ensure that it has been produced in accordance with a system of Quality Assurance." This regulation is currently in the process of being revised, and the draft issued in 2008 for industry comment contains further requirements for vendor audits [19]. Continuing: Section 5.1: "The supplier of software should be qualified appropriately; this may include assessment and/or audit"; and Section 5.3: "Quality system and audit information relating to suppliers or developers of software and systems implemented by the manufacturing authorisation holder should be made available to inspectors on request, as supporting material intended to demonstrate the quality of the development processes."

The GAMP guide version 5 [7] recommends that a vendor audit be undertaken for category 4 software to ensure that the system was developed in a high-quality manner, is supported adequately by the vendor, and is a reasonable interpretation of the proposed update of Annex 11. The vendor audit should take place once the product has been selected but before the system has been purchased, in case issues discovered during the audit affect the purchase decision. The purpose of the audit is to see if an adequate quality management system is in place and operated effectively for LIMS development and support. The evaluation and audit process is a very important part of the life cycle, as it ensures that the design, build, and testing stages (which are under the control of the supplier) have been checked to ensure compliance with regulations. The audit should be planned in advance and cover items such as the design and programming phases, product testing and release, and documentation and support; a report of the audit should be produced after the visit, and if the EU GMP update is unchanged, will be available to inspectors [19].

Many LIMS suppliers are certified to the International Standards Organisation (ISO) 9001 [21] or ISO 90003 [22] and offer a certificate that the system conforms to their quality processes. This is adequate for supporting the development phase of the life cycle, but remember that there is no requirement for product quality in ISO 9000, and product warranties do not guarantee that the system is either fit for the purpose or error-free [8]. If the system is critical to GMP operations, it is better to consider a vendor audit. For further reading on vendor audits, readers are referred to the relevant chapter in McDowall [8].

6.4 Selecting and Purchasing the System

Selection and purchase of a new LIMS should be a formal process to see if an application matches the main requirements of the URS. The outline tests can be used to screen and select the system. An in-house test of a system is strongly advised if there are sufficient time and resources, as this will give increased confidence that a system can undertake the laboratory's work. A selection report would be the outcome of this phase of the work and would form part of the supporting evidence for LIMS validation.

The company's internal procurement processes should be followed to write the capital expenditure justification to be circulated for approval. At the same time, the vendor's contract terms and conditions should be reviewed and changed to correct any issues, such as payment terms, before the purchase order is placed. Once the LIMS request is approved, the purchase order can be placed. This provides the first phase of the configuration management of the system, as it defines the components of the system to be delivered.

6.5 Controlling the Work: Validation and Project Plans

Note that writing the validation plan appears relatively late in the life cycle, typically after the decision to purchase the LIMS. This is because there may be issues discovered during selection of the initial system or in training to use the system selected, which may affect the way that the system is implemented or rolled out. Therefore, the controlling document for the validation work is brought in here to avoid the need to issue amendments or a new version of the plan. However, it is important that the LIMS project team remember that the earlier phases of the project should be documented adequately to ensure that the selection process is not documented retrospectively. The name used for this document varies widely from laboratory to laboratory: validation plan, master validation plan or validation master plan, or even quality plan. Regardless of what it is called in an organization, it should cover the work to be done so that the system provides business benefits as well as meeting regulatory compliance.

The validation plan should define:

* The system to be validated (name and version number), including its scope and boundaries
* The roles and responsibilities of the people involved in the project
* The life cycle to be followed and the documented evidence to be produced when this is followed
* How to deal with and document any deviations from the plan

There will typically be a separate project plan with further breakdown of tasks in a project planning application (e.g., MicroSoft Project). This plan should be referred to in the validation plan but does not include the dates and tasks in the validation plan itself. The reason for this is that some plans can be rather optimistic and therefore do not include the time scale in the validation plan, as this gives the project manager the flexibility to update the plan separately without the need to update and reauthorize the validation plan.

6.6 Refining the User Requirements Specification During Implementation

It is important to realize that a URS is a living document that must be updated as system requirements change and evolve; for example, a URS should be written to select a system. Then the document will be reviewed and updated to reflect the LIMS and version selected that is to be validated, and the functions specific to the laboratory and the systems interfaced to it. The reason for this is that the system selected may have more or fewer features than are contained in the original URS, and therefore the document must be updated to reflect the system to be validated. It is also not unknown for the URS to be updated at least once more, especially after piloting requirements and during writing the user acceptance tests (PQ) to reflect further changes in the users' understanding of the system, in the way that the laboratory wants to operate, and to correct requirements that were written incorrectly or not fully understood earlier in the life cycle. This is normal and expected.

6.7 Piloting the Requirements

Although a URS is essential for defining the requirements for system selection, it is often not sufficiently detailed or the best way of working with the system purchased. Therefore, an excellent way of refining and defining requirements further is through a pilot phase of the LIMS. However, this needs to be structured and managed well so that the project and the specification documentation benefit. From a practical perspective, one should have only two pilot phases, so that the piloting does not go on forever. Each phase needs to be defined carefully in terms of the scope of functions to be prototyped, the time to be spent on the work, and how requirements and test documents will be written or updated.

Piloting requirements are usually informal, in that any documents written will not be checked by quality assurance personnel, and this fact should be stated in the validation plan. However, URS documents need to be updated and outline test scripts written at the end of each phase to reflect that the final prototyped functions work as intended. This approach enables the final URS and any configuration specifications to be finalized relatively shortly after the end of the prototyping phase and not delay the overall project.

System prototyping can be conducted on an unqualified or a qualified system. However, the risk of using either approach is that when the prototyping work is completed, management can think that the implementation is completed; this misunderstanding has to be managed accordingly.

6.8 Functional Risk Assessment and Traceability Matrix

Although not a current regulatory requirement, traceability of requirements, from the URS through all subsequent phases, is a regulatory expectation [23,24]. More important, it is a vital business tool in ensuring that all requirements captured in the URS are managed in subsequent phases of the life cycle. To achieve this effectively, it is important that requirements be presented correctly and in a manner that facilitates traceability. It is all very well that the regulations state that users must define their requirements in a URS, but what does this mean in practice? Table 2 illustrates a way in which capacity requirements can be documented; note that each requirement is:

- Numbered uniquely.
- Written so that it can be tested, if required, in the PQ or verified later in the life cycle.
- Prioritized as either mandatory (essential for system functionality) or desirable (nice to have but the system could be used without it). This prioritization can be used in risk analysis of the functions and also for tracing the requirements through the rest of the life cycle.

Each requirement must be written so that it is either testable or verifiable. *Testable* means that based on the requirement as written, a test or tests can be devised to show that the LIMS delivered does or does not meet the requirement. *Verifiable* means that a requirement is met by carrying out an activity (e.g., the installation of a component, writing a procedure, auditing a vendor). To write requirements that are testable or verifiable, URS writers should follow the process defined in IEEE standard 1233 [25]. This states that a well-defined requirement must address capability and condition, and may include a constraint. Remember, as shown in Figure 6, the URS requirements are the input to the tests carried out in the PQ phase of the life cycle. If the requirements are not specified in sufficient detail, they cannot be tested.

The next stage in the process is to carry out a risk assessment of each function to determine if the function is or is not business and/or regulatory risk critical.

TABLE 2 Some Capacity Requirements for a LIMS

Req. No.	Requirement	Priority[a]
3.2.1	LIMS has the capacity to support up to 20 named users using the system at the same time from any laboratory network location.	M
3.2.2	LIMS has the capacity to hold 800 gigabytes of live data on the system.	M
3.2.3	The system can print to any of four network printers.	M
3.2.4	The LIMS will be available for a minimum of 98% per month of downtime for maintenance and backup per month (e.g., 3%).	M
3.2.5	The LIMS response time is ≤30 seconds for a successful login, and this must not be degraded by fluctuations in network traffic.	D

[a]M, mandatory; D, desirable.

TABLE 3 Functional Risk Assessment and Traceability Matrix

Req. No.	Requirement	Priority[a]	Risk[b]	Test Req.
3.2.1	LIMS has the capacity to support up to 20 named users using the system at the same time from any laboratory network location.	M	C	TS05
3.2.2	LIMS has the capacity to hold 800 gagabytes of live data on the system.	M	C	IQ
3.2.3	The system can print to any of four network printers.	M	C	IQ
3.2.4	The LIMS will be available for a minimum of 98% per month of downtime for maintenance and backup per month (e.g., 3%).	M	C	SLA
3.2.5	The LIMS response time is ≤30 seconds for a successful login, and this must not be degraded by fluctuations in network traffic.	D	N	

[a]M, mandatory; D, desirable.
[b]C, critical; N, not critical.

Table 2 for URS now has two additional columns added on the right, as shown in Table 3. This approach allows priority and risk to be assessed together. Only those functions that are classified as both mandatory and critical are tested in the qualification phase of the validation [8,15]. Therefore, in Table 3, requirement 3.2.5 will not be considered for testing, as it does not meet the selection criteria. Of the remaining requirements, some are traced to the installation of system components to meet requirements 3.2.2 and 3.2.3, and the service-level agreement (SLA) to meet requirement 3.2.4. Requirement 3.2.1 will be tested in a capacity test script, which in this example is called test script 05 (TS05). In this way, requirements are prioritized and classified for risk, and the most critical ones can be traced to the PQ test script.

6.9   Specifying the Configuration of the System

The way that the LIMS will be configured and/or customized must be documented, and for a LIMS application a configuration specification is the best way to do this. The typical configuration elements of a LIMS covered in such a document will encompass:

- A definition of user types and their access privileges
- The configuration of any system policies: functions and any settings turned on (e.g., password length, use of electronic signatures, audit trail configuration)
- A definition of context-sensitive default entries for the audit trail, such as the reason for data changes
- Information about the instruments interfaced within the laboratory
- Identification of any instruments and systems for which training records, maintenance, and qualification status will be maintained

There will need to be links between the URS and the configuration specification to aid traceability. For example, a URS requirement could state that five user types would be set up in the LIMS application and name them with a reference to the configuration specification. In the latter document there would be a reference back to the URS; the access privileges would be defined for each user type. Typically, the functions available in a LIMS will vary with each version of the software from a vendor. Therefore, it is important that the configuration specification be linked to the specific LIMS version being validated. The configuration specification is another example of a living document that has to be kept current.

6.10   Writing the Technical Architecture

This document, which can also be called a *technical specification*, is typically written by the IT department, taking into consideration the recommendations of the supplier in terms of server sizing (e.g., minimum processor power, memory and disk sizing) and the organization's corporate standards. A technical architecture will document the servers and their operating systems; for example:

- Database server
- Application server
- Terminal emulation or Web server for the application; the advantage of this, especially for a large system, is that installation of the application on clients is reduced to a single task
- Use of virtual servers for some LIMS instances
- Operating system and service packs installed on each server
- Other applications, tools, and utilities to be installed on the servers (e.g., antivirus tool, backup agent)

Diagrams are a very useful way of illustrating how the components come together to constitute the overall system and help to collate the individual server specifications.

The number of environments or instances for the LIMS will need to be specified in this document. For example, there could be the following instances for the system:

- Sandbox or development
- Training
- Validation
- Production or operational

At least two, if not three, instances are required for a LIMS. The validation and production instances should be mirror images so that most of the testing can be conducted in the validation rather than the operational instance. If the hardware platforms of these two are identical, the validation instance could be a disaster recovery for the operational instance. User training should be conducted in an instance that is identical to the operational system.

6.11 Installing and Qualifying the System Components

The installation of the LIMS components will be undertaken in a number of layers, starting with the hardware platform and proceeding through the operating system, utilities and tools, and database, and finishing up with the LIMS application software. The work carried out in this phase of the project will be based on the technical architecture that will detail the nature and number of instances that will be established. First, each server will be installed, qualified, and documented by the IT department and there may be an option to install the database that the LIMS will use if the organization has the appropriate license. Then the LIMS vendor will install the database (if not done by the IT personnel) and the LIMS application software. The LIMS vendor should supply the IQ and OQ documentation before execution so that the laboratory staff can review and approve the document before execution.

During the IQ the initial configuration baseline should be established by taking inventory of the entire system, including hardware, software, and documentation. For an LIMS, the IQ should cover:

- Server (for data storage) installation by the IT department, server supplier, or manufacturer
- Interfacing of any instruments to the system (either as part of the initial LIMS IQ or separate from it)
- Installation of bar code readers and printers
- Processing or data review workstations by either the IT department or contractors working on their behalf (typically, with an operating system configured to corporate requirements)

- Connection of the servers and any new workstations to the corporate network
- Installation of the LIMS application software for data processing on the workstations

The IQ work around the LIMS is typically supported by the vendor, system administrators from the laboratory, and IT department, depending on the complexity of the configuration of the system and the LIMS environment. Planning is essential, as retrospective documentation of any phase of this work is far more costly and time consuming. During this phase of the project, analytical instruments and instrument data systems will also be connected and interfaced to the LIMS. There needs to be appropriate qualification documentation to cover the instruments and data systems interfaced. Care needs to be taken to include any configuration of the LIMS or middleware used to communicate between the LIMS and instrument/data system especially, if data are extracted from an instrument data system.

6.12 Determining the Need for an OQ for the LIMS Application Installed

The operational qualification (OQ) is carried out shortly after the IQ and is intended to demonstrate that the application works the way the vendor says it should. Note that the OQ will be carried out on the unconfigured and noncustomized LIMS application that has just been installed. Most LIMS suppliers will supply OQ scripts and, if required, the staff to execute the scripts. The depth and coverage of these OQ packages vary enormously, and the problem, from the laboratory perspective, is deciding if such an OQ package offers value for the money.

The decision process should be risk based and documented. Asking such questions as these will be helpful:

- How close will you operate to the core LIMS being installed?
- Will you be using standard LIMS functionality?
- Will you be configuring the system?
- Will you be customizing the system?
- Am I repeating the vendor's internal testing in the OQ?
- Can I leverage any work in the OQ and reduce my PQ effort?

Subject to a satisfactory vendor audit report, standard LIMS functionality can be assumed to work, as the vendor has tested this, and the standard functionality used by the laboratory will be tested implicitly during the PQ phase of the project. Therefore, why bother to test functions in the OQ that the vendor has already tested? Configured and custom elements cannot be tested in this phase of the work, as they will be specific to the laboratory and will be input into the system after this phase of the project. Therefore, the OQ of a LIMS should be a limited test to indicate that the system that has been installed works.

U.S. GMP regulations (clause 160 [26]) require that before execution the test protocols be approved by the QC/QA unit and also that whatever is written in them needs to be scientifically sound. Here is an example of a warning letter sent by the FDA to Spolana [27], a Czech company, in October 2000: "Furthermore, calibration data and results provided by an outside contractor were not checked, reviewed and approved by a responsible Q.C. or Q.A. official." Therefore, never accept IQ or OQ documentation from a supplier without evaluating and approving it. Check not only coverage of testing but also that test results are quantified (i.e., have supporting evidence) rather than relying solely on qualified terms (e.g., pass/fail). Quantified results allow for subsequent review and independent evaluation of the test results. Further, ensure that personnel involved with IQ and OQ work from the vendor are trained appropriately by checking documented evidence of such training, such as that certificates are current, before the work is carried out.

6.13 Configuring the LIMS Application

Following the installation and qualification of the LIMS application software, it needs to be configured according to the parameters documented in the configuration specification:

- Defining user types and the access privileges for each
- Setting up user accounts and allocating each a user type
- Turning the system policies on or off and inputting the defined settings from the configuration specification
- Entering the default entries for the audit trail

The setup of the system should be documented and traced against the approved configuration specification. This has two advantages: The first is documentation of the application configuration for the validation document suite, and the second is documentation for disaster recovery purposes, as it will enable a rebuild of the system from the application disks.

6.14 Specifying and Coding Custom Software Modules

Custom coding using the scripting language provided by the LIMS vendor also needs to be specified, coded, tested, and integrated within the LIMS application. One control mechanism for this can be documented in the validation plan, and another could be via change control once the system is operational. An alternative approach could be to use an SOP to define exactly what is required in the way of specification, coding, and testing for each custom module that could be used during implementation and when operational. The latter approach may be best from the start, as there is a single way of coding the LIMS.

From both business and validation perspectives, custom coding is the highest-risk software (software category 5), as it is unique and the entire process is in the user's hands. Therefore, unless absolutely necessary, keep custom elements to an absolute minimum unless there is a good business case for taking this path. There have been many cases where too much customization has made it difficult for a laboratory to upgrade from a particular version of a LIMS.

Customization of the LIMS will need software design specifications written for each custom module required. This specification will need to detail:

- Data inputs and how this will occur: manually or automatically and from where in the system
- Handling the data with specifications as to how data will be manipulated, including specification of the equations with ranges of the values being input and how outputs will be presented
- Outputs from the module
- Integration of the custom module with the LIMS application
- Appropriate error handling, including verification of data entry
- Compliance issues such as access security and audit trail

Customization of the LIMS can also include new tables in the database and screens, and these need to be documented as well. The LIMS development environment is where the initial custom code will be written against the software design specification. The writing of the code using the LIMS internal scripting language should follow the standards laid out by the software vendor. There will be testing of the code by the developer that will be informal and not documented. This is a normal part of the software development process by the programmer: an iteration of programming, followed by testing and recoding to enable the software to work correctly. When the software module is ready for formal testing, it will be transported from the development environment to the validation instance and formal testing will be carried out against the software design specification.

6.15 Testing and Integrating Custom Software Modules

Formal testing of the custom modules is carried out in the validation environment against the requirements contained in the software design specification. Depending on the complexity of the custom module there may be white box testing of the algorithms (testing that the inputs, calculations, and outputs are as expected) as well as black box testing (the overall module and its integration with the LIMS). This work will be documented and the documentation suite subjected to QA approval before release of the module into the production environment.

6.16 Population of the Database

Population of the database with the laboratory methods, analytical procedures, and product specifications can be a long process, especially for laboratories that have a large number of products. This phase of the work is often not planned, overlooked, or underestimated. The scope of work will cover the raw materials including active ingredients, intermediates, and in-process testing and primary and secondary finished product testing. In some cases, a LIMS implementation has been phased, based on the population of the database work (e.g., raw materials could be first, primary finished products second, and secondary product testing third—the actual order will be dependent on laboratory and business priorities).

Entry of this material must be controlled and checked before its use, as an incorrect test or specification could result in product being released that was either under or over strength, with potential impact on patient safety. This process will be ongoing after the end of the application validation, and the best way to control this is via an SOP. Note that this is the normal way that a LIMS operates and therefore does not need to be under change control, as the main procedure should have the input specification, testing, and release process contained within it.

6.17 User Acceptance Testing: Performance Qualification

The performance qualification (PQ) stage of the overall validation of the system can also be considered as the user acceptance testing. This should be undertaken by trained users and be based on the way the system is used (including configuration and customization) in a particular laboratory and the surrounding environment. Therefore, a LIMS cannot be considered validated simply because another laboratory has validated the same software: the operations of two laboratories may differ markedly even within the same organization. The functions to be tested in the PQ must be based on the prioritized requirements defined in the URS, so with the numbering of individual requirements can be traced back to the system requirements via the traceability matrix [23,24]. Documentation of the PQ can be done in a number of ways, but the one preferred by the author is to have a controlling PQ test plan that describes the overall approach to testing and a number of PQ test scripts (test cases) that sit underneath the plan. There will be QA approval of the plan, as this is a high-level document, but not of the individual test scripts, as there is little added value that can be input by QA. However, the entire PQ package will be reviewed by QA after execution of the work to check that compliance and standards have been adhered to.

PQ Test Plan A way to document the PQ is to use an overall PQ test plan that outlines the scope of the system to be tested, the features of the LIMS to test, as well as those that will not be tested, with a discussion of the assumptions, exclusions, and limitations of the testing undertaken. A documentation standard for the PQ test plan can be found in the IEEE standard 829–1998 [28], and adaptation of this for practical use is presented here:

1. Introduction
2. Test system
3. Test environment(s)
4. Features to be tested, including description of the test scripts and the test procedures contained in each

5. Features not to be tested, with a rationale for each feature excluded
6. Test approach, including assumptions, exclusions, and limitations of testing for each test script
7. Pass/fail acceptance criteria
8. Suspension criteria and resumption requirements
9. Test deliverables

The key sections of a PQ test plan are the features to test and those that will not be tested, and associated with the features to be tested are the written notes of the assumptions, exclusions, and limitations to the testing undertaken. The assumptions, exclusions, and limitations of the testing effort were recorded in the appropriate section of the PQ test plan to provide contemporaneous notes of why particular approaches were taken. This is very useful if an inspection occurs in the future, as there is a reference back to the rationale for the testing. It is also very important, as no user can fully test a LIMS or any other software application. For example, the operating system and any database used would be explicitly excluded from testing, as the LIMS application software implicitly or indirectly tests these elements of the system.

Release notes for the LIMS version being validated will document the known features or errors of the system and may be a reference document for the overall validation. However, PQ tests carried out in any validation effort should not be designed to confirm the existence of known errors but to test how the system is used on a day-to-day basis. The role of the user in the testing is to demonstrate the intended purpose of the LIMS. It is the role of the software development team to find and fix errors. If these or other software errors were found during the PQ testing, the test scripts have space to record the fact and describe the steps that would be taken to resolve the problem.

PQ Test Scripts   In the same IEEE standard [28] can be found the basis for the test documentation that is the heart of any PQ effort: the test script. This document consists of one or more test procedures for specific requirements testing and each test procedure will contain test execution instructions, collation of documented evidence, and the acceptance criteria for the test procedure as follows:

- Outline one or more test procedures that are required to test the specific LIMS requirements
- Each test procedure will consist of a number of test steps that define how the test will be carried out
- For each key test step the results expected must be defined (not all test steps need to contain expected results, especially if they are instructions to move from one part of the system to another)
- There will be space to write the results observed and note if the test step passes or fails compared with the results expected
- There is a test log to highlight any deviations from the testing

- Sections will collate any documented evidence produced during the testing; this must include both paper and electronic documented evidence
- Definition of the acceptance criteria for each test procedure and if the test passes or fails
- A test summary log, collating the results of all testing
- A sign-off of the test script, stating whether the script passed or failed
- Sections throughout the document for a reviewer to check and approve the work

Testing Overview   One key point is that to ensure that the PQ stage progresses quickly, a test script should test as many functions as possible as simply as possible (great coverage and simple design). Software testing has four main features, known as the 4E's [29]:

1. *Effective:* demonstrating that the system tested both meets the system requirements and finds errors
2. *Exemplary:* can test more than one function simultaneously
3. *Economical:* tests are quick to design and quick to perform
4. *Evolvable:* able to change to cope with new versions of the software and changes in the user interface

It is an abject failure if the PQ testing documentation is written to test one requirement per script, as this does not test the system as it is intended to be used.

Writing the PQ Test Scripts   It is difficult to estimate the number of test scripts that a LIMS implementation requires, as the number of features used within the system can vary, as will the range and extent of configuration, as well as applications and instruments interfaced with the system. However, testing LIMS functionality should consider:

- Analytical process flows as configured by the laboratory: raw materials, in-process analysis, and release
- Specification management
- Stability protocols and reports with alerts
- Interfaces and data transfer between the LIMS and instruments and between the LIMS and applications
- Sample management from registration to disposal
- Unavailability of the network: buffering of data and prevention of data loss
- Custom calculations implemented within the system
- System capacity tests (e.g., analyzing the largest number of samples expected in a batch, the number of users on the system)
- Interfaces between the LIMS and other software applications (e.g., CDS)

Testing should also consider any electronic record or signature requirements (e.g., 21 CFR Part 11) and other regulatory requirements:

- System security and access control, including between departments or remote sites
- Preservation of electronic records (e.g., backup and recovery, archiving and retrieval)
- Data integrity
- Audit trail functions along with date and time stamps, especially if the system is used between time zones
- Electronic signatures
- Identifying altered and invalid records

Some of these compliance requirements can be integrated with some of the main LIMS functionality testing listed earlier. For example, electronic signing of results by the tester and reviewer can be integrated into the analysis of a sample and audit trail entries generated by altering incorrect data entries. The aim of this approach is to test the system as it is intended to be used.

Considerations When Designing Test Procedures Some of the considerations for designing test procedures for a LIMS are discussed here. Note that all aspects of the system that need to be tested must be defined in the URS. The simplest will be to consider a configured aspect of the LIMS, which will be logical security and access control. Although logical security appears at first glance to be a very mundane subject, the inclusion of this topic as a test is very important for regulatory reasons, as it is stated explicitly in 21 CFR, Part 11, and the predicate rules that access to a system should be limited to authorized individuals. Also, when explored in more depth, it provides a good example in the design of a test case.

The test design for access control should consist of the following basic test components as a minimum:

- An incorrect account name fails to gain access to the system.
- The correct account name and password gain access to the system.
- The correct account name but minor modifications of the password fail to gain access to the software.
- The account locks after three failed attempts to access, with an alert sent to the system administrator.

Important considerations in this test design are:

- Successful test cases are not just those that are designed to pass but are also designed to fail. Good test case design is a key success factor in the quality of LIMS validation efforts. Of the test cases above, most are designed to fail, to demonstrate the effectiveness of the logical security of the system.

- The test relies on good computing practices being implemented by the system administrator to ensure that users change or are forced to change their passwords on a regular basis and that these are of reasonable length (minimum six to eight characters).
- Locking an account can also ensure that a requirement of Part 11 is tested (e.g., alerting a system administrator to a potential problem).

Other test case designs that should be used:

- *Boundary test*. Entry of valid data within the known range of a field (e.g., a pH value would be acceptable only between 0 and 14.
- *Stress test*. Test entering data outside designed limits (e.g., a pH value of 15).
- *Predicted output*. Knowing the function of the module to be tested, a known input should have a predicted output.
- *Consistent operation*. Important tests of major functions should have repetition built into them to demonstrate that system operation is reproducible.
- *Common problems*. The operational and support aspects of the computer system should both be part of any validation plan (e.g., backup works; incorrect data inputs can be corrected in a compliant way with corresponding audit trail entries).

The predictability of a system under these tests must generate confidence in the LIMS operations (i.e., trustworthiness and reliability of electronic records and electronic signatures) and the IT support. For more information about the format of the document and more details of PQ testing, see the book by McDowall [8].

6.18 Migrating Data from the Existing LIMS or Another System

Data migration from another system (this could be an existing LIMS or another system) into the new LIMS can often be a component of a LIMS project or a separate project in its own right. This can be a difficult phase of the work, as an understanding of a legacy system may have been lost by the company, especially following reorganization, merger, or acquisition. In essence, the work needs to be planned and the data moved from the originating system to the LIMS, identified, and mapped with the fields in the LIMS database. Do not expect a 1 : 1 relationship between the data, as the two systems were developed independently and therefore work may be needed to ensure a fit or consistent reduction of data. In the latter case, not all data elements may be transferred, and the organization needs to determine what they will do with data that cannot be moved. If scripts are to be written to automate the data transfer, they will have to be specified, developed, and tested so that they are validated before use.

6.19   Training Records of the Users

All personnel involved with the selection, installation, operation, and use of a LIMS should have training records to demonstrate that they are suitably qualified to carry out and maintain their functions. It is especially important to have training records and résumés of installers and operators of a system, as this is a particularly weak area and a system can generate an observation for noncompliance. Major suppliers of LIMS will usually provide certificates of training for their engineers, but the IT department staff responsible for the network and utility operations will also require evidence of the engineers' education, experience, and training.

The types of personnel who could be involved in validation and training requirements are:

Supplier's staff: responsible for the installation and initial testing of the data system software. Need copies of their training certificates listing the products on which they were trained to work. These were to be checked to confirm that they were current and covered the relevant products and then included in the validation package.

System managers: training in the use of the system and administration tasks to be provided by the supplier and documented in the validation package.

Users: either analytical chemists or technicians trained initially by the supplier staff to use the data system. This training will be documented in their training records.

Consultants: must provide a résumé and a written summary of skills to include in the validation package for the system as required by the GXP regulations (e.g., Section 211.25) [26].

IT staff: training records and job descriptions outlining the combination of education, training, and skills that each member has.

Training records for LIMS users are usually updated at the launch of a system but can lapse as a system matures. To demonstrate operational control, training records need to be updated regularly, especially after software changes to the system. Error fixes do not usually require additional training; however, major enhancement or upgrade should trigger the consideration of additional training. The prudent laboratory would document the decision and the reasons not to offer additional training in this event. To get the best out of the investment in a LIMS, periodic retraining, refresher training, or even advanced training courses could be very useful for a large or complex LIMS. Again, this additional training must be documented.

6.20   System Documentation

Vendor Documents   The documentation supplied with the LIMS application or system (both hardware and software), release notes, user guides, and user standard operating procedures will not be discussed here, as it is too specific and also depends on management's approach in an individual laboratory. However, the importance of this system-specific documentation for validation should not be underestimated. Keeping this documentation current should be considered a vital part of ensuring the operational validation of any computerized system. Users should know where to find the current copies of documentation to enable them to do their jobs. The old versions of user standard operating procedures and system and user documentation should be archived.

Standard Operating Procedures Standard operating procedures (SOPs) are required for the operation of both the LIMS applications software and the system itself. SOPs are the main medium for formalizing procedures by describing the exact procedures to be followed to achieve a defined outcome. Procedures have the advantages that the same task is undertaken consistently, is done correctly, and nothing is omitted, and that new employees are trained faster [8]. The aim is to ensure a high-quality operation.

The FDA Guidance for Industry on computerized systems in clinical investigations provides a minimum list of SOPs expected for a computerized system in a GCP environment [30]. This list is presented following editing for a LIMS operating in a GLP or GMP laboratory.

- System setup and installation (including the description and specific use of software, hardware, and the physical environment and the relationship)
- System operating manual (user's guide)
- Archive and restore (including the associated audit trails)
- System maintenance
- System security measures
- Change control
- Data backup, recovery, and contingency plans
- Alternative recording methods (in the case of system unavailability)
- Computer user training, and roles and responsibilities of staff using the system

Note that this is a generalized list of SOPs, and more procedures may be required if the operating environment is more complex. Conversely, some of the procedures above could be condensed into a single SOP with more scope. The key issue is that all areas for the operation and maintenance of the system are controlled by procedure.

6.21   IT Service Level Agreement

In the case of outsourcing the support for the hardware platforms and network that run the LIMS either to the internal IT department or an outsourced IT function, a service level agreement (SLA) must be written to ensure that the IT department does not destroy the validation status of the system. This SLA should cover procedures such as:

- Controlling and implementing changes in the system and the IT infrastructure
- Database administration activities
- Backup and recovery, including media management
- Storage and long-term archiving of data
- Disaster recovery

This SLA will cover the minimum service levels agreed to, together with any performance metrics, so that the IT department can be monitored for effectiveness.

6.22 Reporting the Work: Validation Summary Report

The validation summary report brings together all of the documentation collected throughout the entire life cycle and presents a recommendation for management approval when the system is validated. The emphasis is on using a summary report as a rapid and efficient means of presenting results, as the details are contained in the other documentation in the validation package.

The report should summarize the work undertaken by the project team and be checked against the intention in the validation plan. This gives the writer an opportunity to document and discuss any changes in the plan. A list of all the documents produced during the validation should be generated as an appendix to the report. Finally, there should be a release statement signed by the system owner and QA authorizing release of the system for operational use. Some organizations have a separate release certificate that achieves the same endpoint and is quicker to release than waiting for approval of the validation summary report.

7 MAINTAINING THE VALIDATED STATUS

After operational release of the LIMS comes the most difficult part of computerized system validation: maintaining the validation status of a system throughout its entire operational life. Looking at the challenges that will be faced when dealing with maintaining the validation of a system, some of the types of changes that will affect an operational LIMS are:

- Software bugs found and associated fixes installed
- Upgrading of application software, operating system, and any software tools or middleware used by the LIMS
- Network improvements: changes in hardware, cabling, routers, and switches to cope with increased traffic and volume
- Hardware changes: PCs and server upgraded, or increases in memory, disk storage, etc.
- Interface to new applications (e.g., spreadsheets or laboratory data systems)

- Expansion or contraction of the system for work or organizational reasons
- Environmental changes: moving or renovating laboratories

All of these changes need to be controlled to maintain the validation status of the LIMS. In addition, other factors also affect the system from a validation perspective:

- Problem reporting and resolution
- Software errors and maintenance
- Backup and recovery of data
- Archiving and restoration of data
- Maintenance of hardware
- Disaster recovery (business continuity planning)
- Written procedures for all of the above

In this section, a number of measures are discussed that need to be in place to maintain the validation status of a LIMS.

7.1 Change Control and Configuration Management

Changes will occur throughout the lifetime of the system from a variety of sources:

- Service packs for the LIMS software to fix software errors
- New versions of the LIMS software, offering new functions
- Interfacing of new instruments to the LIMS
- Upgrades of network and operating system software
- Changes to the hardware: additional memory, processor upgrade, disk increases, etc.
- Extension of the system for new users

This is the key item from the release of the system for operational use to its retirement and is essential for maintaining the validation status of the LIMS. All changes must be controlled. From a regulatory perspective there are specific references to the control of change in both the OECD consensus document [31] and EU GMP regulations [20]. Therefore, change control is the primary means of ensuring that unauthorized changes cannot be made to a system, which would cause it immediately to lose its validation status.

A change request form is the means of requesting and assessing a change in the LIMS or in a system in the LIMS environment:

- The change requested was described first by the submitter.
- The business and regulatory benefits should be described along with the cost estimate of making the change.

- The impact of the change should be assessed by the system managers and then approved or rejected by management.
- Changes that were approved were implemented, tested, and qualified before operational release.

The degree of revalidation work to be done is determined during the impact analysis. Changes that affect the configuration (hardware, software, and documentation) are recorded in a configuration log maintained within Excel, for example.

Revalidation Criteria Any change in a LIMS should trigger consideration if revalidation of the system is required. Note the use of the word *consider*. There is often a knee-jerk reaction that any change means that the entire system should be revalidated. One should take a more objective evaluation of the change and its impact before deciding if full revalidation is necessary. First, if revalidation is necessary, is it required to test only the feature undergoing change, the module within the system, or the entire application? There may even be instances where no revalidation would be necessary after a change. However, the decision must be documented, together with the rationale for the decision.

Therefore, a procedure is required to evaluate the impact of any change in a system and act accordingly. One way to evaluate a change is to review the impact that it would have on data accuracy, security, and integrity [2]. This will provide an indication of the impact of the change on the system and the areas of the application affected. This allows the revalidation effort to target the change being made.

7.2 Operational Logbooks

To document the basic operations of a computer system, a number of logbooks are required. The term *logbook* is used flexibly in this context; the actual physical form that the information takes is not the issue but the information that is required to demonstrate that the procedure actually occurred. The physical form of the log can be a bound notebook, a proforma sheet, a database, or anything else that records the information needed, as long as the security and integrity of the records (paper or electronic) are maintained.

Backup and Backup Log The aim of a backup log is to provide a written record of data backup and the location of duplicate copies of the system (operating system and application software programs) and the data held on the computer. The backup schedule for the disks can vary. In a larger system, the operating system and applications software will be separated from the data, which are stored on separate disks. The data change on a fast time scale that reflects the progress of the samples through the laboratory and must be backed up more frequently. In contrast, the operating system and application programs change at a slower pace and are therefore more static; the backup schedule can reflect this.

Some of the key questions to ask when determining the backup requirements for a LIMS are:

- How much time should elapse between backups? Ideally, backups should occur daily.
- What is the nature of the backup: full, incremental, or differential? The best security is daily full backups, but this is very time consuming, so the best compromise is a combination of full once per week and either incremental or differential backups daily.
- Who is authorized to perform backups, and who signs off the log? The laboratory manager, in conjunction with the person responsible for the system, should determine this. The authorization and any review signature required should be defined in an SOP.
- When should duplicate copies be made for data security? This question is related to the security of data and programs. Duplicate copies should be part of the backup procedure at predetermined intervals. The duplicate copies should be stored in a separate location in case of a computer glitch, with the original backups located nearby. Duplicate backups are also necessary to overcome problems that occur in reading the primary backup copies.

Problem Recording and Recovery   During computer boot-up, backup, or other system functions, errors will inevitably occur. It is essential that these errors be recorded and that their solution also be written down. Over time, this can provide a useful historical record of the operation of the computer system and the location of any problem areas in the basic operation. An area where this might be the case may be in peripherals, where a print queue has stalled. This is a relatively minor case; however, there may be cases where the application fails due to a previously undetected error. In the latter case there is a need to link error resolution to the change control system.

Software Error Logging and Resolution   As it is impossible to test completely all the pathways through LIMS software or any software [32], it is inevitable that errors will occur during the system operation. These must be recorded and tracked until there is a resolution. The key elements of this process are to record the error, notify the support group (in-house or supplier), classify the problem, and identify a way to resolve it. Not all reported problems of a LIMS will be resolved; they might be minor and have no fundamental effect on the operation of the system, and may not even be fixed. Alternatively, a workaround may be required, which should be documented; sometimes, even retraining may be necessary. Other errors may be fatal or major, which means that the system cannot be used until fixed. In these cases, the revalidation policy will be triggered and the fix tested and validated before the LIMS can again be operational.

Maintenance Records   All high-quality systems need to demonstrate that the equipment used is properly maintained and in some instances calibrated. Computers are no exception to this. Therefore, records of the maintenance of the

LIMS need to be set up and updated in line with the work carried out on it. The main emphasis of the maintenance records is toward the physical components of a system: hardware, networking, and peripherals; software maintenance is covered under the error logging system described above.

If there is a preventive maintenance contract for the hardware, the service records after each call should be placed in a file to create a historical record. Also, any additional problems that occur that require maintenance will be recorded in the system log and cross-references to the appropriate record will be required. Many smaller computer systems have few, if any, preventive maintenance requirements, but this does not absolve the laboratory from keeping system maintenance records. If a fault occurs requiring a service engineer to visit, this must be recorded as well.

On sites where maintenance of personal computers is maintained centrally for reasons of cost or convenience, maintenance records may be held centrally. The invoice from the central maintenance group may cover all areas of a site or organization, including regulated or accredited as well as nonaccredited groups. It is important for the central maintenance group to maintain records sufficient to demonstrate to an inspector the work they undertake. As defined in EU GMP Annex 11, a third party undertaking this work should have a service agreement and the résumés of its service personnel available and up to date.

7.3 Disaster Recovery

Good computing practices require that a documented *and* tested disaster recovery plan must be available for all major computerized systems, but it rarely is. Failure to have a disaster recovery plan places the data and information stored by major systems at risk, the ultimate losers being the workers in the laboratory and the organization. Here it is not the laboratory personnel who are responsible for the disaster recovery plan but the IT department.

Disaster recovery is usually forgotten, or not considered, as "it will never happen to me." The recovery plan should have several shades of disaster documented: from the loss of a disk drive—how will data be restored from tape or a backup store and then updated with data not on the backup—through to the complete loss of a computer room or building through fire or natural disaster. Once the plans have been formulated, they should be tested and documented to see if they work. Failure to test the recovery plan will give a false sense of security and compound any disaster. This plan needs to be tested and also updated, as the IT technologies used by the organization change over time.

8 SUMMARY

The first stage in a LIMS validation is to map the current business processes inside and outside the laboratory where samples are generated and information used and then simplify and optimize them for electronic working. The time and effort spent on the mapping process is saved as the overall implementation is now simpler and faster. Risk-based validation of a LIMS environment is presented based on a life cycle for GAMP category 4 software. The first stage is to determine if the system needs to be validated, the second stage assesses the risk presented by each requirement to determine whether or not it needs to be tested, and in the third stage as much as possible of the vendor's development work is leveraged to reduce the amount of testing that the users have to perform. However, a LIMS can also have configurable and custom elements that will be specific to the laboratory where the LIMS is being implemented; therefore, the validation effort is directed toward these software elements, as this is where the highest risk is based. Measures to maintain the validation status of the system when the LIMS is operational are also discussed; the major one of these is change control.

REFERENCES

1. R. R. Mahaffey. *LIMS Applied Information Technology for the Laboratory*. Van Nostrand Reinhold, New York, 1990.
2. R. D. McDowall. Risk management for automation projects. *J. Assoc. Lab. Automation*, 9: 72–86, 2004.
3. *U.S. Pharmacopeia*. General Chapter 1058, Analytical Instrument Qualification. First Supplement to USP 31, 2008, p. 3587.
4. *Guidance for Industry: General Principles of Software Validation*. FDA, Washington, DC, 2002.
5. *GAMP Good Practice Guide: Validation of Laboratory Computerized Systems*. International Society for Pharmaceutical Engineering, Tampa, FL, 2005.
6. R. D. McDowall. Validation of spectrometry software: critique of the GAMP Good Practice Guide for Validation of Laboratory Computerized Systems. *Spectroscopy*, 21(4): 14–30, 2006.
7. *Good Automated Manufacturing Practice Guidelines*, Version 5. International Society for Pharmaceutical Engineering, Tampa, FL, 2008.
8. R. D. McDowall. *Validation of Chromatography Data Systems: Meeting Business Needs and Regulatory Requirements*. RSC Chromatography Monographs Series, R. M. Smith, Ed. Royal Society of Chemistry, Cambridge, UK, 2005.
9. R. D. McDowall and D. C. Mattes. Architecture for a comprehensive laboratory information management system. *Anal. Chem.*, 62: 1069A–1076A, 1990.
10. R. D. McDowall. A matrix for a LIMS with a strategic focus. *Lab. Automation Inf. Manag.*, 31: 57–64, 1995.
11. S. Jenkins. Invited Symposium on the Paperless Laboratory. Pittsburgh Conference on Analytical Chemistry and Applied Spectroscopy, 2004.
12. Strategic analysis of US laboratory information management systems markets, Frost and Sullivan 2008, cited by K. Shah. *Pharm. Technol. Eur.*, 20: 31–32, May 2009.
13. R. D. McDowall. Understanding and interpreting the new GAMP 5 software categories. *Spectroscopy*, June 2009.
14. *GAMP Good Practice Guide: Testing GXP Systems*. International Society for Pharmaceutical Engineering, Tampa, FL, 2006.

15. R. D. McDowall. Effective and practical risk management options for computerised system validation. *Qual. Assur. J*., 9: 196–227, 2005.

16. R. D. McDowall. Validation of commercial computerised systems using a single life cycle document (integrated validation document). *Qual. Assur. J*., 12: 64–78, 2009.

17. R. M. Sicnolfi and S. Bishop. RAMP (risk assessment and management process): an approach to risk-based computer system validation and Part 11 compliance. *Drug Inform. J*., 41: 69–79, 2007.

18. Electronic Records; Electronic Signatures Final Rule (21 CFR 11). *Fed. Reg*., 62: 13430–13466, 1997.

19. *Good Manufacturing Practice for Medicinal Products in the European Community*, Annex 11, *Computerised Systems*. Commission of the European Communities, Brussels, Belgium, 2007.

20. Proposed revision for public comment. *Good Manufacturing Practice for Medicinal Products in the European Community*, Annex 11, *Computerised Systems*. Commission of the European Communities, Brussels, Belgium, 2008.

21. *Quality Management Systems: Requirements*. ISO Standard 9001. ISO, Geneva, Switzerland, 2005.

22. *Software Engineering: Guidelines for the Application of ISO 9001-2000 to Computer Software*. ISO Standard 90003. ISO, Geneva, Switzerland, 2004.

23. R. D. McDowall. Validation of spectrometry software: the proactive use of a traceability matrix in spectrometry software validation: 1. Principles. *Spectroscopy*, 23(11): 22–27, 2008.

24. R. D. McDowall. Validation of spectrometry software: the proactive use of a traceability matrix in spectrometry software validation: 2. Practice. *Spectroscopy*, 23(12): 78–86, 2008.

25. *Guide for Developing Software Requirements Specifications*. IEEE Standard 1233–1998. IEEE, Piscataway, NJ, 1998.

26. *U.S. Current Good Manufacturing Practice Regulations for Finished Pharmaceuticals* (21 CFR 211), with revisions as of 2008.

27. Spolana. FDA warning letter, Oct. 2000.

28. *Software Test Documentation*. IEEE Standard 829–1998. IEEE, Piscataway, NJ, 1998.

29. M. Fewster and D. Graham. *Software Test Automation: Effective Use of Test Execution Tools*. Addison-Wesley, London, 1999.

30. *Guidance for Industry: Computerised Systems in Clinical Investigations*. FDA, Washington, DC, 2007.

31. *Consensus Document on Principles of Good Laboratory Practice Applied to Computerised Systems*. Organisation for Economic Co-operation and Development, Paris, 1995.

32. B. Boehm. *Some Information Processing Implications of Air Force Missions: 1970–1980*. RAND Corporation, Santa Monica, CA, 1970.

8

PERFORMANCE QUALIFICATION AND VERIFICATION OF BALANCE

CHUNG CHOW CHAN
CCC Consulting

HERMAN LAM
Wild Crane Horizon Inc.

ARTHUR REICHMUTH AND IAN CIESNIEWSKI
Mettler Toledo Inc.

1 INTRODUCTION

The balance is the most important piece of equipment related to any methodology that will perform quantification. The fundamental concept of all quantification procedures will involve the mathematical computation of data (e.g., absorbance of the analyte) obtained from the sample against the data obtained from the reference. To perform a comparison, a certain quantity of reference standard will be weighed, dissolved in a certain amount of reagent, and the analytical response measured. The same procedure will be applied to the sample in question. These two responses are then compared and computed mathematically to give the desired results. The balance measures the mass of the reference standard, which is the most critical component and the fundamental information needed for this quantitation.

Practical Approaches to Method Validation and Essential Instrument Qualification,
Edited by Chung Chow Chan, Herman Lam, and Xue Ming Zhang
Copyright © 2010 John Wiley & Sons, Inc.

1.1 Working Principles of Laboratory Balances

High-resolving balances, balances with more than 100,000 displayed digits, rely almost exclusively on the electrodynamic compensation principle. A first characteristic of this principle is that the weight force, G, which is the product of the mass, m, of the weighing object and the gravity, g:

$$G = mg \qquad (1)$$

is not measured directly, but is compensated for instead. The same physical transducer principle as that of a loudspeaker is put to use with the balance. The difference between the balance and the loudspeaker is that with the balance, it is not the movement that is of primary interest. Rather, the balance relies on the property of the electrodynamic principle by which a current flowing through a wire in a magnetic field produces a force, F. This force is equal to the product of the amount of the electrical current, I, the strength of the magnetic flux, B, and the length, l, of the conductor:

$$F = IBl \qquad (2)$$

If this compensation force is adjusted such that it is equal to the weight force, the electrical current required is proportional to the mass on the weighing pan, provided that the magnetic flux density, the length of the electric conductor, and the gravitation remain constant. This is certainly true to a large degree; nevertheless, with high resolving balances, it is mandatory to compensate for potential drifts.

The compensation current, carrying the information about the unknown mass, is now fed into an analog-to-digital converter. It produces a number that is the digital equivalent of the mass being weighed. This value is further processed by a digital signal-processing unit to compensate for nonlinearity, sensitivity deviation, and temperature drift, among other parameters, before it is displayed.

Mechanical guidance (solid plates, attached to the body of the weigh cell via flexible bearings known as the upper and lower guides) is required to direct the weighing cell, enabling the weighing platform to support the weighing object. Also, the lever is required to provide mechanical advantage to transform the producible maximum force by the electrodynamic transducer to the weighing capacity of the balance. The upper and lower guides and the lever need to be robust to deal with accidental force impacts (e.g., in shipment); however, they must also not introduce inadmissible forces degrading the weighing value. Figure 1 is a simplified cross-sectional view of the interaction between these pieces in an analytical balance that works on the electrodynamic compensation principle.

1.2 Types of Balance in the Analytical Laboratory

There are three common balances [1,2] that are used routinely in an analytical laboratory:

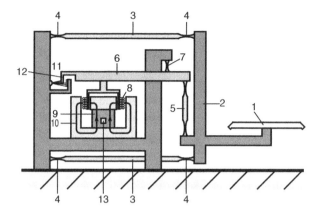

FIGURE 1 Simplified cross-sectional view of an analytical balance, using the electro-dynamic compensation principle. 1, weighing pan; 2, hanger; 3, guide; 4, flexible guide; 5, coupling; 6, lever; 7, lever bearing; 8, compensation coil; 9, permanent magnet; 10, magnetic flux; 11, optical position sensor; 12, position vane; 13, temperature sensor.

1. *Top-loading balance.* This is normally used to weigh higher weights in the gram range. This balance is normally used for weighing buffering salts for buffer solutions for high-performance liquid chromatagraphy (HPLC) analysis.

2. *Analytical balance.* An analytical balance is used to measure mass to very high precision and accuracy. An analytical balance usually has readabilities from 0.01 to 0.1 mg and a capacity from low to approximately the 500-g range.

3. *Microbalance.* As the name implies, a microbalance is an instrument capable of making precise measurements of the weight of objects of relatively small mass: on the order of 1 million parts of a *gram*, the microgram range. Generally, a microbalance is used for very precise low-capacity weighing (e.g., with a capacity of 2 to 5 g and readability of 0.1 to 1.0 µg)

1.3 Basics of Balance Operation

Each balance normally comes with a user's manual from the manufacturer that details step-by-step use of the balance. The laboratory should rewrite the instructions in the form of a standard operating procedure (SOP) for operation of the balance.

1.4 Definitions and Acronyms Commonly Used in the Operation and Qualification of a Balance

In a performance qualification, it is important to define the terms that will be used. The common definitions and acronyms are listed below.

ASTM: American Society for Testing and Materials.

Calibration: a set of operations that establishes the relationship between values indicated by a balance or mass values represented by a weight and the corresponding known values realized by standards.

Corner weighing: eccentricity (or corner load) error is a measure of the magnitude of this effect and is usually expressed as the deviation between the mass displayed when a test weight is placed at the center and then at other specified positions on the pan.

Linearity: a measure of the deviation from the ideal linear behavior of the relationship between applied mass and indicated mass.

NIST: National Institute of Standards and Technology.

OIML: International Organization of Legal Metrology.

Range: a feature of a balance's ability to display a weighing result that is characterized by its readability and minimum and maximum values.

Readability: the mass value of the smallest scale or digital interval displayed by the weighing machine. For example, a balance with a display of four places in grams has a readability of 0.0001 g or 0.1 mg.

Repeatability: a measure of a balance's ability to indicate the same result in replicate weighings under the same conditions. (For modern electronic balances utilizing electromotive force compensation weigh cells, repeatability, or precision error, is the dominant component of measurement uncertainty affecting the weighing, up to approximately 80% of the capacity of the balance.)

1.5   Classes of Weights for Calibration

Pharmacopeial tests and assays require balances that vary in capacity, sensitivity, and reproducibility. Unless otherwise specified, when substances are to be weighed accurately for assay, the weighing is to be performed with a weighing device whose measurement uncertainty does not exceed 0.1% of the reading. Measurement uncertainty is satisfactory if twice the standard deviation of not less than 10 replicate weighings divided by the amount weighed does not exceed 0.001. Unless otherwise specified, for titrimetric limits tests in USP, the weighing shall be performed to provide the number of significant figures in the weight of the analyte that corresponds to the number of significant figures in the concentration of the titrant.

The following class designations are in order of increasing tolerance.

Class 1 Weights   These weights are designated as high-precision standards for calibration [3,4]. They may be used for weighing accurately quantities below 20 mg. Class 1 weights meet the tolerance to be used to calibrate and check balances with a displayed accuracy of very high magnitude (e.g., micro- and ultramicrobalances) as well as high-specification semi-microbalances ($d = 0.01$ mg), where a high safety factor (calibration ratio) is required.

Class 2 Weights   These are used as working standards for calibration, built-in weights for analytical balances, and laboratory weights for routine analytical work. In fact, class 2 weights are of a high enough tolerance accuracy that they are suitable for use with semi-micro and analytical balances for in situ calibration and verification work. This is why they are the default standard in the ASTM procedure (OIML F1 for everybody else) and for daily balance checks. Users buy class 1, as they exceed class 2 and can be used for all lab balances. Class 2 may be used for quantities greater than 20 mg.

Note: Built-in weights inside the analytical balances meet either class 1 or 2 standards, but cannot be classified as such, as their shape cannot meet any known weight standards.

Class 3 and 4 Weights   These are used with moderate-precision laboratory balances. A weight class is chosen so that the tolerance of the weights used does not exceed 0.1% of the amount weighed. Generally, class 3 is used for quantities greater than 50 mg, and class 4 for quantities greater than 100 mg. Weights should be calibrated periodically by a qualified mass metrology recalibration laboratory, traceable to national standards, according to the requirements of the state or country, or as recommended by the qualified mass laboratory.

2   PERFORMANCE QUALIFICATION

Performance qualification requires very basic activities to determine its performance characteristics, but these are essential and critical elements for quantitative work. Potential issues and resolution of these issues are discussed in detail in Section 3. Five basic performance parameters are commonly required for performance qualification: (1) accuracy, (2) repeatability, (3) linearity, (4) calibration, and (5) corner load deviation (eccentric load deviation). All the calibration weights used must be either ASTM class 1 or OIML class E2 weights.

2.1   Performance Qualification Parameters

Accuracy   The accuracy of the balance must be verified once daily (or before use). The verification procedure should comprise a low standard weight and a high standard weight. The low and high weights used will be based on the weight range of the balance. In the case of a 200-g analytical balance, an example weight range for the low and high standard weight will be 100 mg to 200 g standard.

Note: From a metrological standpoint, the accuracy measurement is better verified at the highest weight used. This is because below approximately 80% of the capacity of the balance, repeatability (precision error) is wholly dominant and far exceeds all other uncertainties. Multipoint checking provides good manufacturing practice (GMP) assurance.

Repeatability Perform no less than 10 replicate weighings of the weight desired. (Do not zero between readings, and take the difference between loaded and unloaded weighing results.) The measurement uncertainty is satisfactory if twice the standard deviation of not less than 10 replicate weighings divided by the amount weighed does not exceed 0.001 (0.1%).

Linearity This test verifies the linearity of the weight responses. Perform the weighings to include the extreme end of the range and a reasonable number of weighings within the range. Plot the weight obtained vs. the actual weight used. The linear regression of the plot should give a visual straight line and a coefficient of correlation of greater or equal to 0.9999.

Calibration Most new balances have internal weights built into the calibration. Perform both internal and external calibrations.

Internal Calibration Calibration routine is normally performed when there is no load on the balance. Perform the internal calibration as suggested in the user's manual for the balance. Record the difference between the measured weight and the calibration weight that will normally be displayed. Confirm that the difference is within the acceptance criteria. An example of the acceptance criterion will be 0.1% for a 200-g load.

Technically, the internal calibration feature of a balance is not a calibration process. It is an adjustment process to adjust the results displayed to suit any change in environmental conditions since the last adjustment. External calibration is the method used to characterize the metrology of the instrument. Use of the internal adjustment will allow the user to decrease the frequency of external calibration, because it reduces the repeatability errors over time.

External Calibration External calibration should be performed with a range of calibration weight sets that cover the intended weight range usage of the balance. An example of the weight range that will be used for a 200-g analytical balance may include the following calibration weights: 20, 50, 100, 200, and 500 mg, and 1, 2, 5, 10, 20, 100, and 200 g. Calculate the balance accuracy from these weight sets and compare against the acceptance criteria. An acceptance criterion that is commonly used and accepted is 0.1%.

Corner Load Deviation (Eccentric Load Deviation) Deviation between two weighing values, obtained with the same load, where the load is placed once out of center (eccentrically) and once at the center of the weighing platform with otherwise identical conditions, is

$$y_{ECC}(x, r) = y(x, r) - y(x, 0) \qquad (3)$$

where x is the load, r the off-center distance, and y are the readings. Usually, five weighings are carried out to determine corner load deviation (Figure 2): one

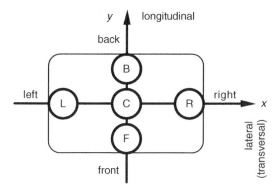

FIGURE 2   Eccentric load determination.

in the center of the platform (C), one each to the left and right of the platform (L, R), and one each to the front and back of the platform (F, B).

Note: The test should be carried out at 100% of capacity (full-scale deflection), as at that point, the eccentricity is approximately 40% of the total uncertainty. The sensitivity accounts for about 50 to 55%.

2.2   Performance Verification: Calibration Practice

A generic template for a performance qualification is provided in the appendix to this chapter.

3   COMMON PROBLEMS AND SOLUTIONS

Influences on balances and weighing samples are physical quantities that inadvertently, and often unknowingly, introduce systematic or random deviations into the mass measurement value. At least three prerequisites are involved when carrying out a weighing: the balance, the weighing object (sample), and the environment. Neither the balance nor the weighing object should be considered isolated objects; they always stay in a state of interchange with their environment.

Although a balance is built to determine the mass, and only the mass, of the weighing object, it is impossible to prevent internal and external influences from disturbing the balance to a certain extent. This is also true for the weighing object; the environment tends to influence it even more. Although there are weighing objects that are fairly immune to external influences, such as dense metals with compact shapes and polished surfaces, most weighing objects are clearly subject to considerable environmental influence. There are samples that are very difficult to weigh, such as hygroscopic or volatile substances, or substances with low density, objects with extending shapes, porous surfaces, or even objects consisting of an electrically nonconductive material.

In the following sections we discuss the most important common influences on weighings [5]:

* Ambient climate
* Air draft and pressure fluctuations
* Radiation
* Mechanical influences
* Electromagnetic influences

3.1   Ambient Climate

Temperature Difference Between Balance and Environment   Electronic balances are usually compensated against the influences of ambient temperature. Even so, a change in ambient temperature may cause a deviation not only of a balance's sensitivity (slope) but also its zero point (display when no weight is in the balance). The remaining temperature coefficient of sensitivity of a properly adjusted precision balance amounts to 3 to 10 ppm/K typically or to 1 to 3 ppm/K typically for an analytical or microbalance. However, with most weighing operations, the zero's temperature coefficient is not of major concern: namely, when taring is allowed before weighing. On the contrary, with all weighings where re-zeroing is not applicable, the zero's temperature drift must be taken into consideration.

What is more, a balance whose temperature differs by a few degrees from its environment undergoes an acclimatization process, during which a change in the displayed value occurs which may amount to a multiple of the value to be expected from the static temperature coefficient. Such temperature transients occur predominantly when the balance is plugged to the power supply, because then the electronic components begin to dissipate power. These effects are especially pronounced when a balance is moved from a cold environment (such as a car's trunk) to a warm one (such as a laboratory). Such a transient interval may extend from one to multiple hours, depending on the type of balance. If an application requires the ultimate in resolution and accuracy from an instrument, it is recommended that a precision balance be powered up 1 to 2 hours before use at the precise location of measurement; for an analytical or microbalance, 6 to 12 hours are recommended. For the same reason, an electronic balance should not be disconnected from the power source during measurement pauses, since reconnecting the balance retriggers the warm-up phase. Most balances provide a standby mode for this reason.

Temperature Difference Between Weighing Object and the Environment or Weighing Chamber   The air inside the draft shield of balances undergoes a temperature rise above ambient temperature because of power dissipated by electronic components. When the draft shield is opened, this warmer air escapes and colder air streams into the weighing chamber. This air draft exerts forces on the weighing pan and the object weighed. As a consequence, the displayed value gets distorted and unstable. Only after 15 to 30 seconds, when the air draft has died out, do these perturbations disappear. If the draft shield has been opened for the first time after a long lapse, the thermal equilibrium will be distorted. As a consequence, the weighing may be distorted (see below). In such cases it is recommended the draft shield be opened and closeed several times (as if when weighing), before performing the first weighing.

For the same reason, one should not reach with hands into the weighing chamber of a high-resolving balance, nor touch the weighing object, nor place the object on the platform with bare hands. A pair of pliers or tweezers should be used to do these operations. If a weighing object is not at the same temperature as the ambient air, an air draft emerges along the surface of the object, especially if the object possesses an extended vertical shape (such as a glass beaker). When the object is warmer than its surrounding air, this gives rise to an upstream, while an object colder produces a downstream current. Because of viscous friction, an upward or downward force acts on the weighing object, making it seem lighter or heavier, respectively. A mere few tenths of a degree are sufficient to offset the weighing value or at least to degrade repeatability and prolong stabilization time. The effect fades away when the weighing object has acclimatized. (*Note:* Acclimatizing a beaker may take half an hour or more.)

Humidity Exchange Between Weighing Object and Environment   Analogous to temperature differences, a humidity change in the environment can cause deviations of the weighing value. The effect caused by a humidity difference between a weighing object and its environment is approximately proportional to its surface. Rough surfaces tend to absorb larger amounts than do smooth ones. A polished, clean metal surface, subjected to a humidity change of the ambient air from 40% to 80% RH, adsorbs a water film of about 0.1 to 0.4 µg per square centimeter of surface area (this applies to calibration weights, for example). In the contrast, anodized aluminum will adsorb up to 40 $\mu g/cm^2$. Glassware is also known to adsorb water on its surface. The amount adsorbed depends on the quality of glass, its cleanliness, and its pretreatment. Glassware from soda-lime glass (soft glass), borosilicate glass (Pyrex/Duran), hard glass, or quartz adsorb, in this order, decreasing amounts of water.

Weighing of Hygroscopic, Volatile, or Reactive Weighing Objects   The body (volume) of a weighing object is capable of exchanging water, too, although at a much slower pace than its surface. This effect often results in a never-ending drift in the weighing value. The situation gets critical if the weighing object is obviously hygroscopic, volatile, or reactive. Alcohols, which evaporate easily, and water belong to this category. Ashes, salts, and other substances are hygroscopic. Many substances react with the surrounding air or with the humidity in the air.

As a rule, such substances should always be kept sealed to minimize any exchange with their environment. Suitable for weighing are containers with narrow necks or sealable containers. Syringes, or containers sealed with (rubber) membranes, also help in reducing the exchange. Volatile fluids can also be kept under an oil layer, provided that the density of the liquid to be protected is higher than the density of the cover substance. Reactive substances should be kept under inert gases or other substances that prevent reactions.

When gravimetrically calibrating pipettes, a moisture trap can lower the evaporation rate. This device consists of a glass cylinder with only a small opening through which the dispensed liquid is dosed into a container. If the exchange is not too pronounced, the repeatability of the weighing may be improved if the weighing operation always follows the same procedure and time pattern, including loading of the balance and reading of the display. However, this method cannot prevent a possible systematic deviation (bias) caused by a reading at an inappropriate instant, either too early or too late.

3.2   Air Draft and Pressure Fluctuations

Air Draft   Any air draft generates, through impact pressure or viscous friction, spurious forces onto the weighing platform and the weighing object; clearly, these forces do not emerge from the mass of the latter. Especially if the weighing object possesses a large surface, one has to be prepared for such influences. Air currents cause not only fluctuations of the weighing value, but may offset them as well. Air conditioners and fume hoods are known for their notorious ability to produce air currents. During cold seasons, heaters and windows with large surfaces may give rise to air currents as well.

Disregarding extreme situations or sensitive weighing objects for the time being, one may expect to weigh trouble free without a draft shield down to a readability of 10 mg. Around a readability of 1 mg it is recommended to use a draft shield, and from 0.1 mg (analytical balance) it is indispensable. Usually, these balances already provide a draft shield that is fully integrated into their design, defining the balance's weighing chamber. From a readability of 10 µg, it is highly recommended to further reduce the volume of the weighing chamber, if possible. To achieve a readability of 1 µg (microbalance), the weighing chamber and the weighing pan area must be kept small. This is even more true for ultramicrobalances with a readability of 0.1 µg.

Pressure Fluctuations   The inside volume of the balance housing, containing the weighing cell and the electronics, is connected by the load transfer mechanism to the weighing platform outside this housing. With balances equipped with a draft shield, the latter further isolates the weighing pan from the outside. Pressure changes in the environment of the balance lead to pressure differences between the outside and inside of the balance. Since the feed-through channel of the weighing platform cannot be hermetically sealed by a membrane (suffice it to mention the barometric distortion, as well as the guiding forces introduced by such a membrane), the balance's exterior and interior remain connected to each other. These pressure differences generate compensation currents, which will exert flow forces onto the weighing pan or the weighing cell, disturbing the weighing value. For a laboratory that will house high-resolving balances, sliding doors are preferred instead of hinged doors. Air conditioners that are used should not produce strong air currents or pressure fluctuations.

3.3 Radiation

Heat Radiation The air contained within a balance's draft shield, and the balance itself, will begin to warm up if exposed to any form of heat radiation. Temperature differences, or varying temperatures of the balance or the weighing object, cause perturbations described earlier. Therefore, a balance should never be placed in the path of a heat source e.g. direct sunlight from a window. The sun is one of the most frequent, especially strong and highly varying heat source.

Operator Radiation An operator's body may radiate sufficient heat during a weighing on an analytical or microbalance to perturb the weighing. With a surface temperature of about $30°C$ (or 300 K), a person emits enough heat to change the conditions in the weighing chamber within about half an hour.

3.4 Mechanical Influences

Vibrations Vibrations, translational as well as rotational, at the site of the balance may introduce deviations into the weighing value. The higher the readability of a balance, the more susceptible it usually is to vibrations. The lower the frequency of a vibration, the more difficult it is for the digital filter of the balance to suppress this disturbance. As a rule of thumb, movements with frequencies below 0.1 to 1 Hz become visible in the display, such that the weighing value cannot be read unambiguously. In these instances, the influence is clearly recognizable.

More treacherous are vibrations in higher-frequency bands, above 10 to beyond 100 Hz. The digital filter is capable of eliminating these fluctuations without difficulty, and the display remains stable. However, the balance may, depending on the weighing load, exhibit resonances at certain frequencies, which may lead to nonlinear effects. Vibrations at the place of installation should be suppressed and, if possible; should be kept below 10 mG (with analytical or microbalances; $G = 9.81$ m/s^2; hence, 10 mG $\approx$ 0.1 m/s^2). If this cannot be realized at the location of installation, a preliminary test with known weights should be carried out to reveal potential deviations caused by vibrations.

Placement of the Weighing Object As described as part of the qualification exercise, moving the load to a different position will result in a deviation of the displayed weighing value, called *eccentric load* (error) [also known as *corner load* (error) or *shift error*]. The magnitude of this deviation increases, usually over-proportional, with the weight of the load and distance between its center of gravity and the center of the platform. It is suggested that the off-center deviation of an approved balance is smaller than one approval step at one-third of the weighing capacity. With an analytical balance, a corner load deviation typically of 0.1 to 0.2 mg may be expected at half-capacity and with the load placed entirely at the edge of the pan.

From the above it should become clear that it is advisable to place the weighing object always in the center of the platform. Should this not be possible, the second-best solution is always to place it at the same location on the platform. Thereby, the influence of eccentric loading can at least be reduced.

3.5 Electromagnetic Influences

Electrostatic Forces   According to Coulomb's law, two electric charges exert a force, F_E, on each other:

$$F_E = \frac{1}{4\pi\varepsilon_0}\frac{Q_1 Q_2}{r^2} \tag{4}$$

which is proportional to the product of both electric charges, Q, and the square of the distance, r, between them. A balance or weighing object carrying such electric charges will cause an electrostatic attraction force. The vertical component of this force is being measured by the balance. Because the force is unexpected, it will mostly go undetected and wrongly be interpreted as weight force equivalent to a mass $\Delta m = F_E/g$, where g is the value of the local gravity.

How can an object get electrically charged in the first place? If two objects touching each other are separated, electrons from one object are left behind on the other. In that way both objects get electrically charged: the electron donator positively, the electron acceptor negatively. According to this mechanism, a glass container can get electrically charged, for example, by gripping it with tweezers, by rubbing it dry with a cloth, or by shoving it on a surface. This charging mechanism is often called *triboelectricity*, although no friction is required per se for this process; separation is sufficient. Liquids and gases may become charged as well. For example, a liquid that is poured from one container into another can get charged by this process.

For charges to exist on bodies over an extended time without dissipating, the bodies (in this case, the balance and the weighing object) need to be electrical insulators. Electrical charge can flow through the body (bulk) of a body, or it can flow along its surface. Many weighing objects possess virtually no body conductivity and hence are isolators. Containers and other glassware made from borosilicate glass are excellent electrical isolators. Even soda-lime glass ("window glass") is a good insulator. Many components of laboratory supplies are manufactured from plastic materials which are fairly good to excellent insulators.

Even though the bulk of a body is from electrically insulating material, electrical charge can still dissipate if the surface of the body provides sufficient surface conductance. Although the electric bulk conductance of an object may be small, its surface conductance can still be substantial. Because of the adsorbed water layer on the surface at normal air humidity and ubiquitous contamination (by salt and other chemicals), as a rule of thumb most labware possesses sufficient surface conductance to dissipate electrical charges within a short time.

This surface conductance depends largely on air humidity, especially when contamination is present. However, if the air humidity falls below 40 to 45% relative humidity (RH) (during wintertime, when heating is on), chances are that electric charges will barely dissipate, if at all. This is especially true for clean equipment. Charges will manifest themselves such that repeated weighings of the very same object will produce different weighing values or that a weighing value will never settle [i.e., it drifts for a long time (seconds, minutes)]. Due to the lack of an obvious reason, most often the balance gets blamed for this behavior.

If the ambient air humidity is extremely low, around 10 to 20% RH, as can be the case in a laboratory glove box with a controlled atmosphere, for example, electrical charges may stay for hours on objects or on the balance. If recognized correctly, the effect may be overcome with simple measures. First, a normal atmosphere with sufficient humidity (50% RH) should be provided, and the distance between attracting or repulsing charges should be increased. If this is not possible, or the suppression is not sufficient, the countermeasures described next will help.

Weighing Object Electrically Charged In this case the weighing object should be raised in the center between the platform and the top of the draft shield (where present). A spacer can be used to achieve this. Because it also gets weighed, it should not be too heavy, and there should be no doubt about the constancy of its mass. This method increases the distance between the electric charge and the base of the balance.

Another countermeasure is to enclose the weighing object with a Faraday cage, which is electrically as well as mechanically connected to the weighing pan. Thereby, all electric field lines—the innate cause for forces—will end inside the weighing object, the cage, or the pan; no field line will leave this structure. Consequentially, the electric charges cannot exert any force to the outside, hence producing no weighing force either. A remedy can be ionized air. It neutralizes the electrical charges. It is important that the ionized air current is electrically neutral (i.e., carries an equal number of positively and negatively charged ions). Ionized air can be produced with handheld piezo ionizers. Such devices are well suited for occasional applications. If the electrical charges occur systematically, and if a high throughput is required, electrically operated ionizers that produce a steady stream of ions are the choice. The electrode of such ionizers can be installed next to the balance such that the weighing object passes through the operating zone of the ionizer and gets discharged while loading it on the balance. In persistent cases the ionizer may be mounted directly inside the weighing chamber. The ionizer must be switched off during the weighing, and the decay of the air currents inside the draft chamber must die out, before the balance can be read.

Balance Electrically Charged Usually, the balance gets grounded automatically by connecting the power plug to the electrical outlet. If the power plug has only two prongs (no earth prong), or if the balance is powered from a battery, it is recommended that a bare spot of the balance (a screw, for example) be connected to earth with a wire. The glass windows of the draft shield are also capable of accumulating electrical charges, thereby causing a force on the weighing object or the weighing pan. To alleviate this situation, an analogous strategy as described above may be used, with the difference that the Faraday cage now needs to be connected electrically and mechanically to the frame of the balance (not its weighing pan).

Magnetostatic Forces   Magnetic forces emerge between two magnetized bodies or between a magnetized body and a magnetically permeable body. Although magnetic phenomena are more difficult to describe than those in the case of electrostatics, we still may write approximately

$$F_M \approx \frac{M}{r^2} \tag{5}$$

where M is the magnetization of a body and r is the distance between the magnetically active objects. Again, the force decreases with the square of the distance between the magnetized and the attracted body. A magnetically active weighing object will be affected by repelling or attracting magnetic forces. The vertical component of such a magnetic force, F_{mag}, acts on the weighing object. Because the force is unexpected, it mostly remains undetected and is interpreted as a weight force equivalent to a mass $\Delta m = F_{mag}/g$, where g is the value of the local gravity.

Magnetically highly permeable elements (elements with a large relative magnetic permeability, $\mu r \gg 1$) are iron, nickel, and cobalt, as well as their alloys. Metallic glasses and ferrites are also permeable. Magnetically "hard" materials can be magnetized: for example, those from hard steel, and permanent magnetic materials, designed expressively for that reason. Magnetic fields also emerge from electromagnets or cables carrying high electric currents. Typical sources of magnetic fields, for example, are permanent magnets, also (mostly inadvertently or unknowingly) magnetizable objects such as screwdrivers, knives, tweezers, or other tools or objects manufactured from steel. Furthermore, equipment containing electromagnetic components such as motors, valves, and transformers should not be overlooked. In suppressing the influences of magnetic effects, we have to keep apart the cases described below.

Weighing Object Magnetized   Most often, the magnetized weighing object attracts magnetically permeable elements in its vicinity, including those of the balance. Although much thought is given to design balances without making use of permeable materials, it cannot be done without. The electrodynamic transducer, by its nature, must contain magnetically highly permeable material, as well as other components of the balance, such as calibration drive motors or magnetic screens. It may also happen that the balance stands on a partially or fully magnetically permeable support (iron table, iron-reinforced stone table, or iron-strengthened wood table), next to permeable equipment (steel enclosings) or other iron construction.

The effect can be alleviated by placing the weighing object onto a nonmagnetic, nonpermeable (and not too heavy) body to gain distance. This is the preferred method employed with balances that have no draft shield. Alternatively, the weighing object can be screened with a cage made from magnetically permeable material (soft iron or alloys such as mumetal), which itself rests on the weighing pan. Thereby, magnetic field lines are caught within the case (at least to the extent of the cage's screening capability) and cannot reach the outside, which reduces forces onto the weighing pan.

Analogous to the countermeasures against electrostatic forces, the cage need not be closed entirely in every case. Depending on the situation, an iron sheet metal between a weighing object and a weighing pan may sometimes help. However, contrary to electrostatic field lines, where a single layer of electrical conductor, independent of its thickness, provides a perfect screen, a magnetically permeable layer may not fully screen against magnetic fields. A thicker layer, or multiple separate plies, may be required to suppress the magnetic field entirely. If none of the methods mentioned is applicable, or if they are not efficient enough, the weighing object should at least always be placed in the same position and with the same orientation.

Weighing Object Magnetically Permeable   In this case, sources of external magnetic fields should be removed, or if this is not possible, the balance should be screened with magnetically permeable material that is supported by the balance or its rest, not by the weighing pan. Here, as well as above, the screen need not always cover the entire balance; it may suffice to use a permeable iron mesh.

REFERENCES

1. *U.S. Pharmacopeia*, General Chapter <41>, Weights and Balances. USP, Rockville, MD. 2006.
2. *U.S. Pharmacopeia*, General Chapter <1251>, Weighing on an Analytical Balance. USP, Rockville, MD. 2006.
3. *Troemner Mass Standard Handbook*. Troemner, Thorofare, NJ.
4. *Standard Specification for Laboratory Weights and Precision Mass Standards*. ASTM E617-97. ASTM, West Conshohocken, PA, 2008.
5. A. Reichmuth. *Adverse Influences and Their Prevention in Weighing*, Vol. 1.3. Mettler Toledo, Columbus, OH, 2001.

Appendix: Example Qualification Protocol for ABC Analytical Balance, Equipment No. XXX

Title:	Example Qualification Protocol for ABC Analytical Balance, Equipment No. XXX		
Document No.:	XXX	Version No.:	XXX

Prepared by:

_____   _____

XYZ   Date

Reviewed by:

_____   _____

XYZ   Date

Approved by:

_____   _____

XYZ   Date

Table of Contents

1. Introduction
2. System Description
3. Scope
4. Approach to Qualification
5. Roles and Responsibilities
6. References
7. Operational/Performance Qualification
8. Protocol Deviations
9. Revision History

1 Introduction

This protocol outlines the steps for the qualification of the ABC analytical balance, Equipment No. XXX. The balance is used for weighing analytical samples, reagents, and standards in laboratory ABC.

2 System Description

The ABC analytical balance has a readability of 0.1 mg and a load capability of XX grams. The balance has a built-in calibration weight for internal calibration. A printer is attached to the balance for weight measurement recording.

3 Scope

Documented qualification testing will be carried out to confirm that the balance is installed properly and is performing satisfactorily prior to routine use in support of GXP studies.

4 Approach to Qualification

A holistic approach, which tests the balance and the printer as a single functional unit, will be used for the qualification. The installation qualification (IQ) requirements, along with the operational qualification (OQ) and performance qualification (PQ) tests, which are essentially the same, will be combined into this single qualification protocol.

5 Roles and Responsibilities

The equipment owner or delegate from XYZ Department will be responsible for the qualification of the balance.

6 Reference

1. *Operation Manual for the ABC Analytical Balance.*

7 Operational/Performance Qualification

All the calibration weights used in qualification testing must be either ASTM E617-97 class 1 or OIML class E2 weights.

1. Perform an internal calibration before executing the qualification tests.
2. The balance should be tared before test measurement.
3. *Weighing accuracy test.* This test verifies that the balance is capable of accurate weight measurements from 20 mg to 200 g.
   a. Measure the following calibration weights: 20 mg, 50 mg, 100 mg, 200 mg, 500 mg, 1 g, 2 g, 5 g, 10 g, 20 g, 50 g, 100 g, 200 g.

Acceptance Criteria The weight difference between the measured weight and the certified calibration weight should ≤0.1% of the calibration weights.

Example Results of the Weighing Accuracy Test:

Standard Weight	Measured Weight	Acceptance Range		Pass or Fail
20 mg		20.02	19.98	
50 mg		50.05	49.95	
100 mg		100.1	99.90	
200 mg		200.2	199.80	
500 mg		500.5	499.50	
1 g		1.0010	0.9990	
2 g		2.0020	1.9980	
5 g		5.0050	4.9950	
10 g		10.0100	9.9900	
20 g		20.0200	19.98	
50 g		50.0500	49.95	
100 g		100.1000	99.90	
200 g		200.2000	199.80	

Testing executed by/Date: _____

Verified by/Date: _____

4. *Example repeatability test for 100-g weight.* This test verifies the precision of the weight measurement.
   a. Take 10 measurements of a 100-g weight.
   b. Calculate the standard deviation of the 10 measurements.

Acceptance Criteria The measurement uncertainty is satisfactory if twice the standard deviation of not less than 10 replicate weighings, divided by the amount weighed, does not exceed 0.001 (0.1%).

Results of the Repeatability Test

Measurement Number	Weight
1	
2	
3	
4	
5	
6	
7	
8	
9	
10	
Standard deviation	

$2 \times$ standard deviation$/100$ g $=$ _____ (Pass or Fail)

Testing executed by/Date: _____

Verified by/Date: _____

5. *Linearity test.* This test verifies the linearity of the weight responses.
   a. The linearity check procedure comprises of weighings of an XX-g, YY-g, and ZZ-g standard weight.

Acceptance Criteria The plot of actual weight vs. observed weight should be a linear curve. The coefficient of correlation should be within at least 0.99999.

Results:

Testing executed by: _____

Date: _____

6. *Example Corner-Weighing Difference.* This test verifies the accuracy of the weighing process when the load is off center.
   a. The corner-weighing difference of the balance is checked using a 100-g standard weight.
   b. Measure the weight at the center and close to the corners of the weighing pan.

Acceptance Criteria The weights recorded should be within 0.1% of the standard weight.

Results of the Corner Weighing Test

Pan Position	Measured Weight	Acceptance	Pass or Fail
Center Corner 1 Corner 2 Corner 3			

Testing executed by/Date: _____

Verified by/Date: _____

7. *Example printer verification test.* This test verifies the accuracy of the weight printing process.
   1. Take a measurement of a 100-g weight.
   2. Print the weight record and verify the printed record with the weight on the display panel.
   3. Repeat the procedure with the following weights: 50 g, 10 g, and 1 g.

Acceptance Criteria The weights recorded must match the weight on the display panel.

Example Results of the Printer Verification Test

Standard Weight	Measured Weight on Display	Printed Weight on Paper	Pass or Fail
100 g 50 g 10 g 1 g			

Testing executed by/Date: _____

Verified by/Date: _____

8 Protocol Deviations

All protocol deviations, including test results that fail to meet the acceptance criteria, must be investigated. In consultation with department management and QA (as necessary), resume testing when the problem that caused the failure has been resolved. The problems and their resolutions must be documented, along with any other protocol deviation, as an attachment to this executed protocol.

9   Revision History

Version	Reason for Change
1	New

9

PERFORMANCE VERIFICATION OF NIR SPECTROPHOTOMETERS

HERMAN LAM AND SHAUNA ROTMAN
Wild Crane Horizon Inc.

1 INTRODUCTION

Near-infrared (NIR) spectrophotometry is a measurement technique that can be used for both qualitative and quantitative analyses. The NIR region extends from 780 to 2500 nm (12,820 to 4000 cm^{-1}) as defined by the IUPAC [1], the *United States Pharmacopeia* (USP) [2], the *European Pharmacopoeia* (EP) [3], and the *British Pharmacopoeia* (BP) [4]. The NIR spectra are composed of overtones and combination bands originating from the fundamental vibrational transitions in the mid-infrared region. The extinction coefficients of the NIR bands are much weaker by about one to four orders of magnitude than the bands in the midinfrared range. The low molar absorptivity allows deeper penetration of the NIR radiation into the samples, which enables direct spectral measurement without sample preparation. Many materials, such as glass, are relatively transparent in the NIR region. Together with the additional advantages of being a fast and nondestructive technique, NIR spectroscopy has been implemented in many areas, including research and development, receiving (raw materials identification), manufacturing (process analytical technology), and quality control.

For many NIR applications, the typical spectral features are very board and highly overlapped. Quantitative analysis based on univariate (single-wavelength)

Practical Approaches to Method Validation and Essential Instrument Qualification,
Edited by Chung Chow Chan, Herman Lam, and Xue Ming Zhang
Copyright © 2010 John Wiley & Sons, Inc.

177

calibration is seldom used in the NIR region. Chemometric methods based on multivariate calibrations, which involve predictive model development using spectral information from multiple wavelengths over the spectral regions of interest or full spectrum, are often required [5,6]. Even for qualitative applications such as raw materials identification, chemometric methods are preferred over the simple direct comparison of the sample spectrum against the reference spectrum. Unlike a typical ultraviolet–visible application, where the spectral data of the samples are compared with the reference samples within a very short period of time, the NIR applications usually involve modeling the spectral data of the samples against a calibration model developed days or months ago. The reliable and consistent performance of NIR spectrophotometers is vital for successful modeling and hence the robustness and transferability of the applications.

To assure the robustness and long-term applicability of the chemometric models developed for NIR applications, the performance of NIR spectrophotometers have to be verified regularly over time. The intent of this chapter is to elucidate the tests pertinent to the performance verification of NIR spectrophotometers, using the relevant pharmacopeial documents guidelines. The performance verification tests include wavelength accuracy, wavelength repeatability, photometric noise, photometric linearity, and response stability. Attention is focused on the testing required by the various pharmacopeias and to the standards appropriate for verifying the performance of the instrument in both reflectance and transmittance modes. Finally, some operating tips pertaining to the performance verification of NIR instruments are described. Which tests are required by each pharmacopeia is summarized in Table 1.

To obtain spectra from a sample using NIR, one of three different modes of operation can be employed: reflectance, transmittance, and transflectance. The choice between these three modes depends on the applicability of the operating mode to the task at hand [7]. In transmittance mode, the radiant power of the light passing entirely through a sample (or standard) is assessed (Figure 1). The power of this light has been diminished by sample attenuation. The measurements of transmittance are carried out by dividing the radiant power of the light transmitted through the sample (I) by the radiant power of the incident light (I_0). In reflectance mode, light originating from the source is directed at a

TABLE 1 Performance Tests Required for NIR by Various Pharmacopeias

Test	Pharmacopeia[a]		
	USP 31	BP 2008	EP 7 Ed.
Wavelength accuracy	√	√	√
Wavelength repeatability	—	√	√
Photometric noise	√	√	√
Photometric linearity	√	√	√
Response stability	√	√	√

[a]USP, *United States Pharmacopeia*, General Chapter ⟨1119⟩ [2]; BP, *British Pharmacopoeia*, Appendix IIA [4]; EP, *European Pharmacopoeia*, Section 2.2.40 [3].

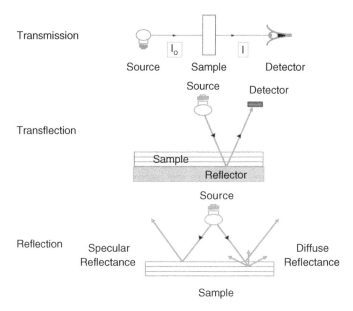

FIGURE 1 Mode of NIR measurements.

sample or standard. The light contacts the sample (or standard) and penetrates it, to some extent, before being reflected. The radiant power of the reflected light is then assessed and quantified by a detector. A measurement of reflectance is then calculated by dividing the attenuated radiant power of the light reflected off the sample (R) by that of the incident light (R_0) reflected off a reference material such as Spectralon [sintered poly(tetrafluoroethylene) (PTFE)]. Transflectance is frequently described as a combination of the reflectance and transmittance modes. In transflectance mode, light is detected at a point where it has twice traveled the length of the sample (or standard), thereby attaining a pathlength double that of the transmittance mode. This mode of operation is achieved by placing a reflector behind the sample (or standard) so that after light has been transmitted through the sample (or standard) once, it may be reflected and transmitted back through the sample or standard (again being subjected to reflectance and attenuating factors).

2 PERFORMANCE ATTRIBUTES

2.1 Wavelength Accuracy

Wavelength accuracy is defined as the deviation of the wavelength reading at an absorption or emission band from the known wavelength of the band. The wavelength deviation can cause significant errors in applications that are based on chemometric models to calculate the qualitative or quantitative results. The test of wavelength accuracy is known by a number of names, including test of wavelength uncertainty, verification of wavelength scale, and abscissa test [2–4,8]. Wavelength accuracy tests serve to ensure that the spectra collected by an NIR system exhibit peaks at wavelengths and wavenumbers that are consistent and within a given tolerance. By verifying that a spectrophotometer is performing within the tolerances given for wavelength accuracy, the consistency of the spectra generated by a frequently verified spectrophotometer with respect to time can be maintained. Additionally, the utilization of traceable standards serves to link the spectra obtained by the spectrophotometer of interest with those obtained by other systems that have also been verified using traceable standards.

While spectra generated from grating instruments have an approximate linear wavelength scale λ (nm), the spectra from Fourier transform instruments are linear in wavenumber υ (cm^{-1}). Most instruments can interconvert the spectral scale from wavelength λ (nm) to wavenumber υ (cm^{-1}), and vice versa. The conversion between wavelength and wavenumber is give by the relationship $\lambda = 10^7/\upsilon$. The relationship between wavelength and wavenumber scale is nonlinear. Care should be taken in selecting the proper wavelength accuracy standard to suit the instrument and the application needs. Some standards are certified in both wavelength and wavenumber scale.

Generally, standards appropriate for this use demonstrate both spectral repeatability and uniform spectral coverage throughout the NIR region [9]. The former of these properties can be achieved by selecting standards that are relatively unaffected by changing the instrument resolution [8]. Several types of material have been tested for their ability to assess the wavelength accuracy of an instrument performing in the NIR region [10]. Table 2 contains a summary of some of these materials and their advantages and disadvantages. A list of some commercially available standards suitable for the assessment of wavelength accuracy test is included in Appendix 1.

Rare earth oxide–based standards have found considerable acceptance as the material of choice for the verification of wavelength accuracy, due to the fact that they can be combined to provide adequate coverage of the NIR region (up to 2000 nm). In fact, all of the wavelength uncertainty and accuracy standards currently recommended by USP ⟨1119⟩ incorporate rare earth oxides [2]. A summary of the wavelength accuracy standards mentioned by the relevant pharmacopeial documents is presented in Table 3.

Standards for Transmittance Mode of Measurement in USP, EP, and BP

NIST SRM 2035 (Glass Standard)   NIST (National Institute of Standards and Technology) SRM 2035 was initially developed as a result of discussions held in 1994 to lay out the desired parameters for a new x-axis performance verification and calibration standard for the NIR region [15]. Two of the main specifications laid out in these discussions were that the peaks of the new standard should be both broad and symmetric (to minimize the peak location dependence on resolution) and that the standard should cover, at a minimum, the range from 1000 to 2000 nm [15]. NIST SRM 2035 is composed of holmium, samarium, ytterbium,

TABLE 2 Wavelength Accuracy Standards and Their Advantages and Disadvantages

Standard	Advantages	Disadvantages
Water vapor	Spectral peaks throughout the NIR region Inexpensive	NIR spectra are complex; very high resolution is required to resolve bands Values cited for some peaks in BP and EP are incorrect [10]; it is advisable to check the HITRAN database [11]
Polystyrene	Some sharp bands in the NIR region Inexpensive	Lacks adequate spectral features, especially at shorter wavelengths of the NIR region Bands are asymmetric and thus dependent on spectral resolution Single thickness generally doesn't have enough intensity to cover the NIR range
Gas standards (acetylene and hydrogen cyanide)	Suitable for high-resolution applications (resolved spectra can be obtained at resolutions of 0.25 cm^{-1} and higher) [11]	Provide traceable peaks only over small wavelength ranges (1513–1541 nm for acetylene and 1528–1563 for HCN) [12,13]
Trichlorobenzene [10,14]	Some sharp peaks or bases in the NIR region	Bands asymmetric Usable features exist only between about 1650 and 2500 nm
High-pressure mercury arc emissions [14]	Sharp bands throughout the NIR region	Generally impractical for routine operation
Methylene chloride	Sharp bands Can be used for transmission and diffuse reflectance (in conjunction with titanium dioxide) Two prominent peaks for wavelength calibration above 2000 nm	Bands less than 1600 nm generally weak; bands used in calibration do not cover the entire NIR region
Rare earth oxides (REOs)	Can be made to have good coverage of the NIR region (up to 2000 nm), depending on the REO mixture Talc is being incorporated in USP NIR Calibrator RS to provide peaks above 2000 nm Stable Smooth and relatively symmetric bands	Most standards currently available do not have the band above 2000 nm

TABLE 3 Summary of the Wavelength Accuracy Standards Mentioned by Various Pharmacopeias

Standard Category	Specific Standard	Suitable for:			Mentioned in:		
		Reflectance	Transmittance	Transflectance	USP 31 [2]	BP 2008 [4]	EP 7 Ed. [3]
Rare earth oxides	NIST 1920(a)[a]	✓					✓
	NIST SRM 2035[b]	✓	✓		✓		
	NIST SRM 2036[c]	✓			✓		
	USP Calibrator RS[d]	✓			✓		
Miscellaneous	Water vapor		✓			✓	✓
	Methylene chloride R		✓			✓	✓
	Methylene chloride R + titanium dioxide	✓				✓	✓

[a]Discontinued; replaced with NIST (National Institute of Standards and Technology) SRM 2036.
[b]Certificate of analysis certifies NIST SRM 2035 for transmittance; however, one of the footnotes in USP 29 says that it can be used in transflectance.
[c]Included in USP 29 as a transmittance standard because if the backing is removed, it can theoretically be used as such, despite the fact that the certificate of analysis says that it is a diffuse reflectance standard.
[d]USP item number 1457844.

TABLE 4 Band Locations of SRM 2035 at 4, 8, and 16 cm⁻¹

Band Location	4-cm⁻¹ Bandwidth		8-cm⁻¹ Bandwidth		16-cm⁻¹ Bandwidth	
	nm	cm⁻¹	nm	cm⁻¹	nm	cm⁻¹
1	1,945.8	5,137.9	1,945.7	5,138.0	1,945.7	5,138.2
2	1,469.6	6,803.9	1,469.4	6,803.4	1,469.3	6,804.0
3	1,366.4	7,316.5	1,366.8	7,315.2	1,366.9	7,314.0
4	1,222.5	8,178.0	1,222.5	8,177.3	1,222.5	8,178.0
5	1,151.7	8,680.6	1,151.7	8,680.9	1,151.6	8,681.6
6	1,075.7	9,294.0	1,075.7	9,294.1	1,075.7	9,293.9
7	975.8	10,245.6	975.8	10,245.4	975.8	10,245.0

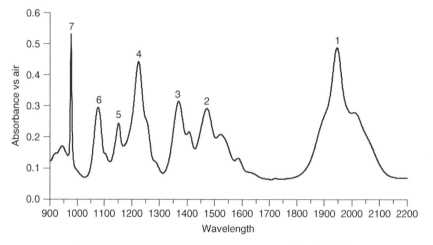

FIGURE 2 NIST SRM 2035 spectrum; 8-nm SSW.

and neodymium rare earth oxides in a base glass [16,17]. The standard is certified for the location of seven absorbance bands ranging from 10,300 to 5130 cm⁻¹ (971 to 1949 nm) at six spectral bandwidths from 4 to 128 cm⁻¹. A summary of the band locations at 4, 8, and 16 cm⁻¹ is given in Table 4. A spectrum of SRM 2035 is shown in Figure 2. SRM 2035 is one of NIST's current transmission standards for the assessment of wavelength and wavenumber accuracy in the NIR spectral region. It is worth to mentioning that SRM 2065 was designed to expand on the work done for SRM 2035, extending the operational range of the standard into the ultraviolet–visible region. It is certified for 13 additional absorption band locations from 805 to 334 nm [17]. Both SRM 2035 and 2065 are composed of holmium, samarium, ytterbium, and neodymium rare earth oxides in a base glass.

Methylene Chloride (Liquid Standard) According to EP and BP, methylene chloride at an optical pathlength of 1.0 mm can be used for wavelength verification [3,4]. Methylene chloride has eight characteristic bands: at 1155, 1366, 1417, 1690, 1838, 1894, 2068, and 2245 nm (8658.0, 7320.6, 7057.2, 5917.2, 5440.7,

5279.8, 4835.6, and 4454.3 cm^{-1}) for comparison. The bands at 1155, 1690, and 2245 nm are used for calibration.

Water Vapor Water vapor contains prominent peaks with good signal-to-noise ratios in the spectral region between 5000 and 5600 cm^{-1} and 7000 and 7425 cm^{-1} [8]. For resolution higher than 4 cm^{-1}, the absorption band water vapor at 7299.86 cm^{-1} (1369.9 nm) can be used for wavelength accuracy check [3,4]. The wavelength value cited in EP and BP is slightly different from the value cited (7299.43 cm^{-1}, 1360.59 nm) in the HITRAN 2004 database [10]. The resolution required to resolve the water vapor peaks does not represent routine normal usage.

Standards for Reflectance Mode of Measurement in USP, EP, and BP

NIST SRM 1920a Since the early 1980s, NIST (formerly the National Bureau of Standards), has been developing and/or producing rare earth oxide–based standards suitable for the assessment of wavelength accuracy in the near infrared spectral region [18]. The first such standard, NIST SRM 1920a, was a reflectance standard that contained a mixture of dysprosium, erbium, and holmium rare earth oxide powders that were mixed and pressed into an aluminum cavity and sealed off using a sapphire window and a retaining ring [19]. Sapphire was employed because of its lack of absorption features in the near infrared. The standard is used for wavelength verification from 740 to 2000 nm for spectral bandwidths of 2, 3, 4, 5, and 10 nm. Thirty-seven band locations have been reported in the certificate. The uncertainty of the peak location is no greater than 1 nm. Three bands at 1261, 1681, and 1935 nm are generally used for wavelength accuracy comparison [2].

Because NIST SRM 1920a was introduced at a time when there were virtually no other standards available for wavelength validation of the near infrared, it became (somewhat by default), the industry standard for wavelength validation [18]. As time went on, numerous commercial standard vendors began to produce secondary standards with compositions similar to SRM 1920a. NIST ceased its production of SRM 1920a around 2005, citing the commercial acceptance of the standard and the fact that highly similar standards were available from many secondary standard vendors as reasons for its discontinuance [18]. Those still requiring NIST-produced and NIST-certified wavelength standards were instructed to use NIST SRM 2036, the near infrared wavelength/wavenumber reflection standard [18].

NIST SRM 2036 Despite the success and widespread acceptance of SRM 1920a, it was apparent early that the standard had numerous shortcomings. Many industries employing SRM 1920a (including the pharmaceutical industry) require lower uncertainties, certification at higher resolution than the current 2-nm slit spectral width limitation, less dependence of the band position on instrument resolution, and for Fourier transform instruments, certification of the band locations in wavenumbers rather than wavelengths. SRM 2036 addressed some of the problems inherent to SRM 1920a. Most notably, it minimizes the dependence of band position on resolution.

SRM 2036 is made by contacting a rare earth oxide glass (composed from a formulation initially designed and used for transmittance measurements), with a piece of sintered poly(tetrafluoroethylene) [18]. It is worth noting that the glass portion of SRM 2036 is actually made from the same formulation as SRM 2035 and is identical to the formulation found in SRM 2065 [18]. SRM 2036 is suitable for the verification and calibration of the wavelength/wavenumber scale of NIR spectrometers operating in diffuse reflectance mode [20]. The standard is certified for the location of seven absorbance bands ranging from 971 to 1949 nm (10,300 to 5130 cm^{-1}) at three spectral bandwidths of 3, 5, and 10 cm^{-1}. A summary of the band locations is given in Table 5. A spectrum of SRM 2036 is shown in Figure 3.

TABLE 5   Band Locations of SRM 2036 at 3, 5, and 10 cm^{-1}

Band Location	3-cm^{-1} Bandwidth (nm)	5-cm^{-1} Bandwidth (nm)	10-cm^{-1} Bandwidth (nm)
1	1945.7	1945.8	1945.6
2	1469.6	1469.6	1469.5
3	1367.1	1367.2	1367.3
4	1222.1	1222.1	1222.1
5	1151.4	1151.2	1151.0
6	1075.7	1075.7	1075.8
7	976.0	976.0	975.9

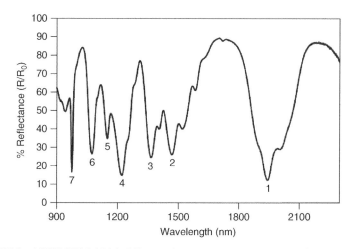

FIGURE 3   NIST SRM 2036 diffuse reflectance spectrum; 5-nm SSW. (From [18].)

USP NIR System Suitability Reference Standard   The USP NIR system suitability reference standard consists of a mixture of dysprosium, erbium, and holmium rare earth oxides pressed into a proprietary housing with a mechanical seal that mitigates moisture exposure [21]. This standard is similar to NIST 1920a but with the addition of talc (acid metasilicates of magnesium) to the rare earth oxide mixture to give an additional absorbance band at 2319 nm [22]. The standard issued for wavelength verification from 700 to 2500 nm for two spectral bandwidth classes (≥ 5 or <5 nm). Eight band locations have been reported in the certificate. A summary of the band locations is given in Table 6. A spectrum of the USP NIR system suitability reference standard is shown in Figure 4.

TABLE 6   Band Locations of the USP NIR System Suitability Reference Standard

Band	Bandwidth Not Greater Than 5 cm^{-1} (nm)	Bandwidth 5 cm^{-1} or Greater (nm)
Dy$_2$O$_3$	1065.4	1066.3
Dy$_2$O$_3$	1261.1	1261.8
Er$_2$O$_3$	1536.2	1534.4
Dy$_2$O$_3$	1612.2	1612.2
Dy$_2$O$_3$	1682.7	1681.4
Ho$_2$O$_3$	1847.6	1847.3
Ho$_2$O$_3$	1931.9	1935.5
Talc	2391.8	2391.8

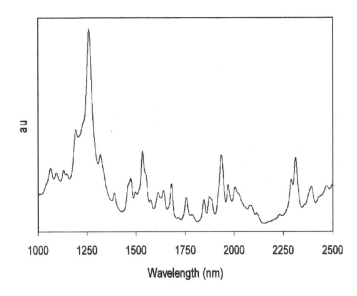

FIGURE 4   Spectrum of USP NIR system suitability reference standard.

Methylene Chloride Plus Titanium Dioxide   The *European Pharmacopoeia* has described a suspension of 12 g of titanium dioxide in about 4 mL of methylene chloride for wavelength verification in reflectance mode [3]. The mixture is shaken vigorously and allowed to settle for 2 minutes before a spectrum is measured. Titanium dioxide has no absorption in the NIR region and provides a diffuse reflectance matrix. Methylene chloride has eight characteristic bands: at 1155, 1366, 1417, 1690, 1838, 1894, 2068, and 2245 nm (8658.0, 7320.6, 7057.2, 5917.2, 5440.7, 5279.8, 4835.6, and 4454.3 cm^{-1}) for comparison.

Standards for Transflectance Mode of Measurement   Even though transflectance measurements are fairly commonly used in NIR, there is still the lack of a well-established standard. The task to develop a suitable reference standard that can accommodate the optical configuration and path length variations for the transflectance measurement can be very challenging. The use of NIST SRM 2036 for wavelength verification in transflectance mode has been mentioned [18]. The EP and BP have not explicitly provided a reference for the wavelength verification in transflectance mode. However, the methylene chloride and titanium dioxide mixture used for reflectance measure can also be used for transflectance measurement according to Burgess and Hammond [10].

Tests   Wavelength accuracy is verified by measuring a known wavelength reference standard with well-characterized absorption peaks and comparing the recorded wavelength of the absorption peak(s) against the value(s) listed in the certificate of that reference standard. All three of the relevant pharmacopeial chapters prescribe wavelength accuracy tests that involve the comparison of a standard's known wavelength/wavenumber values (obtained from the appropriate certificate of analysis) to those obtained by scanning the same standard using the instrument to be verified.

Set the NIR spectrophotometer to the same resolution as that used to certify your standard (see the certificate of analysis that corresponds to your wavelength/wavenumber standard). Next, obtain the spectrum of the wavelength standard and compare the positions of at least three peaks located across the spectral regions (see Table 7) to the certified values in the certificate of analysis for the peak chosen.

Acceptance   Each peak measured should be within the tolerance of the value certified. Ranges of tolerance may be obtained by applying the tolerance of the nearest wavelength/wavenumber to the values you have obtained. The tolerances in wavelength and wavenumber in various spectral ranges outlined in the BP, EP, and USP are shown in Table 7.

2.2   Wavelength Repeatability

While it is important to maintain wavelength accuracy, it is also important for the spectrometer to have long-term wavelength reproducibility for successful

TABLE 7 General Acceptance for Wavelength Accuracy Verification

Wavelength (nm)		Wavenumber (cm^{-1})	
Range	Acceptance	Range	Acceptance
EP and BP			
1200	±1	8300	±8
1600	±1	6250	±4
2000	±1.5	5000	±4
USP 31			
700–2000	±1	below 5000	±8
2000–2500	±1.5	5000–14,000	±4

NIR applications. Fourier transform NIR systems that use an internal HeNe laser for reference generally have good wavelength reproducibility. The wavelength reproducibility for dispersive systems that use a mechanical wavelength selection device needs to be monitored regularly.

Tests Tests of wavelength repeatability assess the precision of the abscissa (*x*-axis) scale. Both the British and European pharmacopeias have specified that "suitable" standards should be used to acquire spectra appropriate to test an instrument's wavelength repeatability [3,4]. The PASG (Pharmaceutical Analytical Sciences Group) NIR subgroup and the ASTM (American Society for Testing and Materials) provide further guidance as to which standards are acceptable for this type of testing. PASG notes that polystyrene and rare earth oxides are examples of appropriate standards, whereas ASTM names high-pressure mercury, trichlorobenzene, and SRM 2035 [8,14]. Since all of the aforementioned standards are used for the testing of wavelength accuracy, it is acceptable to use a wavelength accuracy standard, as it should provide well-defined peaks.

To adequately assess the wavelength repeatability of an instrument, a sufficient number of scans of the standard chosen should be obtained to ensure that the data set used to assess this property is large enough to encompass the normal variability for the wavelength measurement of the standard. Instructions regarding the number of scans required of the standard are lacking from most of the documents examined here; however, the ASTM specifies that the standard should be scanned a total of 10 times [14].

Acceptance The British and European pharmacopeias, as well as PASG, have specified that wavelength repeatability should be assessed by ensuring that the standard deviation of a wavelength is consistent with the specifications laid out by the NIR manufacturer [3,4,8]. Conversely, the ASTM essentially leaves the selection of tolerances up to the instrument's user, but specifies that the standard deviation should be collected at two wavelengths [14]. These two wavelengths should ideally be certified, well defined, and should surround the region of interest.

The standard deviation of a peak should be calculated as:

$$S = \sqrt{\frac{\sum (\lambda_i - \lambda_{avg})^2}{n - 1}} \qquad (1)$$

where S is the standard deviation, λ_i the wavelength of the ith wavelength observed, λ_{avg} the average peak wavelength at the peak location, and n the number of sample spectra taken (according to the ASTM guideline, $n = 10$). Ideally, to meet all the guidelines laid out for wavelength repeatability by the PASG, ASTM, and British and European pharmacopeias, one would take 10 spectral scans of the same standard and assess the standard deviations of two certified peaks spanning the region for which NIR is being employed [3,4,8].

Typical values given by NIR manufacturers for wavelength/wavenumber reproducibility (wavelength repeatability) vary in value. For example, PerkinElmer's Spectrum 100N FT-NIR spectrometer specifications note that the machine should yield results that are better than 0.004 nm at 1390 nm (or 0.02 cm^{-1} at 7200 cm^{-1}) [23], whereas Bruker Optics' multipurpose analyzer (NIR) specifications note that results should be better than 0.01 nm at 1250 nm (or 0.05 cm^{-1} at 8000 cm^{-1}) [24].

2.3 Photometric Noise

Photometric noise can originate from a number of sources within an instrument. The detector noise is usually a key contributor to the overall noise in NIR measurements. Assessment of photometric noise is crucial to ensuring that data integrity is maintained, as the effect of noise can obscure the true data encompassed within the raw data. A good signal-to-noise ratio is desirable to differentiate spectral data associated with the analytes from the background interferences from the sample matrixes. The noise problem is more prominent in a high absorption situation (low flux), where there is less light reaching the detector. Consequently, the signal-to-noise ratio decreases, and noise plays a larger role in obscuring the data.

Tests The guidelines for the assessment of photometric noise laid out by the USP differ from those of the British and European pharmacopeias. In short, the USP chapter delves into the details of the test by setting out parameters for testing as well as performance tolerances. The British and European pharmacopeias defer largely to the noise guidelines laid out by the individual spectrophotometer's manufacturer.

A test of photometric noise for NIR involves an assessment of the noise contained within a spectrum that was generated by scanning an appropriate standard. The USP guidelines actually stipulate that two test settings should be made for photometric noise: one at high flux (using a high flux standard, e.g., 99% reflectance) and one at low flux (using a low flux standard, e.g., 10% reflectance) [2].

The U.S., British, and European pharmacopeias make some further recommendations as to which standards should be employed for testing instrument noise. Specifically, the USP notes that the standards should be "traceable," whereas the British and European pharmacopeias mention that the reflectance standards should be "suitable" and that they could be made of white reflective ceramic tiles or reflective thermoplastic resins (e.g., PTFE) [2–4]. It has been suggested that the ideal standard should be homogeneous, spectrally flat, and nonglossy [25]. A list of some commercially available standards suitable for the assessment of noise is included in Appendix 2 to provide readers with a starting point in their search for appropriate standards. Information about which standards are appropriate for testing noise in the transmittance mode is obtained somewhat less easily.

Both the British and European pharmacopeias state that the test should be performed over the wavelength range specified by the manufacturer. The previous USP general chapter on NIR used to recommend that the test should be performed in "successive nominal 100 nm segments" across the instrument's range, from 1200 to 2200 nm [2]. Noise is generally assessed by measuring the root mean square or peak-to-peak noise of a spectrum. The USP used to recommend that the root mean square should be measured at successive nominal 100-nm segments and should be calculated using the equation [2]

$$\text{root mean square} = \sqrt{\frac{1}{N} \sum_{i}^{N} (A_i - \overline{A})^2} \tag{2}$$

where A_i is the absorbance at a data point, $\overline{A}$ the average absorbance for the spectral segment being evaluated, and N the number of data points in the spectral segment. The noise evaluation section in the NIR general chapter was revised in 2008 to mention the use of built-in software to determine the system noise in the operation range. As no specific has been given as to how to determine the noise in the new NIR general chapter, the stipulations in previous versions of the NIR general chapter will be a useful reference in evaluating the automated routines for noise determination.

In contrast, the British and European pharmacopeias specify that the peak-to-peak noise over a spectral range recommended by the instrument manufacturer should be calculated [3,4]. The peak-to-peak noise can be obtained by subtracting the minimum peak from the maximum peak for the wavelength range being assessed. According to EP, the peak-to-peak noise is approximately twice the standard deviation. (*Note*: Another way to estimate the standard derivation of the noise, as one-fifth of the peak-to-peak noise, has been mentioned elsewhere [26].)

Acceptance
USP Requirements Prior to 2008, the acceptance for both high and low flux (separately) was as follows: Calculate the root mean square for each 100-nm spectral segment from 1200 to 2200 nm. Next, the average of the root mean

squares for each flux is calculated. The high-flux root mean square average should be less than 0.3×10^{-3}, whereas the low-flux root mean square average should be less than 1×10^{-3}. Additionally, determine which 100-nm spectral segment yields the highest root-mean-square value for high and low flux. For the high-flux segment having the largest root mean square, ensure that its value is less than 0.8×10^{-3}, and for the low-flux sample, less than 2.0×10^{-3}. The recommended acceptance for the noise testing procedures was revised in 2008. The tolerance for the noise-testing procedures should demonstrate a suitable signal-to-noise ratio for the intended application. The previous acceptance can still serve as a useful reference.

British and European Pharmacopeial Requirements   Measure the peak-to-peak noise for the manufacturer-suggested wavelength range. Ensure that this measurement is consistent with the noise specification laid out by the instrument manufacturer. It should be noted that typical values for noise vary by manufacturer and instrument model.

2.4   Photometric Linearity and Response Stability

Linearity can be defined as the directly proportional relationship between concentration and response. The assessment of photometric linearity is an important test of performance verification, since it helps to ensure that compositional changes are reflected in the absorbance measured in a directly proportional manner. Ideally, it is assuring to demonstrate absolute photometric accuracy for NIR measurements. However, the spectra obtained from the reference standards using the NIR system to be tested are subjected to unavoidable differences in experimental conditions from the provider of reference standards. Hence, the use of a set of calibration standards for photometric linearity may not be used for a photometric accuracy test. The percent reflectance and absorbance values simply allow the user to know that an acceptable range of sample-driven responses can be made. Instead of demonstrating absolute photometric accuracy, a less vigorous test to demonstrate response stability will be verified over time. The response stability can provide some assurance that the photometric responses will be consistent over time. Good photometric linearity and stability are necessary to assure consistency in the responses to enable robust chemometric applications.

According to the EP, as long as the standards have not changed and the same reference background and identical experimental conditions, including the positioning of the standard, have been used, consistency of response through both reference and observational scans can be assessed [2–4]. Since the detector responses differ over the NIR spectral range, it is desirable to conduct the photometric linearity test at different spectral regions. Before 2008, USP specified the wavelengths at which linearity testing should be performed (1200, 1600, and 2000 nm). The wavelength recommendation in the USP was removed in 2008. The British and European pharmacopeias do not specify the wavelength locations where the photometric linearity should be verified. However, it should be noted that the Pharmaceutical and Analytical Sciences NIR subgroup (based in the UK) has published a rather comprehensive paper on the development and validation of NIR methods and has provided wavelengths and tolerances that are consistent with those provided previously by the USP for linearity testing [2,8]. Consequently, it is desirable to perform the tests at the wavelengths specified by the USP, as doing so will ensure compliance with the U.S., British, and European pharmacopeias.

Tests

Photometric Linearity   Standards should be obtained in accordance with the pharmacopeial guidelines, which stipulate that a minimum set of four standards with reflectance varying between 10 and 90% should be acquired [2–4]. Additionally, standards having lower reflectances (e.g., 2%, 5%) should be added to the set if the instrument is being employed for the measurement of low reflectance (less than 10% reflectance) samples [2–4]. Use an instrument setting similar to those used to obtain the certified values.

Tests of photometric linearity essentially compare the difference in response between reference and observed spectra. These response values are obtained from scans on a set of standards of varying relative reflectance or transmittance. The reference spectra should only be acquired once (unless there are special circumstances) and should be obtained at a time when the instrument is known to be in good operating condition. Assuming that the reference spectra have already been obtained, rescan the set of standards to obtain the spectra observed during current verification and, subsequently, plot the observed reflectance against the reference reflectance for various wavelengths (Figure 5). This will yield a number of plots, one for each wavelength. A linear regression is then performed on each plot, the results of which are compared to the tolerances for the test.

Response Stability   Compare the reflectance (or absorbance) responses from the spectra obtained from the previous verification time point with the responses obtained from the current time point.

Acceptance

Photometric Linearity   Plot R_{obs} vs. R_{ref} at 1200, 1600, and 2000 nm [2,8]. Perform a linear regression for the plot of each wavelength; the slope should be within the range 1.0 ± 0.05, and the intercept should be in the range 0.0 ± 0.05 for each of the three regressions.

Response Stability   A tolerance of $\pm 2\%$ is acceptable [3,4].

3   PRACTICAL TIPS IN NIR PERFORMANCE VERIFICATION

1. It should be noted that instrument vendors frequently include software that, when executed, leads the user through tests to assess the instrument's performance

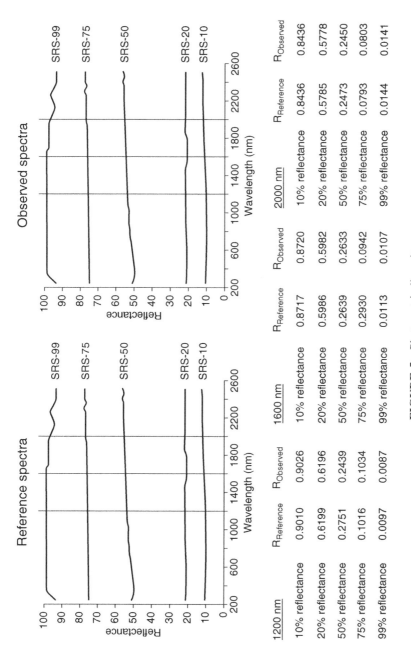

FIGURE 5 Photometric linearity.

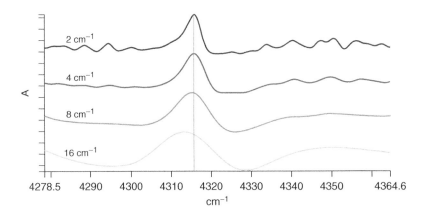

FIGURE 6 Visual peak location using methane spectra obtained under varying resolutions; peak at approximately 4314 cm^{-1}.

with respect to a specific pharmacopeial guideline. These built-in routines are highly useful.

2. Each of the three relevant pharmacopeial chapters lists a number of factors that can affect the quality of the spectra obtained by an NIR instrument. The factors noted are sample temperature, moisture and residual solvent, sample thickness, sample optical properties, polymorphism, and the age of the samples [2–4]. Attempts to control those factors to improve the consistency of the test results are a worthwhile investment of time.

3. The appropriate care and handling of standards plays a critical role in ensuring their reliability. Instructions pertaining to these protective measures and to the maintenance of standards are generally issued by the standard's manufacturer and accompany the standard when shipped. Some common preventive techniques include the wearing of gloves (to prevent surface contamination) as well as the use of protective cases at all times when the standard is not in use.

4. Spectral resolution often plays an important role in determining the sharpness of the spectral features obtained. For example, visual assessment of the location of a peak maximum may vary with resolution (see Figure 6). An improper resolution setting may lead to a merging of adjacent peaks (lower resolution than specified) or emerging of an extract peak of the shoulder near the peak of interest (higher resolution than specified). Consequently, it is important to compare spectra that have been obtained using the same resolution.

5. When using Spectralon [sinter poly(tetrafluoroethylene)] as a reflectance standard, care must be taken to ensure that the Spectralon is free of contamination. Spectralon is highly porous to hydrocarbons and thus prone to hydrocarbon contaminations. The hydrocarbon contaminations can cause changes in the reflectance where the materials absorb. Refinishing the Spectralon materials can be done by washing or mild sanding with no loss of reflectance properties [23].

6. Appropriate reference material must be used for reflectance measurements.

7. A number of standards exist for the purpose of assessing wavelength accuracy in the NIR region, and care must be taken to select the appropriate standard for your instrument's configuration and the methods used for wavelength separation. For a dispersive instrument, wavelength accuracy is best compared on a nanometer scale. Wavenumber scale comparison is better for a Fourier transform type of instrument.

8. Performance verification standards should be obtained from commercial vendors that have demonstrated traceability to one or more governmental measurement institutes (including the National Institute of Standards and Technology, the National Physical Laboratory, and the National Research Council of Canada). Since the availability and cost of NIST standards may be prohibitive to some users, abbreviated lists of vendors and information about the standards that they provide for the verification of wavelength accuracy and photometric noise evaluations in the NIR region has been included in Appendixes 1 and 2 for ease of reference.

9. Even though there is no mention of the frequency of verification tests, it is advisable to conduct the tests at least twice a year, to provide assurance and maintain robustness of the NIR application running on the instrument.

REFERENCES

1. N. Sheppard, H. A. Willis, and J. C. Rigg. International Union of Pure and Applied Chemistry: names, symbols, definitions and units of quantities in optical spectroscopy. *Pure Appl. Chem.*, 57(1), 1985.

2. *U.S. Pharmacopeia*, 31 Chapter ⟨1119⟩, Near-Infrared Spectrophotometry. USP, Rockville, MD.

3. *European Pharmacopoeia*, 6th ed., sec 2.2.40, Near-Infrared Spectrophotometry. European Directorate for the Quality of Medicine, Strasbourg, 2005.

4. *British Pharmacopoeia*, Appendix IIA, Infrared Spectrophotometry: Near-Infrared Spectrophotometry. BP, Norwich, UK, 2008.

5. R. Kramer. *Chemometric Techniques for Quantitative Analysis*. Marcel Dekker, New York, 1998.

6. J. Miller and J. Miller. *Statistics and Chemometrics for Analytical Chemistry*, 5th ed., Pearson Prentice Hall, Upper Saddle River, NJ, 2005.

7. G. Reich. Near infrared spectroscopy and imaging: basic principles and pharmaceutical applications. *Adv. Drug Deliv. Rev.*, 57:1109–1143, 2005.

8. *Guidelines for the Development and Validation of Near Infrared (NIR) Spectroscopic Methods*. Pharmaceutical Analytical Sciences Group, NIR Sub Group, UK, 2001 (http://www.pasg.org.uk/).

9. S. J. Choquette, J. C. Travis, C. Zhu, and D. L. Duewer. Wavenumber standards for near-infrared spectrometry. In *Handbook of Vibrational Spectroscopy*. Wiley, Hoboken, NJ, 2002.

10. C. Burgess and J. Hammond. Wavelength standards for the near-infrared spectral region. *Spectroscopy*, 22(4):40–48, 2007.

11. L. S. Rothman, D. Jacquemart, A. Barbe, D. C. Benner, M. Birk, L. R. Brown, et al. The HITRAN 2004 molecular spectroscopic database. *J. Quant. Spectrosc. Radiat. Transfer*, 96, 2005.

12. S. L. Gilbert and W. C. Swann. *Acetylene* $^{12}C_2H_2$ *Absorption Reference for 1510 nm to 1540 nm Wavelength Calibration*. SRM 2517a. NIST Special Publication 260–133. NIST, Gaithersburg, MD, 2001.

13. S. L. Gilbert, W. C. Swann, and C.-M. Wang. *Hydrogen Cyanide* $H^{13}C^{14}N$ *Absorption Reference for 1530 nm to 1565 nm Wavelength Calibration*. SRM 2519a. NIST Special Publication 260–137. NIST, Gaithersburg, MD, 2005.

14. *Standard Practice for Describing and Measuring Performance of Ultraviolet, Visible, and Near-Infrared Spectrophotometers*. ASTM E 275–01. ASTM, West Conshohocken, PA, 2007.

15. S. Choquette, J. Travis, L. O'Neal, C. Zhu, and D. Duewer. A rare-earth oxide glass for the wavelength calibration of near-infrared dispersive and Fourier-transorm spectrometers. *Spectroscopy*, 16(4):14–19, 2007.

16. *Near Infrared Transmission Wavelength Standard from 10300 cm^{-1} to 5130 cm^{-1}*. Certificate of Analysis. SRM 2035. NIST, Gaithersburg, MD.

17. *Ultraviolet–Visible–Near-Infrared Transmission Wavelength/Vacuum Standard*. Certificate of Analysis. SRM 2065. NIST, Gaithersburg, MD.

18. S. J. Choquette, D. L. Duewer, L. M. Hanssen, E. A. Early. Standard Reference Material 2036: Near-Infrared Reflection Wavelength Standard. *Appl. Spectrosc.*, 59(4):496–504, 2005.

19. *Near Infrared Reflectance Wavelength Standard from 740 nm to 2000 nm*. Certificate of Analysis. SRM 1920a. NIST, Gaithersburg, MD.

20. *Near-Infrared Wavelength/Wavenumber Reflection Standard*. Certificate of Analysis. SRM 2036. NIST, Gaithersburg, MD.

21. B. Koch. RE: Annual recertification not required for USP NIR system suitability RS. USP, Rockville, MD, Apr. 30, 2008.

22. USP certificate: USP NIR system suitability RS. Catalog No. 1457844. USP, Rockville, MD, May 2008.

23. Technical specifications for the Spectrum 100N FT-NIR spectrometer. PerkinElmer, Waltham, MA.

24. Specifications and options: Bruker Multi Purpose Analyzer. Bruker Optics, Billerica, MA.

25. A. Springsteen. Standards for the measurement of diffuse reflectance: an overview of available materials and measurement laboratories. *Anal. Chim. Acta*, 380:379–390, 1999.

26. G. M. Hietje. *Anal. Chem.*, 44(6):81A–88A, May 1972.

Appendix 1: Standards for Wavelength Accuracy Testing

| Company | Part/Description | Suitable for: | | Made of: | | | | Approximate Price in 2007 | Traceable to: |
		Reflectance	Transmittance	Dysprosium Oxide	Holmium Oxide	Samarium Oxide	Erbium Oxide		
Avian Technologies, Inc.	Multicomponent wavelength calibration standard (250–850 nm)	✓		✓	✓	✓			NRC, NIST
	WC-FW-NIR-01 (1.25-in. diameter)							$560.00	
	WC-FW-NIR-02 (2.0-in. diameter) In durable and inert matrix, in delrin holder							$625.00	
	Packed powder wavelength calibration standard (identical to NIST SRM 1920a)	✓		✓	✓		✓		Calibrated by NRC (doesn't say "traceable" in documents)
	WC-PP-1920-01 (1.25-in. window)							$1400.00	
	WC-PP-1920-02 (1.5-in. window)							$1500.00	
	Holmium oxide filter		✓		✓				NRC
	WC-HO-01(1-in. filter)							$375.00	
	WC-HO-02 (2-in. filter)							$550.00	
	WC-HO-CUV(10-mm cuvette filter)							$375.00	
	Didymium oxide filter		✓						NRC
	WC-DD-02(2-in. square filter)							$475.00	
	WC-DD-CUV(10-mm cuvette filter)							$375.00	
	Mixture of praseodymium and neodymium oxides								

(Continued overleaf)

Appendix 1: (*Continued*)

Company	Part/Description	Suitable for:		Made of:				Approximate Price in 2007	Traceable to:
		Reflectance	Transmittance	Dysprosium Oxide	Holmium Oxide	Samarium Oxide	Erbium Oxide		
Labsphere	Spectralon UV-Vis-NIR Wavelength calibration standards sets	✓		✓	✓		✓		NIST
	WSS-03-010 (1.25-in. diameter)							$475	
	WSS-03-020 (2.00-in. diameter)							$525	
	Contains 3 standards, each composed of a different rare earth oxide								
	Spectralon multicomponent wavelength calibration standard	✓		✓	✓		✓		NIST
	WCS-MC-010 (1.25-in. diameter)							$650	
	WCS-MC-020 (2.00-in. diameter)							$705	
	Three oxides combined into one standard								
McCrone Scientific	McCrone standard Neodymium-doped YAG		✓						NPL
Nelson Gemmological Instruments	Nelson M-42 standard Monocrystalline erbium-doped YAG		✓				✓		NPL

Appendix 2: Standards for Linearity and Noise Testing

Company	Part/Description	Suitable for: Reflectance	Transmittance	Approximate Price in 2007[a]	Traceable to:
Avian Technologies, Inc.	Fluorilon gray scale standard sets (2.0 in. in diameter)	✓			NRC
	FSS-04-02 (4 standards: 2%,10%, 50%, and 99% reflectance)			$1400.00	
	FSS-08-02c (8 standards: 2%, 5%, 10%, 20%, 40%, 60%, 80%, and 99% reflectance)			$2700.00	
	Inconel on optical glass neutral density filter sets calibrated from 400–2000 nm		✓		NRC and/or NIST
	TS-ND06-02C (2.0 in. square, 6 filters); nominal densities are 0.03, 0.1, 0.3, 0.5, 1.0, and 2.0 OD			$1800.00	
	TFS-06B (10-mm cuvettes, with 6 standards); includes 4 different absorbance standards (0.5, 1.0, 1.5, and 2.0 A), a blank, and a didymium oxide wavelength calibration filter			$1250.00	
Labsphere	Spectralon diffuse reflectance standard sets[b]	✓			NIST
	RSS-04-010 and RSS-04-020 (1.25- and 2.00-in. diameters, respectively; contains 4 standards: 2%, 50%, 75%, and 99% reflectance)			$1320, $1465	
	RSS-08-010 and RSS-08-020 (1.25-in. and 2.00-in. diameters, respectively; contain 8 standards: 2%, 5%, 10%, 20%, 40%, 60%, 80%, and 99% reflectance)			$2585, $2810	
SphereOptics	Zenith diffuse reflectance standards, typical reflectance UV/vis-NIR gray scale standards sets (calibrated from 250 to 2500 nm with calibration data in 50-nm steps)	✓			NIST, PTB
	SG 3089 and 3088 (1.25- and 2-in. diameters, respectively; 25%, 50%, 70%, and 99% reflectance)			$1680, $1885	
	SG 3043 and 3044 (1.25- and 2-in. diameters, respectively; 2%, 5%, 10%, 25%, 50%, 75%, and 99% reflectance)			$3165, $3365	

[a]Standards are available individually as well (and possibly at different percentage reflectances than the sets).
[b]Approximate 2007 prices; subject to change.

10

OPERATIONAL QUALIFICATION IN PRACTICE FOR GAS CHROMATOGRAPHY INSTRUMENTS

WOLFGANG WINTER
Matthias Hohner AG

STEPHAN JANSEN, PAUL LARSON, CHARLES T. MANFREDI, AND WILLIAM H. WILSON
Agilent Technologies Inc.

HERMAN LAM
Wild Crane Horizon Inc.

1 INTRODUCTION

Gas chromatography (GC) is a powerful analytical technique used widely for pharmaceutical, chemical, food, cosmetic, and environmental applications. High-resolution capillary GC techniques are capable of separating and quantitating components with sufficient volatility in very complex mixtures with high sensitivity. To provide a high level of assurance that the data generated from the GC analysis are reliable, the performance of a GC system should be monitored. Unlike the very well established performance verification guidance for high-performance liquid chromatography (HPLC) systems [1,2], discussions on performance verification for GC systems are relatively scarce [3,4]. In this chapter, the rationale and tests to verify some key performance attributes for a typical GC system and troubleshooting suggestions for verification tests and routine operations are discussed.

Practical Approaches to Method Validation and Essential Instrument Qualification,
Edited by Chung Chow Chan, Herman Lam, and Xue Ming Zhang

TABLE 1   Typical Device Certifications

Mark	Certification	Description
$C\epsilon$	CE	European Union EMC and safety
c(UL)us LISTED	UL/CSA	North American product safety
FC	FCC	North American EMC

2   PARAMETERS FOR QUALIFICATION

2.1   Test Design and Rationale

Instrument performance verification is an integral component of the instrument life cycle [5–10]. In heavily regulated environments, these efforts can be quite substantial and typically must demonstrate operation to a wider standard. The majority of vendors can provide very sophisticated and highly automated systems to meet user needs. Off-the-shelf gas chromatographic equipment should conform to industry design standard, build, and test practices, and it should meet appropriate manufacturing industry certifications. Typical certifications are listed in Table 1.

Performance testing of a GC system frequently employs a combination of modular and holistic tests. *Modular testing* consists of direct measurements made against a system or a component of a system. These measurements typically employ independently calibrated measurement tools, such as precision thermometers (electronic or mechanical), flow measurement devices (dynamic flow meters), and manometers. *Holistic* (or *inference*) *testing* comprises qualitative and/or quantitative evaluations that are performed as an instrument is used in its typical operational modes. If well-characterized material is analyzed using a precisely defined method, the output can be evaluated against expected criteria. This is a very useful approach because it evaluates the device in its fully connected state (some variables required by the test methodology, such as an evaluation column, must be carefully controlled). The premise is that all components in the system must be working as expected to provide the predetermined output.

The two testing methods should be employed in concert wherever possible, because each has limitations. Holistic testing typically fails to provide the independent influence of an externally indexed measurement device. Pure modular testing fails to test the whole with respect to the connection points and potential for the cascading effects of series components. Measurements made against established standards allow for comparison of the results across vendors, geographies, and environmental influences. This rigor provides confidence that methods are transferable.

2.2   Risk Assessment

Test selection should be based on a combination of risk assessment and scientific rational. Scientific rational must always be used to temper decisions that result

TABLE 2 Risk Probability Classification

Probability Code	Probability of Occurrence
A	Likely to occur under most operating conditions or frequency of use
B	Likely to occur under certain conditions or over time
C	May occur with time or operating conditions
D	Unlikely to occur

TABLE 3 Failure Severity Classification

Severity	Impact If Not Detected
High	Renders data unusable and/or difficult to pick up by secondary instrument readiness testing (system suitability/check testing)
Medium	Affects data but is likely to be detected by secondary instrument readiness testing
Low	May affect data and is obvious to secondary instrument readiness testing
Minimal	Unlikely to affect data

TABLE 4 Risk Assessment Ranking[a]

Failure Severity	Probability of Occurrence			
	A	B	C	D
High	1	1	2	3
Medium	1	2	3	4
Low	2	3	4	5
Minimal	3	4	5	5

[a] 1, Critical; 2, serious; 3, moderate; 4, minor; 5, negligible.

from simple risk analysis exercises. When developing a test suite, always ask whether a test provides value or is just being run because it can be. Risk-based assessment allows the user to assign a risk value to components, operations, or processes within a system. Risk assessments are based on the probability or potential of failure (Table 2) and the criticality or severity if a failure does occur, as well as the ease of failure detection (Table 3). A numerical ranking of the combination of probability and severity assessment is given in Table 4. This model is particularly useful for determining preventive maintenance intervals and part replacement choices.

Sections 2.3 to 2.6 provide examples of the GC performance test selection based on operation function and risk assessment.

2.3 Thermal Precision

Include the Test?
Failure mode. The temperature is not reproducible.

Probability of occurrence. Code C; may occur with time or operating conditions.

Subsystem failure potential
- Irreproducible physical leakage (e.g., faulty seals, missing or poorly positioned insulation, venting restrictions, feedback issues)
- Control failure (e.g., weak or leaking control elements, triacs, switches, sensor intermittent)
- Control failure (e.g., intermittent temperature)
- Logic failure (e.g., algorithmic issues and timing)

Failure severity. Medium to high, depending on the frequency or duration between failures.

Qualitative risk. Serious to moderate.

Include in test suite? Yes.

Holistic or Modular Testing?

Modular: Series temperature measurements with an external thermometer. Depends on a representative oven location being accessible:
- The integrity of the space must not be compromised by the introduction of a probe.
- The probe must be of sufficient accuracy and known calibration status.

Modular: Series measurements with instrument's own temperature sensor. Depends on being able to read the sensor or capture its output to a format that can be evaluated.

Holistic: Retention of peaks from a series of injections. Depends on the following:
- Flow accuracy of the system
- Physical integrity (i.e., leaks, restrictions, etc.)
- Column performance
- Peak shape
- Integration quality

    Use any of the modes described above as long as the dependencies can be controlled. Choosing a test model with the fewest number of dependencies (or with dependencies that are most likely to be met) reduces the inherit risk. When the risk is comparable, the choice can be made based on efficiency or cost.

2.4 Thermal Accuracy

Include the Test?

Failure mode. The oven is functioning but the temperature is not accurate.

Probability of occurrence. Code C; may occur with time or operating conditions.

Subsystem failure potential

- Physical leaks (e.g., faulty seals, missing or poorly positioned insulation)
- Control failure (e.g., weak heater elements, sensor calibration)
- Logic failure (e.g., algorithmic issues and timing)
- Environmental effects (e.g., room or area too hot or, less likely, too cold)

Failure severity. High to medium, depending on whether the system is compared to other systems or is subject to method transfer.

Qualitative risk. Serious, assuming that the system is used for method transfer.

Include in test suite? Yes.

Holistic or Modular Testing?

Modular: Temperature measurement(s) with an external thermometer. Depends on a representative oven location being accessible:
- The integrity of the space must not be compromised by the introduction of a probe.
- The probe must be of sufficient accuracy and known calibration status.

Modular: Measurements with instrument's own temperature sensor. Depends on whether the probe can be calibrated against an external source.

Holistic: Retention of peaks from a series of injections. Depends on the following:
- Column performance or condition
- Temperature or flow dependency
- All of the conditions being replicated on a reference system of known fidelity

In this case, direct measurement with an external source is the only sufficiently reliable option.

2.5 Injection Precision

Include the Test?

Failure mode. Sampling is occurring, but the amount sampled is not precise.

Probability of failure. Code B; likely to occur with time or operating conditions.

Subsystem failure potential

- Sampler automated piston or plunger drive movement not reproducible
- Needle or sampling probe plugged or restricted
- Syringe, plunger, or metering device seal not tight
- Connecting tubing leaking or poorly attached
- Logical or mechanical mismatch between sample viscosity and draw speed

Failure severity. High, due to the potential difficulty in picking up a failure with readiness testing if internal standards are not used. Medium to minimal if internal standards are used, but very poor reproducibility could lead to difficulty in establishing a well-characterized lower limit.

Qualitative risk. Serious (assuming the use of external standards).

Include in Test Suite? Yes.

Holistic or Modular Testing?

Modular: Gravimetric measurement of aspirated volume. Depends on the following:
- Access to balance of appropriate resolution and known precision
- Knowledge of equipment sampling method compatibility with this measurement method
- Operator with sufficient skills to make precision weight measurements
- Statistical calculator to determine weight precision

Holistic: Area of peaks from a series of injections. Depends on the following:
- Column condition
- Instrument conditions (e.g., temperature, flow, detection consistency)
- Chromatographic data system to perform integration and data reduction
In this case, either technique would work, but direct measurement relies heavily on external skills and equipment that may not be present or of a known state. If skills or equipment is an issue, use a well-defined holistic test.

2.6 Injection Linearity

Include the Test?

Failure mode. Changing the sample volume does not reflect a linear change in the sample detected.

Probability of occurrence. It could be considered to be "code D—low probability" when the narrow range of acceptable sample size is factored in, or code C if we just look at the potential to fail.

Subsystem failure potential
- Sampler automated piston or plunger drive movement not reproducible
- Needle or sampling probe plugged or restricted
- Syringe, plunger, or metering device bore wear not consistent throughout
- Syringe or metering device not correct for loop or sample holding volume
- Logical or mechanical mismatch between sample viscosity and draw speed

Failure severity. It depends! If an injection precision test was run before this one, most failures will have been picked up already. Furthermore, most failures described here are due to setup issues and do not occur spontaneously. For example, changing the injection volume of a method necessitates some

TABLE 5 Sample Attributes in HPLC vs. GC

Parameter	LC Samples	GC Samples
Volatility	Nonvolatile	Volatile
Diluent transparency	Transparent to detection technique	Frequently detected or affects the detector
Diluent effects	Negligible (environment of the mobile phase typically remains the same or similar)	May be dramatic (inert-gas mobile phase vs. vaporized solvent cloud)
Sample expansion	Not a factor	Must be factored into any change in sample volume
Sample size	Large dynamic range	Typically maintained constant except for inlets designed to accept and contain large samples prior to vaporizing (pressure/temperature programming)

level of method validation. Lifting larger samples that worked fine at lower sample sizes may suffer from the weight of the sample column, the viscosity, and the potential for volatilizing the sample in the sampling device. Additionally, the physics of sampling volatile materials requires different considerations than those for nonvolatiles. Adding flash vaporization changes the dynamics as well.

Injection linearity commonly appears in a GC qualification protocol if protocol development began with LCs. However, this test is typically not indicated for GCs. Sample attributes in HPLC are compared against those in GC Table 5.

Under most circumstances, a 1-μL injection of solvent expands sufficiently to fill the inlet liner volume of a typical split–splitless detector. A larger injection typically results in the sample overflowing the liner and being vented or contaminating the flow path farther upstream.

Qualitative risk. The risk is very low.

Include in test suite? No.

Holistic or Modular Testing? This test is not recommended for GC. It depends on the following:

- The sample characteristic is too variable to make any generalizations about injector or inlet performance.
- Large sample injection inlets should be tested according to their intended use.
- The solvent front must be managed.

2.7   Leveraging the Manufacturing Testing

Instrument manufacturers pursue independent agency certification for their instruments for a number of safety and environmental compliance attributes. Vendors also perform countless hours of testing to verify that their instruments meet their design specification goals. Some of these characteristics are functions of the design (e.g., geometry and induced air movement in a GC oven) and are not variable provided that the instrument is not modified or damaged. You should work with the vendor to accept some testing performed in their test facility and only run tests that prove that the bases for these measurements has not changed at your site. For the oven, for example, the vendor probably has test data showing the thermal profile across a representative sampling matrix.

3   OPERATIONAL QUALIFICATION

3.1   Basic Safety and Operation

A basic safety and operation test should assess whether the instrument is suitable for qualification and can include the following checks:

- *A power-on self-test that does not report errors*. If available, this is one of the simplest and most generic checks.
- *Inspection for physical damage*. Focus on power connections and other environmental considerations.
- *Verification that all gases required are present and of suitable quality*. This is often done prior to testing as part of site or instrument preparation.
- *Validation of safety shutoffs*. Vendors are introducing features to provide safety checks that cover periods of unattended operation. One notable example is the hydrogen-pressure integrity check.

3.2   Oven Temperature Accuracy and Precision

GCs typically employ an air-bath oven with effective temperature regulation. The heat is generated by an electrical heating wire, and the air is distributed by a rotating fan. Cooling usually involves some means of exposing the oven's interior to ambient air through controlled ducts or automatic door openers. Much lower temperatures and greater cooling speed can be accomplished with cryogenic cooling. Liquefied nontoxic gases are sprayed into the ovens, and their natural expansion speeds the cooling process. A small number of vendors use Peltier cooling to supplement the cooling ramp or allow some degree of subambient operation.

After column selection, the most significant factor in gas-phase separation is temperature. It is important to know that the temperature is accurate, precise, and stable uniformly in the oven or at least where the column is positioned. Mixing fans are used to provide the desired homogeneity. In principle, only one place in any heated zone will have the exact temperature that is being used for temperature control, and that place is the location of the controlling thermometer. Because the control circuitry of the GC is based on the temperature returned from the controlling thermometer, accuracy should be measured using an external probe and meter. The number of readings depends on the equipment and operation. Multiple parallel measurements can be made in several locations simultaneously if a multiprobe meter is used.

Temperature accuracy should be recorded at a minimum of two setpoints: one in the low range (but not subambient) and one in the medium-to-high range. Optionally, one can test in the higher range, but probe selection, described later, becomes even more significant. Even though the vendor's acceptance for temperature accuracy may vary from vendor to vendor, with a tighter limit for lower temperature, in general a limit of $\pm 3^{\circ}C$ should be achievable. In the subambient range, the cryogenic cooling option further drives probe selection and placement. If the cooling only speeds cycle times, it may not make sense to test the cryogenic operation based on temperature. Instead, testing based on cycle-time expectation may be more appropriate.

As is the case in many types of testing, the act of measuring should not affect the measured environment. Most ovens were not designed for intrusion from external measurement devices that can result in heat leaks. The most common place to feed the sensor into the oven is via the oven door, which is further complicated when measuring higher temperatures. Even when this does not influence temperature accuracy directly, it may still have a deleterious effect on the physical components of the GC that were not designed for exposure to elevated temperatures. Temperature probes come in a myriad of shapes, sizes, ranges, resolutions, and environmental tolerances, and the physical connection between the probe and the meter is a factor as well. Make sure that the probe chosen fits the dimensional requirements (the probe end and probe shaft length and thickness), has a calibrated range that extends above and below the expected setpoints (covered in the tools section), and has the required precision for the measurement (inversely proportional to the temperature range covered).

Fortunately, temperature meter selection is usually less complicated. Most meters have settable or automatic ranging, which allows them to work with multiple probe types and working ranges. Purchasing a high-quality meter that has been researched to have the performance envelope required will provide a high level of assurance of the data it supplies. Readings can be taken directly from the meter's display and recorded in the qualification record, or data logging can be used, which includes distinct advantages:

- No transcription errors
- Capture at precise intervals
- Unattended operation

Temperature precision or stability measurements can be achieved in multiple ways:

- Direct measurements can be taken and recorded over a fixed time range with fixed intervals. External or internal sensors can be used.
- Precision of the oven temperature profile can be inferred from analyzing retention precision of peaks that result from multiple injections of a known compound. Imprecise oven temperature may result in a detectable difference in retention. Careful test design is crucial to assure that small changes will be seen in the system under test.

For direct measurements, the readings' duration, intervals, and number depend on a matrix of considerations about the system under test:

- The type of oven (i.e., air bath, direct-contact heater, etc.)
- The reading lag of the temperature probe
- The requirement imposed by the intended use of the chromatograph (a long isothermal analysis may require a different temperature profile analysis than that required by a short thermal gradient analysis)

External probes work best for instruments that do not provide a method of capturing an oven's native temperature sensors' output. They are particularly convenient if data logging was used for the accuracy reading and the probe is still in place. A valid alternative is to use the oven's native sensors if they are available. The use of external sensors and a data logger is valid only when the combined hysteresis of the meter, probe, and logging system have demonstrated an appropriate capture rate.

An internal sensor that was established as accurate during an accuracy test can demonstrate its ability to report that value repeatedly throughout a time-controlled sequence. If the GC is capable of sending the sensor values to the user's data station as an analog representation of the oven temperature profile, the signal can be stored and later analyzed for stability as a function signal fluctuation.

The model described above can be expanded to analyze any profile that the oven is capable of producing. This opens up the possibility of testing ramp efficiencies. Ramping the oven with the maximum ramp rate gives information about the system's ability to reach a maximum temperature within a specified time. This is often referred to as an oven leak test because issues with the oven insulation or other options can affect the oven's integrity. In general, a temperature stability of $0.2°C$ at a particular setpoint should be achievable.

3.3 Inlet Pressure Accuracy and Integrity

Of all chromatography system components, GC inlets differ the most from their LC counterparts. Pumps provide the mobile-phase flow in LCs, but GCs rely on pressure to produce flow through the column, and that pressure is applied at the inlet. Sample is introduced in a substantially different way as well. Although many different types of sampling techniques have existed in the GC world, the most prevalent is the syringe-based liquid sampler. Unlike LC, where the sample is introduced into the carrier stream by rerouting the stream through a preloaded loop, samples are typically introduced by piercing a soft seal at the head of the inlet. This seal (or septum) allows the sample to be introduced into the temperature-, flow-, and pressure-controlled environment of the inlet without interrupting flow.

There have been many schools of thought when it comes to managing carrier flow. The high flow volumes required for packed column chromatography can be managed through mechanical flow controllers, and the simplest controllers are high-accuracy forward-pressure regulators that maintain a constant pressure downstream. These controllers suffer from inconsistent flow rates through the column as the temperature changes. This is due primarily to viscosity changes, which are particularly pronounced when the temperature gradient is programmed. The increase in column backpressure reduces flow as the temperature increases. The flow returns to the set flow upon return to the original set temperature. This inconsistent flow, although thermally repeatable, can result in substantial differences in resolution.

The solution to inconsistent flow is to use mass flow control. In mass flow control, the controller sacrificed pressure to maintain the flow against varying backpressure. As with most control mechanisms, the mass flow controllers suffered from inherit resolution and hysteresis issues. These effects are aggravated by large injection volumes and result in pressure "ringing" or cyclic overcompensation of flow. Even with these limitations, mass flow proved to be a very good method of controlling the types of flows required for packed column chromatography. With the introduction of capillary columns, the flows required for appropriate linear velocities could not practically be provided with forward pressure and mass flow control. Vendors resorted to exploiting the physics of the pressure–flow relationship. For any given restrictor, the flow through the restrictor is directly related to the pressure applied once a threshold pressure is overcome. This is typically accomplished using a backpressure regulator that controls the pressure maintained behind it in the inlet space. Clearly, the pressure integrity of the inlet is a prerequisite for this type of control. This is not only mandated by the need to control the pressure accurately, but to prevent uncontrolled sample loss.

Pressure integrity tests measure decay and typically involve interrupting the inlet gas flow after the pressure has been increased to the operational level. The incoming flow must be occluded completely by a mechanical means (e.g., a shutoff valve placed just before the gas supply enters the GC). Failure to do so can result in a positive change in pressure caused by "normal" flow bypass or masking of an actual leak in the inlet. An unanticipated pressure change can result, due to the tubing between the source and the GC acting as a reservoir (the greater the distance, the more magnified the effect).

Pressure accuracy tests should obtain measurements while the system is actively controlling the pressure after a stabilized pressure has been achieved.

Even though the location of the measurement is partially a function of the inlet geometry, because the effect of that pressure must be felt at the head of the column, measurements taken at this point result in the most meaningful information. Acceptance of the inert pressure test may vary from vendor to vendor. As a reference, if the inert pressure is set at 25 psi, a ±1-psi acceptance should be achievable. Pressure sensors are available in a variety of sizes. Newer sensors have diameters that make them perfect candidates for insertion into the inlet by the same mechanism (nuts and ferrules) as that used for column connection.

3.4 Flow Accuracy

Carrier and support gas flows are significant contributors to consistent and accurate chromatography in GC systems. As described in Section 3.3, packed or capillary flow systems dictate whether flow or pressure is measured. In the following discussion, packed flow control is assumed.

Flow measurement on a manual system is a bit obscure. Flows are not set, achieved, and validated. Instead, they are monitored and adjusted until the setpoint is achieved. Therefore, flow is only as accurate as the external measurement device. Flow measurement on an electronically controlled flow system is more straightforward. A setpoint is established via the keyboard or controlling system, and the GC is left to its own devices to achieve that flow. Therefore, this is a true external check on the system's internal ability to control flow.

When choosing a flow meter, the gas type being measured, flow ranges, and accuracy requirements must be considered. There is no lack of choice in flow meters, although not all flow meters can be used for all applications. For example, the most basic flow meters (e.g., bubble flow meters) cannot be used for all gas types. Very light gases such as hydrogen are too permeable to the soap film and can result in very variable results. Electronic flow meters can be based on simple pressure vs. resistance meters. These can typically be recognized by their different ports for different gas types. Care must be taken with any gauge to make sure that its internal working backpressure does not affect the system under test. Recall from our discussion of mass flow controllers that changes in backpressure will be interpreted as a need to increase flow. Flow meters that are based on thermal conductivity can be very accurate, but some require gas-specific calibration. Mass flow detectors are the best all-around choice, although they are also the most expensive option. Mass can be estimated by the gas's temperature change when a fixed amount of heat is applied.

Another complicating factor in flow measurement may be the difficulty in determining the actual composition of the flow. For example, what is reported as total flow in many capillary systems is a combination of column flow, septum purge flow (if present), and split flow (if applicable). Additionally, the reported "flow" may simply be a calculation based on pressure, or it may be a true measurement based on flow sensing. Further, the column flow displayed by the system is an inferred flow based on the column characteristics specified, which may not be accurate. When the actual geometry of an installed column differs from that of the configured column, there will be a mismatch between the displayed and measured flows. For a capillary system, column flow is typically the lowest of all the combined flows, and the total flow is likely to be similar to the total flow displayed. For packed systems, the carrier flow is usually measured at the detector's outlet. In some detectors, an adapter may be required to connect the detector vent to the flow meter tubing or inlet.

Support gas flows in detectors also affect proper GC operation. Although most detectors tolerate some variation in flow, improper flows can cause increased noise or lower response or in extreme cases, a detector that will not run (e.g., flame ionization detector flameouts, nitrogen–phosphorus detector bead not igniting). As discussed previously, due to the nature of the setup process, manual flow control produces a final result that differs from the setpoint only to the extent the user chooses to allow. If the flows are controlled by electronic feedback systems, the ability of the system to achieve a specific setpoint can be verified by an external meter.

The steps for measuring detector flows differ by detector type and vendor-specific design. In general, the detector column inlet is typically blocked or capped and gases are turned on one at a time and measured at the detector outlet. Combining the gases is not recommended because the composite gas may not be read properly by the flow meter, and the gas combination may be explosive (e.g., FID support gases). The flow meter should be purged for a sufficient time with the gas being measured before the flow is recorded.

For some detector models, dismantling the detector may be required to access the vent. A decision should be made about whether the impact on the system is worth the information obtained from the test. For these models, it might be better to do indirect or inferred testing to qualify the effectiveness of the detector as a whole. In general, an acceptance of $\pm 10\%$ of the setpoint flow should be achievable.

3.5 Flow Rate Precision

Flow rate accuracy is related to flow rate precision. Clearly, flow rate accuracy aids in method transfer and simplifies method setup, but flow rate precision is what makes the chromatography successful between runs. Because the GC analysis (without mass detectors) is principally quantitative, there is a heavy reliance on retention time to identify the peaks of interest. By calculating the relative standard deviation (RSD) of the retention times of eluting peaks, we can determine the interrun precision. This is another example of how a holistic test is predicated on the proper workings of other systems. Flow precision is typically a function of retention time: The shorter the retention time, the smaller the allowable standard deviation must be. A peak with a retention time of 20 minutes can have a standard deviation 10 times higher than that of a test with a retention time of 2 minutes if the relative retention limit is the same for both. A peak with a 0.2-minute retention time that is subjected to a 1% RSD means that the retention time is stable to within 0.002 minute. In general, an acceptance of $\pm 0.5\%$ RSD in retention time should be achievable.

3.6 Signal Noise and Drift

The goal of chromatography is to separate and detect peaks. Drifting, wandering, or otherwise misbehaving baselines make accurate peak detection very difficult. Noise also contributes to the detection of peaks and may even obscure small peaks in extreme cases. A stable baseline with minimized noise is a vitally important attribute for any chromatographic system. Ultimately, it defines the limit on system sensitivity (signal to noise) and affects the reproducibility of results, as signal integration becomes more difficult with smaller peaks and noisy baselines.

Detector noise typically has a high-frequency component (electronic in origin) and lower-frequency components, referred to as wander and drift. *Wander* is random in direction but at a lower frequency than that of short-term noise. *Drift* is a monotonic change in signal over a longer period of time than those for wander and short-term noise (see Figure 1). Noise can be calculated in various ways. Many procedures for detector noise and drift estimation are based on ASTM (American Society for Testing and Materials) Method E 685 [11]. Most chromatographic software is capable of calculating detector noise and drift. A simple way to estimate noise involves determining the standard deviation of the baseline values over the range selected and then multiplying by 6 to get the entire population of values that contribute to the noise. This six-sigma noise model, although useful, is easily affected by baseline wander and spiking. To minimize this effect, noise calculations are applied to subsets of the noise range. The results of those segments are then averaged. This has a tendency to report lower but more representative noise values. The segmented model is also the most effective if the noise is being calculated manually. Use of the segmented method to estimate noise and draft is covered in Method E 685 [11]. (*Note:* One simple way to estimate the standard derivation of the noise, as one-fifth of the peak-to-peak noise, has been described by Hietje [12].)

Noise and drift diagnose many problems that may emanate from the detector or are merely reflections of upstream issues that will affect chromatographic performance. Most often, noise represents a local detector problem, whereas drift can usually be traced back to contamination of the inlet or conditioning issues with the column. As described previously, noise and drift can reflect the overall condition

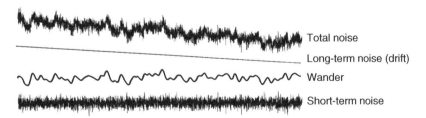

FIGURE 1   Noise, drift, and wander. (From Agilent Technologies, Inc., with permission.)

TABLE 6 Typical Values for Noise and Drift for Common GC Detectors Using ASTM Methods

Detector	ASTM Method	Test	Units	Range
Flame ionization	E 594-96 [13]	Noise	A	$10^{-14} - 10^{-13}$
		Drift	A/h	$10^{-13} - 10^{-12}$
Thermal conductivity	E 516-95a [14]	Noise	μV	$5 - 50$
		Drift	$\mu V/h$	$20 - 250$
Electron capture	E 697-96 [15]	Noise	A	$10^{-12} - 10^{-11}$
		Drift	A/h	$10^{-11} - 10^{-10}$
Flame photometric	E 840-95 [16]	Noise	A	$5 \times 10^{-12} - 5 \times 10^{-10}$
		Drift	A/h	
Nitrogen/phosphorus thermionic	E 1140-95 [17], nitrogen	Noise	A	$10^{-14} - 10^{-13}$
		Drift	A/h	$5 \times 10^{-13} - 5 \times 10^{-12}$
	E 1140-95 [17], phosphorus	Noise	A	$10^{-14} - 10^{-13}$
		Drift	A/h	$5 \times 10^{-13} - 5 \times 10^{-12}$

of a system at rest. Therefore, noise and drift are measured routinely with the system under normal ready-to-run conditions. Using this holistic approach, a system will pass only if the inlet and column are not eluting excessively, and the detector and its electronics are not introducing excessive noise. If the system fails due to these upstream influences, the detector must be isolated to allow measurements independent of these influences. Substitute a plug for the column in the detector inlet and evaluate the output of the detector. If the detector passes under these conditions, one should still attempt to understand how the system operates under normal conditions. If only the detector modularity is tested, some critical influences on performance may not be evaluated. Typical values for noise and drift for common GC detectors using ASTM methods are listed in Table 6.

3.7 Signal-to-Noise Ratio

The signal-to-noise ratio is an extension of the noise test in that it relates a detector's output to an external source. Like any signal emanating from a detector, noise can be artificially low for the wrong reasons. Noise alone cannot be relied on to assure that a detector is sensitive and amplifying appropriately the signal that it experiences. For this, a sample must be introduced that makes possible a comparison of signal output to baseline noise. This is significant in qualification testing, but it has even greater significance for minor-component analysis. Typically, the lowest signal that is detectable with any confidence must be at least three times the amplitude of the noise.

Signal-to-noise test design can vary, but the essence is to have a peak of known response compared to system noise. This can be done with a single run, where the run time chosen allows for a noise range that does not encroach on the peak. Alternatively, one can determine the minimum detectable limit (MDL) by calculating the response factor and then extrapolating what that would be at three times noise. This is an estimate only; to truly detect this limit, decreasing quantities of sample should be injected until the peak can no longer be differentiated from noise. MDL is calculated per compound and applies only to that compound. A good discussion on MDL and signal-to-noise estimation on mass flow detector (response to the mass of the analyte in the detector at a given time) and concentration detectors (response to the concentration of the analyte in the detector volume) is available in the literature [18].

3.8 Detector Response Linearity

Detector response may or may not be directly proportional to the amount of compound detected. To determine this, most chromatographic methods perform standard curve analysis. To produce a curve for GCs, the injection volume is typically maintained and standards of varying concentrations are used. Concentrations should produce results that span the expected detectable range. If the detector is linear with respect to concentration, peak areas different from those obtained by the standard can be extrapolated as to their respective concentration. This is of little significance to users, who only interpret data that fall within their calibrated curve. The most significant piece of information is the point at which a system goes nonlinear, which is compound dependent and difficult to extrapolate.

Response linearity can be determined by running a sequence of five or six injections of increasing sample concentration. From the plot of response vs. the concentration of the solutions, the correlation coefficient between sample concentration and response can be calculated to determine the linearity. The maximum and minimum concentrations above which the response starts to go nonlinear can be estimated by plotting the response factor (response in absorbance unit per unit concentration) vs. the log of concentration. The response factors should be fairly constant if the responses are within the linear range. The plot of response factors vs. log concentration should track along a line parallel to the x-axis within the linear range. A line that traces the almost constant response factors is drawn and set as 100%. Two additional lines at 95% and 105% of the response factor values are drawn. Response factors that fall outside the region between the 95 and 105% lines are considered nonlinear [11].

Using increasing injection volume is not suggested for GC (although it is sometimes done for LC), due to the restrictions imposed by the fixed volume of inlets and the tremendous expansion experienced when flash vaporization is employed. For mass flow detectors such as FID, a straightforward choice of straight-chain hydrocarbons works well. For highly specialized detectors, the choice is more complex and the column type also becomes a factor.

3.9 Injection Precision

Most automatic liquid samplers (ALSs) on the market amount to little more than robot emulators of the manual injection process: Instead of a human hand pulling up a plunger of a syringe, a mechanical hand works the syringe. But ALSs offer higher precision not only in the consistency of picking up the sample but, more important, in its delivery. The dynamics of sample volatilization and delivery in a needle-based injector could fill a chapter of its own. Suffice it to say that the more consistently—and in some cases, rapidly—the process can be performed, the more precise the delivery will be.

A series of injections is made and the areas and/or heights are determined though integration. The peak injection precision calculates the RSD for area (or height) and retention time. Area is preferred over height because it is more immune to baseline and peak shape variations and therefore relates better to the amount injected. In the injection precision test, consumables can play a major role in perceived precision. Factors such as system cleanliness, inlet liner activation, and syringe condition contribute directly to the success or failure of this test. Subtleties in syringe condition (e.g., a partially occluded needle) may only show up under certain conditions (e.g., viscous samples, high draw volumes, tightly sealed sample vials). The syringe must also be appropriate for the sampling range under investigation (or the investigation range must be appropriate for the syringe). Expecting a 250-µL syringe to return good results when tested at 0.5 µL may prove to be disappointing.

Sample choice, column selection, and injection conditions must all be factored into the test design as well. For example, the sample should be stable, have one to four compounds that can clearly be separated on a commonly used column, and be dissolvable in a solvent appropriate for GC analysis. Commonly used GC solvents are 2,4,5-trimethylpentane (*iso*-octane) or *n*-hexane. Solvents such as water or dichloromethane are less suitable for GC testing. Sample and column choice must reflect the expectation that the challenge is on sample delivery, not on the data system's ability to integrate poorly shaped peaks. Peaks that are not well separated increase the impact on integration. A certified standard with compounds of similar polarity is the best choice. In general, an acceptance of ±2% RSD in peak area should be achievable.

3.10 Injection Linearity

Injector linearity testing is not recommended, for the same reasons as those given in Section 3.8 for detector response linearity. With the possible exception of programmed temperature and pressure detectors, GC inlets have a very narrow range of acceptable injection volumes that they can accommodate. Injection linearity commonly appears in a GC qualification protocol if protocol development began with LCs. For LCs it is very common to vary the injection volume (within the limits of the sampler). For GCs it is not a common practice to run different sample volumes within the same injection sequence. Consider the fact that a typical volume of an inlet liner is less than 1 mL and that 1 µL of most solvents expands to nearly this volume when vaporized. Any additional volume will be lost though split vents (for split injection) or potentially will contaminate the flow path in any inlet.

3.11   Temperature Accuracy of Heated Zones

Inlets and detectors are normally heated with an electric heater, present in a metal block that includes a sensor and electronics to control the temperature. Every metal block heated this way generates a temperature profile over the block. During GC design, the heated zone is developed such that the temperature profile is minimized, or at least tailored to the inlet's needs. As stated earlier, a test must be rugged, in that it can be performed by different operators on different GCs in a consistent way with similar results. With this test, the sensor's position is the key factor for consistent test execution and results.

A common approach is to let the sensor enter the top of the inlet, through the septum, and then search for a position in the liner where the measurement is closest to the setpoint. However, this hunting for the expected answer goes against the spirit of qualification testing, and it works only for a limited set of inlets that can be accessed from the top, like a split–splitless or packed inlet. A much more consistent way would be to access the inlet or detector in the same way that the column is accessed. New temperature sensors with dimensions similar to capillary columns dimensions are now available. These sensors can be fed into the inlet and/or detector as if they are columns, fixed with a column nut and column ferrule to a predefined distance, and then measured after the temperature stabilizes. This method can provide improved accuracy and precision.

Because there is always a temperature profile over an inlet or detector, position the sensor in an area where the temperature is relatively consistent, as opposed to searching for the temperature that is closest to the setpoint. With the former approach, the test is much more rugged and measures the inlet's capability instead of the user's finesse. This approach should not be used for electron capture detectors, due to the possibility of contamination.

Take great care not to contaminate MSD transfer lines. Disassembling the MSD components to take measurements or using a dirty temperature probe can contaminate the MSD. For these reasons, holistic testing is preferred.

3.12   Temperature Accuracy of Headspace Heated Zones

The temperatures of headspace heated zones are important in preparing samples for injection into a system. Heated zones include an oven, a sample loop, and a transfer line or syringe, depending on the type of headspace. All heated zones are tested in detail during instrument design, and some tests are difficult to repeat, because of accessibility issues, after the instrument is installed in a lab. Therefore, it is important that sensors be located in representative, easy-to-access locations within the zone. Some vendors provide such preconfigured locations for installing sensors. Depending on the headspace's sampling type, a carryover test may be indicated. Loop systems are good candidates for carryover tests, and syringe-based systems may better be tested as part of system preparation or suitability testing.

3.13 Carryover

Carryover is normally expressed as the percentage of sample that can be detected in a blank run that immediately follows the sample run. Carryover can come from many sources, but the injector itself is usually the culprit. Most other sources of carryover tend not to result in distinct peak formation; instead, they tend to contribute to increased baseline noise and wander. Active adsorption of compounds by liners and other inlet/sampler components can be released on the next injection, which can result in ghost peaks.

Unlike LC or other loop fill samplers (e.g., many headspace samplers), which include a large amount of plumbing in which sample can be retained, GC samplers depend largely on a consumable syringe. In many cases, carryover testing is better left for ongoing suitability testing. Sample type, viscosity, polarity, wash solvent, number of washes, and syringe condition all contribute to the potential for sample carryover. Because of the variety of conditions and settings and the disposable sample path, including a carryover test in an operational qualification (OQ) suite is not suggested.

4 PREVENTIVE MAINTENANCE

Preventive maintenance (PM) is performed to ensure that a system performs properly over a specified time interval. The goal of PM is to check an instrument to make sure that it will perform as expected in the upcoming period of time. Preventive maintenance usually includes replacing septa, liners, and seals, the parts that wear out over a certain period of time and can easily be replaced. The scope of the preventive maintenance should be based on the instrument's usage, sample type, and operating conditions. Standard operating procedures (SOPs) usually specify that septa must be replaced weekly or after a specific number of runs. New instruments often provide counters for these data and issue warnings when limits are reached.

The interval between PMs depends on the operating conditions (higher temperatures wear out parts more easily), the number of samples run, and the condition of the samples. For high-utility instruments that operate 24/7 with dirty samples, maintenance operations should be conducted more frequently. Regular PM prevents unscheduled downtime of systems and reduces the number of unscheduled repairs due to a lack of maintenance. In many organizations, preventive maintenance is a mandatory prerequisite before performing an operational qualification.

The OQ that follows the PM verifies that the GC is in a qualified state. Technically, one could execute an OQ without performing maintenance beforehand. If the instrument passes the OQ acceptance criteria, the OQ retrospectively confirms that the system was performing well before. The issue with this approach is how to manage the risk of failing the OQ. In this case, a practical mitigation of compliance risk is the system suitability tests conducted prior to sample analysis. For analyses conducted between the last OQ and the current OQ, successful system suitability tests demonstrated that the system had met the performance requirements required by the particular analytical methods to execute that particular analysis. The reliability of the analytical results could be assured. A good suite of system suitability tests sets off early triggers when chromatographic performance degrades. The practical difficulty is that system suitability tests provide important data points during routine analyses, but they typically do not challenge the system sufficiently to uncover marginal performance of an individual module.

5 COMMON PROBLEMS AND SOLUTIONS

For this section it is assumed that the gas chromatograph has passed an initial OQ. This means that gases with sufficient purity were initially connected correctly and that the heated zones worked properly. The discussion that follows describes difficulties that can arise while executing performance tests or during routine operation of a GC. Problems can be categorized into three basic groups:

1. The chromatographic behavior (e.g., resolution, efficiency, retention time) has changed from its expected performance.
2. The instrument is unable to initiate or complete an analysis.
3. The results of an analysis, such as area precision or measurement accuracy, do not meet the requirements defined for the analytical application or test method.

Breaking the system into components simplifies the troubleshooting process. Before beginning any troubleshooting, though, it should be confirmed that the system is running the correct method with the correct conditions on a known sample. Many problems are caused by running the wrong method, method parameters, or sample.

5.1 Injector and Sampler Systems

Injector systems encompass liquid autoinjectors and gas-phase autosamplers such as headspace, purge and trap, and thermal desorption systems. With modern GC systems, many of the injector failures that can occur are identified by the system. These types of failures should be covered in the user's manual together with a prescription for repair.

Chromatographic Problems   Usually, the injector does not affect the chromatographic performance (i.e., peak resolution or column efficiency) of a system. Loss of efficiency or resolution can be caused by leaks and contamination, and the user will normally see this more clearly in the degradation of analytical results.

For gas-phase samplers, the same holds true. Alterations in the plumbing of these devices can certainly affect the chromatographic performance; consequently, be sure to determine what has been done to a system since it last worked correctly. The wrong tube diameter or length installed during a previous repair attempt can easily broaden peaks and compromise performance.

Instrumentation Problems Most systems indicate where an instrumentation fault exists, and the supporting documentation suggests solutions. If the instrument is in a "not ready" state, consult the status information to determine what is not ready. For liquid autoinjectors, the plunger and injector carriage are the two likely places where an error has occurred. Inspect the syringe while resetting the injector to ensure that the needle and plunger are intact and that the syringe barrel has not been cracked. Another type of failure is more subtle and is caused by wear in the syringe. As the syringe wears out, it affects area repeatability. If possible, run an initial study using a known standard to determine the number of injections possible before analytical performance is compromised. This type of suitability testing is done routinely in the pharmaceutical industry, and it offers value to other industries as well. With time and resource limitations, however, these types of studies may not always be possible. In such cases, the user can set a modest number of cycles (e.g., 500) and change the syringe before problems occur.

Another part of the injector system to consider is the sample vial. It is possible to cut the vial septum with the needle and introduce these fragments into the sample or into the inlet. Typically, this produces additional unknown peaks in the chromatogram. Inspect the syringe to ensure that there are no particles in it, and make an injection from a clean solvent vial. Depending on vial dimensions and sampler setpoints for vial access, the needle may be hitting the bottom of the vial. This is possible when only a small amount of sample is available for analysis and the needle needs to go deep to take up the sample. When the needle hits the bottom of the vial, the sampler does not necessarily report an error, but the (partially) blocked needle may result in differences in area. An injection precision test with a sufficient amount of a known sample and known conditions can rule out sampler problems.

If area precision varies only after a high number of runs with all injections made from one (rather than two) vials, the cause is the cap. When a cap is very tight and closes very well, each time sample is pulled out of the vial, no air comes in, resulting in a lower gas pressure on top of the sample. When the syringe filled with sample is moving up, the lower pressure above the liquid causes some sample to come out of the syringe again, resulting in problems with RSD.

For gas-phase samplers, errors could be the result of thermal or pneumatic zones not being ready. First review the thermal and pneumatic setpoints to confirm that they are correct. If the system is unable to reach the correct temperature, inspect the heated area for missing or incorrectly installed insulation and covers. If this does not address the problem, call the instrument supplier for service. If the system is unable to reach the setpoint pressure, inspect the flow path and test for leaks with an electronic leak detector. Do not use liquid leak detectors because they can quickly contaminate the sample pathway and create much bigger problems. Replace fittings and tubing if necessary to address the leaks. When the system is able to achieve a "ready" state, run a known sample to confirm the performance.

Analysis Problems   This is the area where injector system problems are most likely to be seen first. From the analysis, the usual problems are no peaks present, reduced peak areas, or distorted peaks. For example, a system is working well and meeting specifications. All of a sudden, the peaks disappear and the resulting analysis is a flat baseline. In this particular example, the clue to what happened is that no peaks are detected. With a liquid autoinjector, this means that no sample was available to be drawn or no sample was drawn. Verify that there is sample in the vial, and then inspect the syringe. With no peaks observed, the needle is either broken or plugged. If there is no visible damage to the syringe, attempt to aspirate solvent and confirm that liquid can be withdrawn and injected into a waste bottle.

When reduced peak areas are observed, be sure to compare the solvent peak areas between an acceptable analysis and the problem analysis. A very useful piece of information is whether the peak area for the solvent is reduced as well. First examine the syringe in the autoinjector. There will be a loss of area if the clamping mechanism of the plunger has loosened and the plunger is no longer drawn upward to pull the sample into the syringe. If the plunger is still connected, the next step is to remove the syringe. Test the syringe by attempting to draw solvent. If solvent cannot be drawn into the syringe, it is plugged and needs to be replaced or cleaned. At this point, the user will have to follow the SOP to return the GC to use as a system suitable for analysis.

For gas-phase samplers, similar problems can be encountered. The first step in troubleshooting a gas-phase sampler is to separate its contribution from that of the gas chromatograph. If possible, inject a sample of the components directly into the GC to determine if the problem lies with the sampler or the chromatograph. Assuming that the problem is still with the sampler, evaluate the results as before and determine if the peaks are missing, too small, or distorted. If all the peaks are missing, it is likely that the sampler flow path is plugged. Start from the sampling point and work backward toward the GC interface and confirm that flow is possible. If some of the peaks are missing, determine whether these compounds are likely to adsorb to the sampler flow path. Active compounds can be lost in the flow path as the deactivation ages or the system accumulates dirt. Run a known sample and compare the recovery of the analytes to results previously acceptable. At this point, unacceptable recoveries suggest that the sampler flow path needs service (either cleaning or replacement).

5.2 Inlet Systems

The inlet of the gas chromatograph is defined as the interface between the injector/sampler and the column. Many varieties of inlets have been developed to address specific application needs, and in general, there are three basic types: split–splitless, on-column, and direct [19]. Other variations of these inlets offer temperature programmability [e.g., a programmed temperature vaporizer (PTV)] but are pneumatically similar to these three basic types. All inlets have thermal and pneumatic components, both of which can be sources of problems. A well-designed method has been characterized so that the user can easily tell when the measurement may be questionable. As in the case of the injector, confirm that the correct method is being run with the correct setpoints on a known good sample before undertaking more involved troubleshooting.

Chromatographic Problems   The purpose of the inlet is to introduce the sample to the column without affecting the separation performance of the column. Chromatographic problems that may be encountered are distorted peaks, lost resolution, and poor retention-time precision. Peak distortion can be traced to a poor injection, dirty liner, or incorrect column installation. Depending on the type of sample analyzed, the trapped sample matrix in the liner can contribute to poor peak shape. This dirt acts as an adsorbent and may slowly release trapped analytes, broadening them. Band broadening leads to a loss of peak resolution, thus compromising the separation and subsequent quantitation. Poor retention-time precision can be the result of dirt in the liner as well as of an unstable pneumatic control system or undesired cold spots in the inlet. Monitor the inlet pressure, preferably with the data system, to see if there are variations during an analysis. Verify the temperature setpoint and check that all insulation and covers have been installed correctly.

Instrumentation Problems   Most inlet problems are related to replaceable components such as liners, septa, O-rings, or seals that are in contact with the sample. For inlets with liners (e.g., split–splitless, PTV), the liner's purpose is to serve as a disposable trap for nonvolatile materials and as a sample homogenizer for split injections. As such, these need to be replaced on a regular basis. A dirty liner can cause a variety of problems, such as loss of peaks, distortion of peaks, discrimination of compounds, ghost or contaminant peaks, poor retention-time precision, and poor area precision. Most chromatographers running dirty samples replace the liner routinely, and they also cut a portion of the inlet side of the column off and replace the septum. These steps normally return the inlet to service. For inlets without a liner (e.g., on-column), cutting the column usually remedies any sample-induced problems, such as contamination. For direct inlets, the inlet volume itself may require cleaning to restore performance.

The sealing components of the inlet are also potential sources of problems. Most inlets use an elastomeric septum to seal the inlet and serve as the syringe interface. The septum is comprised of low- and high-molecular-weight components that can contaminate the inlet. If the needle shears off some of the septum, these particles can enter the inlet and create ghost or contamination peaks. These peaks show up during sample runs as well as solvent blank runs and no-injection runs. In addition, they can act as a trap for the sample components, resulting in the loss of peaks. Most inlet liners use O-rings or gaskets to make a gastight seal. These materials are also elastomeric and can add spurious peaks to the chromatogram. As in the septum contamination case, these peaks show up even with a blank run. In addition, these peaks grow, the longer the oven sits at low temperature, since this concentrates these compounds on the head of the column. Contamination can also collect within the inlet but outside the liner. Although this is much less likely, if all other sources of contamination have been addressed, it may be necessary to clean the inlet shell to restore performance.

Electronic pneumatic control systems are used widely in GC instrumentation. Depending on the operational mode, these systems can employ pressure and/or flow measurements to achieve control. First review the pneumatic setpoints to confirm that they are correct. If the system is unable to reach the setpoint pressure, inspect the flow path and test for leaks with an electronic leak detector. Do not use liquid leak detectors because these can quickly contaminate the sample pathway. While leak checking an inlet or even the external gas supply, be sure to focus on those areas where the user can make or break connections, as shown in Figure 2, because these the areas are the most failure-prone.

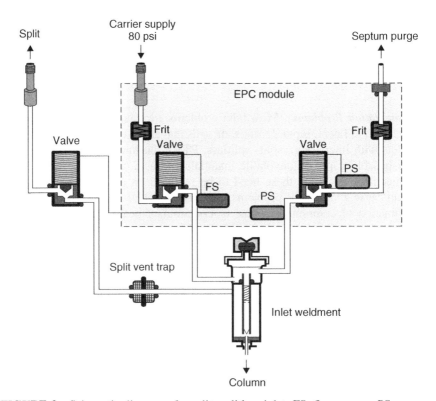

FIGURE 2 Schematic diagram of a split–splitless inlet. FS, flow sensor; PS, pressure sensor. (From Agilent Technologies, Inc., with permission.)

Although leaks can occur anywhere along the flow path, and particularly at component interfaces, the actual inlet is the most likely location for leaks. For a split mode with the inlet design shown in Figure 2, the total flow to the inlet is regulated by the flow sensor (FS), while the inlet pressure is maintained via the pressure sensor (PS) and the split valve. The FS branch is almost always able to reach its setpoint, even with leaks in the system. The PS, on the other hand, will not be able to control pressure if there is insufficient flow or a substantial leak from the system. A quick leak check can be done manually as follows:

1. Put the inlet into splitless mode and turn off the septum purge flow. The control system regulates pressure only in this configuration and opens the supply valve enough to reach the desired pressure setpoint. Although the flow sensor is not being used, its value is still read by the system in real time.
2. Compare the flow sensor reading to the column flow setpoint. These should be nearly identical. If the flow sensor is reading higher than the column flow by more than 1 mL/min, the system is leaking. Other inlet types can have the same type of quick leak checks, and in newer instrumentation this is built into the system. Consult the instrument documentation to see if such features are available.

Thermal problems are also possible with inlet systems. First review the thermal setpoints to confirm that they are correct. If the system is unable to reach temperature, inspect the heated area for missing or incorrectly installed insulation and covers. Newer instrumentation may provide on-board diagnostics to help the user determine the error, so consult the instrument documentation for more information. If this does not clarify the problem, call the instrument supplier for service.

Analysis Problems   In addition to the problems already described (e.g., no peaks, reduced peak areas), analysis problems can also include poor area precision, poor retention-time precision, and reduced analyte recovery. Inlet temperature can have a large impact on all these results, because an insufficiently heated inlet can give irreproducible results and thermally trap high-boiling analytes. Confirm that the inlet is at the original method setpoint. A thermocouple or other temperature probe can quickly show if the inlet is not reaching the desired temperature. Pneumatic systems can also affect the analytical results. If the system is reaching the method setpoints, external measurement of split flow, septum purge flow, and column flow can shed more light on pneumatic problems. Pressure instabilities can lead to both retention time and area irreproducibility. Monitoring these values during the analysis can provide useful information (newer instrumentation can do this internally). Reduced analyte recovery is almost always due to activity or dirt in the inlet. Replacing the septum, liner, and trimming the column usually repairs these problems.

5.3 Thermal Zones

Newer gas chromatographs should have the ability to monitor thermal zone performance and determine whether it is operating within the expected performance band. This diagnostic capability notifies the user of heating problems within the heated zones. Status and pop-up messages are generated if a heated zone is not working properly. The instrument's user's manual should list the messages and may give troubleshooting information. These errors can be caused by a problem with the power to the instrument or a shorted temperature sensor or heater. If the heated zones have shorted or open sensors and/or heaters, the modules require replacement. If the problem was caused by the power delivered to the instrument, restarting the instrument may remedy the problem.

For the oven, further investigation may be required. Inspect any ducting mechanism (e.g., a flap) physically to see if it is operating properly. Normally, peaks tail because of column or inlet liner activity. When this occurs, increases in injected concentration can reduce the peak tailing. If tailing is observed while increasing the amount of the component, it can be an indication of a cold spot in the sample path. If this occurs, check to see if insulation is missing. The inlets and detectors can be supplied with insulation cups, which can reduce the thermal loss at the column connection. Reinstalling these devices should help this type of tailing problem.

5.4 Detectors

Again, let us assume that a GC has initially passed an OQ. The OQ verified that the gases were initially connected correctly and that the heated zones are working properly. The discussion that follows involves detector-related issues that can arise while operating the GC.

There are several different types of difficulties for detectors. Wearout occurs for the nitrogen–phosphorus detector (NPD) bead, and contamination can be a problem for all detectors. Contamination can be introduced from several sources, including the gases, the sample, the sample's solvent, the septum, and the column. For example, chlorinated solvents can saturate an electron capture detector (ECD) and affect the NPD bead performance and lifetime. The column bleed can plug a flame ionization detector (FID) and NPD jets since these detectors share the jet design.

The saturation of the ECD signal by chlorinated materials requires that the detector be heated and purged at higher flow rates. Monitor the signal until it returns to an acceptable level. To prevent this problem, change the solvent, dilute the sample (if the sample itself is the source of the saturation), or use solvent venting devices to dump the solvent before it reaches the detector.

Plugging of the flame jet occurs when silicon dioxide builds up in the jet tip. This is caused by the slow removal of stationary phase as the GC is used. This problem can be more prevalent when carrying out high-temperature analysis. There can be subtle and abrupt indications that this is occurring. The subtle indication is that the retention times of the sample components begin to get longer. This is opposite of the expected shift to shorter retention times, which occurs as stationary phase is lost over time. The shift to longer retention times is caused by the increased pneumatic resistance from the plugged jet. Another symptom can be that the flame goes out while the solvent elutes and then reignites. In the FID design, the hydrogen and makeup gas are mixed with the carrier gas before entering the detector. The jet has to be removed and replaced with another jet. The orifice could also be cleaned and the plugged jet could then be returned to service. If the user replaces the jet, occasionally the electrometer does not contact the collector when the castle assembly is replaced. This results in zero offset and no response when flame ignition occurs. The electrometer then has to be loosened and repositioned to establish contact at the collector. At this point, the user has to follow the SOP to return the GC to use as a system suitable for analysis.

Ironically, the absence of contamination can also cause problems. The availability of air cleaners, hydrogen generators, and gas scrubbers can reduce the low level of hydrocarbons in compressed air and the other gases significantly in terms of the FID's baseline offset. The reduction in offset can be a factor of 3 or more, depending on the flow setpoints for the FID. Gas chromatographs with autoigniting FIDs have a signal offset value which indicates whether or not the flame is on. The autoignition algorithm then compares the measured offset with the signal offset value to determine whether the flame is on. If the value is not exceeded even though condensation is visible on a shiny surface placed above the detector, the system does not recognized the flame as lit. The resulting low offset can require a change in the signal offset value for the detector, which is typically expressed as a current in picoamperes. The signal offset value needs to be changed to a lower value for the detector to determine the "lit" state correctly. Follow the SOP for changes in parameters such as the lit offset.

The flame photometric detector (FPD) can also become contaminated. Since the FPD has an optical path, the windows in the path can become coated with contamination over time and reduce the signal at the photomultiplier. A simple test is to try to ignite the flame. If the ignition uses a glow plug, there should be an offset in the signal from the glow plug turning on. If this signal cannot be seen, the optical components may be contaminated or the glow plug may be open. Check to see if the FPD flame light is on by placing a shiny surface near the exit tube. If condensation is observed, the optical path needs to be examined. This requires disassembling the detector. Replace the windows, O-rings, and filters if present. If this does not remedy the problem, the photomultiplier tube needs to be replaced. If a high offset is measured after changing the components in the optical path, check to see if the optical filter was installed properly. As with the FID, the presence of cleaner gases may require the signal offffset value to be lowered for the detector to determine correctly whether the flame is on. Track any changes as required by the SOP.

Quenching of the FPD signal occurs when there is enough carbon present to reduce the sulfur emission. This can be tested by using two different sets of flow conditions for the FPD and then comparing the areas for each of the components in the sample. The ratio of the areas is reasonably constant for all measured peaks if there is no quenching. The ratio changes for peaks that have quenching.

The response mechanism for the NPD detector causes gradual depletion of the bead material. Eventually, the bead needs to be replaced (follow the manufacturer's instructions). Contamination of other parts of the detector can be determined by having the flows on, the electrometer on, but the bead power off. The detector offset should remain low if contamination is not present while heating the detector to a temperature at or above the desired operating temperature. If this offset with the bead off is a significant portion of the operating offset value desired, the rest of the detector needs to be cleaned or replaced. The amount of contamination from the solvent can be determined by programming the flow for hydrogen to OFF while the solvent peak is eluting or by diverting the solvent peak. As the detector is returned to service, follow any SOP that may apply.

Since the thermal conductivity detector (TCD) is based on the change in resistance over the sensing element, the change in the detector tends to occur over a period of time. One strategy that can be used with the TCD is to track the response of standards over time. Newer gas chromatographs may include diagnostic routines that monitor the resistance and issue an exception message when the filament resistance is too high or open. If the GC in question does not offer this capability, check the user's or service manual for ways to measure the resistance.

The ECD is a detector based on a beta radiation source and as such will be under regulatory scrutiny for the handling of the detector. This limits what can be done on site. For contamination, thermal cleaning is the only type of cleaning permitted. If this does not work, the user is required to exchange the detector cell. For low background, one solution is to change the makeup gas to a methane–argon mixture. This should increase the offset.

Mass Selective Detectors   The quadrupole mass selective detector uses an autotune to calibrate the mass axis using specific masses from a well-known standard. The electron multiplier voltage (EMV) is changed during this tuning process to achieve a certain response for the masses that are used. Because analytes are introduced into the MSD and ionized in the ion source, the ion source gets contaminated first. When the source gets dirty, the EMV increases when a new tune is performed, and this is often the first indication that the source is getting dirty. After the source is cleaned, the EMV is lower than before cleaning, but a bit higher than after the previous cleaning, as a result of an aging multiplier.

5.5   Carrier Gases

One problem that can occur with electronically controlled inlets, secondary pressure sources, and detectors is that a gas type can be configured improperly. This is because of the viscosity differences between the gases typically used for GC. If secondary pressure sources are being used, check the gas type, because nitrogen is generally set as the default gas type. If helium is used but configured as nitrogen, control problems occur.

Another potential problem with the carrier gas is the level of purity. This manifests itself mostly as a shift in the baseline level for the detector. If 99.999% pure carrier gases are not available, selective traps for certain gases are available. Getters are available for helium. Hydrogen generators can be used to provide both hydrogen carrier and fuel gas.

With the introduction of electronically controlled secondary pressure sources, it is easier to use techniques such as Deans switching and backflushing with capillary columns. However, there are a couple of problems that may occur when using these techniques. For techniques that do not reverse the flow, such as heart cutting and comprehensive GC, the pressure in the system is determined primarily by the inlet. The sensor at the secondary source indicates a certain pressure even when the secondary source is not turned on. The setpoint for the secondary pressure must be set high enough so that it controls. Typically, adding a 0.5- to 1-mL/min flow to the column flow at the secondary pressure source accomplishes this.

For backflushing, the reversal of flow can also cause the column and makeup gas flow to increase. For flame-based detectors, this may cause the flame to go out. Care must be taken in setting the method's parameters so this does not occur. For the NPD, care must be taken not to cool the bead too much. This should be investigated while setting up the method. The user can consult information available with the capillary flow technique modules for guidance in selecting appropriate setpoints for the inlet and secondary pressure control devices.

Acknowledgment

This chapter could not have been completed without the dedication and scrutiny of Karen Riley, our proofreader and style editor of many years.

REFERENCES

1. H. Lam. Performance verification of HPLC. In *Analytical Method Validation and Instrument Performance Verification*. Wiley-Interscience, Hoboken, NJ, 2004.

2. J. Crovrther, J. Dowling, R. Hartwick, and B. Ciccone. Performance qualification of HPLC instrumentation in regulated laboratories. *LC-GC North Am.*, 26(5), 2008.

3. J. V. Hinshaw. Flow, pressure and temperature calibration: 1. *LC-GC Eur.*, 18(1): 22, 2004.

4. J. V. Hinshaw. Flow, pressure and temperature calibration: 2. *LC-GC Eur.*, 18(3): 138, 2005.

5. *U.S. Pharmacopeia*, 31, General Chapter <1058>, Analytical Instrument Qualification. USP, Rockville, MD.

6. *General Requirements for the Compliance of Testing and Calibration Laboratories*. ISO, Geneva, Switzerland, ISO/IEC 17025. 1999.

7. L. Huber. *Validation and Qualification in Analytical Laboratories*. Interpharm Press, Boca Raton, FL, 1999.

8. H. Lam. Validation of analytical instrument. In *Pharmaceutical Manufacturing Handbook: Regulation and Quality*. Wiley-Interscience, Hoboken, NJ, 2008, Chap. 8.3.

9. S. K. Bansal, T. Layloff, E. Bush, M. Hamilton, E. Hankinson, J. Landy, S. Lowes, M. Nasr, P. St. Jean, and V. Shah. Qualification of analytical instruments for use in the pharmaceutical industry: a scientific approach. *AAPS PharmSciTech*, 5(1):article 22, 2004.

10. P. Coombes. *Laboratory Systems Validation Testing and Practice*. PDA, Bethesda, MD, 2002.

11. Standard Practice for Testing Fixed Wavelength Photometric Detectors Used in Liquid Chromatography. *Annual Book of ASTM Standards*, Vol. 03.06, E 685-93. ASTM, West Conshohocken, PA,

12. G. M. Hietje. *Anal. Chem.*, 44(6): 81A–88A, May 1972.

13. Standard Practice for Testing Flame Ionization Detectors Used in Gas or Supercritical Fluid Chromatography. *Annual Book of ASTM Standards*, Vol. 03.06, E 594-96. ASTM, West Conshohocken, PA,

14. Standard Practice for Testing Thermal Conductivity Detectors Used in Gas Chromatography. *Annual Book of ASTM Standards*, Vol. 03.06, E 516-95a. ASTM, West Conshohocken, PA,

15. Standard Practice for Testing Electron Capture Detectors Used in Gas Chromatography. *Annual Book of ASTM Standards*, Vol. 03.06, E 697-96. ASTM, West Conshohocken, PA,

16. Standard Practice for Testing Flame Photomertic Detectors Used in Gas Chromatography. *Annual Book of ASTM Standards*, Vol. 03.06, E 840-95. ASTM, West Conshohocken, PA,

17. Standard Practice for Testing Nitrogen/Phosphorus Thermionic Detectors Used in Gas Chromatography. *Annual Book of ASTM Standards*, Vol. 03.06, E 1140-95. ASTM, West Conshohocken, PA,

18. *A Guide to Interpreting Detector Specifications for Gas Chromatographs*. Technical note. Agilent Technologies, Wilmington, DE, 2005.

19. M. S. Klee. *GC Inlets: An Introduction*, 2nd ed. Agilent, Technologies, Wilmington, DE, 2005.

Recommended Reading

D. Rood. *The Troubleshooting and Maintenance Guide for Gas Chromatographers*, 4th rev updated ed. Wiley, Hoboken, NJ, 2007.

K. J. Hyver, Ed. *High Resolution Gas Chromatography*, 3rd ed. Hewlett-Packard Co., Palo Alto, CA, 1989.

11

PERFORMANCE VERIFICATION ON REFRACTIVE INDEX, FLUORESCENCE, AND EVAPORATIVE LIGHT-SCATTERING DETECTION

RICHARD W. ANDREWS

Waters Corporation

1 INTRODUCTION

The differential refractive index detector (dRI), the evaporative light-scattering (ELS) detector, and the fluorescence (FLR) detector are used for analyses where either the analyte(s) do not absorb in the visible or ultraviolet or are at sufficiently low concentrations that absorbance detection is unable to provide an acceptable signal-to-noise ratio. They are used either for their universality (differential RI and ELS) or for their sensitivity (FLR). Because their operational principles are different, the qualification tests for these detectors will be somewhat different from those generally employed for absorbance-based detectors. The proportionality between detector response and concentration, as well as system precision and control, must be demonstrated.

There are two approaches to qualification of high-performance liquid chromatography (HPLC), UPLC, or ultrahigh-performance LC (UHPLC) systems.

Practical Approaches to Method Validation and Essential Instrument Qualification,
Edited by Chung Chow Chan, Herman Lam, and Xue Ming Zhang
Copyright © 2010 John Wiley & Sons, Inc.

The first is module based and focuses on direct measurements of individual performance specifications such as flow rate accuracy and temperature accuracy. The second approach focuses on the chromatograph as a system and infers the performance of individual modules. The latter approach has the advantage of using the chromatograph and its associated data system in a holistic manner. This is particularly important with detectors such as the differential refractive index detector and evaporative light-scattering detector, which require a stable flow rate and thermal equilibration for proper performance. Fluorescence detectors are most commonly used for high sensitivity analysis and are best qualified as part of a system. This minimizes the potential for contamination of a system dedicated for trace analysis, especially when marker compounds are used for determining compositional accuracy and precision.

The differential refractive index detector is a bulk property detector; that is, it nonselectively detects changes in the refractive index of the column effluent as peaks elute from the column. Since the chromatographic peaks are dilute, a reference cell is used to subtract the contribution of the mobile phase. The reference cell contains static mobile phase which should be purged at the beginning of a series of analyses to ensure that it accurately represents the mobile phase in use. Several factors influence the refractive index of liquids and must be controlled: the composition of the mobile phase, the temperature of the flow cell and column, the amount of dissolved air (or sparge gas) in the mobile phase, and the backpressure on the flow cell.

A schematic for a deflection-style differential refractive index detector is shown in Figure 1. In this type of dRI detector the angular deflection of the light beam is determined by the difference in refractive index between the reference and sample sides of the flow cell. The detector is aligned by adjusting the angular deflection so that the two photodiodes have equal currents when the contents of the reference and sample sides of the flow cell are identical. When the refractive index of the column effluent is different from that of the reference half of the flow cell, the beam is deflected as shown in Figure 1b. This off-balance signal is proportional to the difference in refractive index between the reference and sample sides of the flow cell. The operation of the dRI detector is straightforward. The detector must be purged with the mobile phase to be used for the separation, allowed to reach thermal equilibrium, and balanced to ensure that the full dynamic range of the detector is available for measurements.

Figure 2 shows a plot of the change in refractive index for water–methanol solutions as well as the first derivative of the change in RI with weight percent for water–methanol solutions. The slope of the Δ RI vs. weight percent curve varies from $+300$ to -330 µRIU/percent change in methanol. This change in refractive index with composition is sufficiently large that only premixed mobile phases are likely to provide stable baselines. For example, a solvent delivery system that has a compositional ripple of 0.25% would show a change in refractive index at the baseline of approximately ± 60 µRIU at 25% methanol. Consequently,

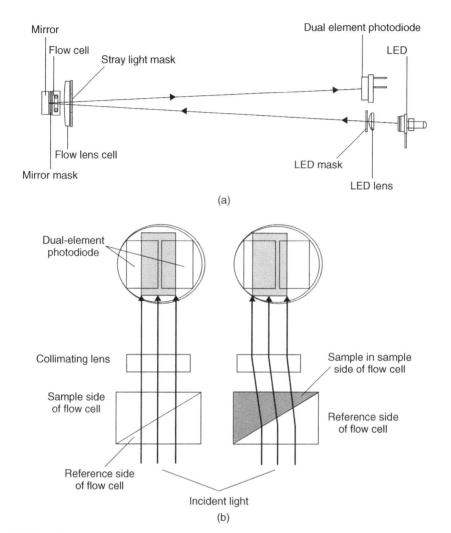

FIGURE 1 (a) Optical schematic of deflection-style differential refractive index detector; (b) deflection of the light beam.

"dial-a-mix" or pump-proportioned solvents as well as gradient separations are not commonly used with refractive index detection.

For dilute solutions, the solute molecules behave independently and the relationship between the refractive index of the solution and the concentration of solute is usually linear. However, the variation of refractive index with concentration may not be linear in all cases, as illustrated in Figure 2. Figure 3 plots the change in refractive index of water–sucrose solutions vs. concentration of sucrose, and is typical of the behavior of dilute solutions of small molecules. It should be noted that 8000 μRIU exceeds the maximum change in refractive

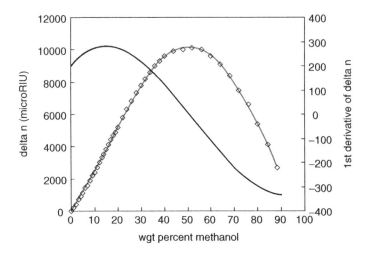

FIGURE 2   Refractive index of water–methanol solutions.

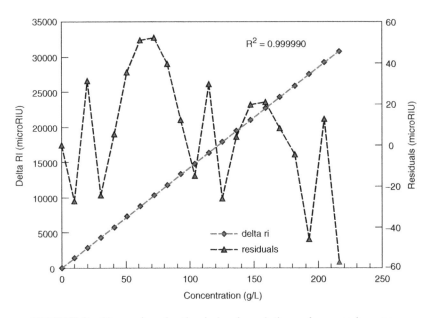

FIGURE 3   Change in refractive index for solutions of sucrose in water.

index that can be measured by most refractive index detectors (typically, 500 to 5000 μRIU). The linearity plot includes both the linear regression and the residuals, which are randomly distributed about zero. The residual pattern is a strong indication of linearity.

Differential refractive index detectors usually require an extended warm-up time to ensure that the detector's internal temperature is sufficiently stabilized to maintain a stable baseline at high sensitivity. When a chromatograph fitted with a refractive index detector is to be qualified, either upon installation or as part of a laboratory standard operating procedure (SOP), it must be set up to run the specific separation chemistry selected to demonstrate that the detector meets its design specifications. In most cases this will require a change in column and mobile phase: for example, from size-exclusion chromatography with a pure organic solvent such as stabilized tetrahydrofuran, to a premixed aqueous–organic solvent and reversed-phase column such as 75 : 25 water–methanol with a bonded C18 reversed-phase column. As a practical matter, such conversions may require a full day for the detector and chromatograph to become sufficiently equilibrated with respect to temperature to ensure that baseline drift does not compromise high-sensitivity analysis. Consequently, verification of detector drift is likely to be unreliable within the short period of time available for detector and system qualification. Detector noise and drift are best measured within the context of system suitability SOPs, which address the full analytical requirements of the assay.

Evaporative light-scattering (ELS) detectors differ from refractive index detectors in that they respond to the nonvolatile components in the chromatographic effluent. Figure 4 is a schematic diagram of a typical ELS detector. In ELS detection, an aerosol is created in the nebulizer, the aerosol droplets are desolvated by evaporation in the drift tube, and the resulting stream of particles is finally detected as the particles scatter light. Each step must be controlled carefully to ensure reproducible results, with the primary considerations being the gas flow

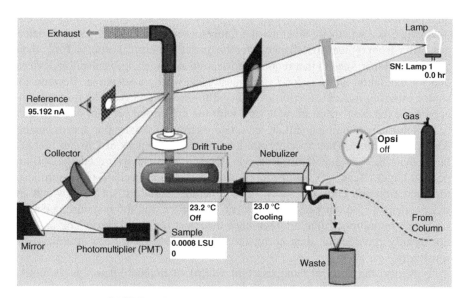

FIGURE 4   Evaporative light-scattering detector.

rates and temperatures, which determine the distribution of droplet sizes produced by the nebulizer and the efficiency of the desolvation process. The desolvation temperature is chosen to minimize the background scattering signal from the mobile phase while reducing the loss of semivolatile analytes. The time required for a droplet to travel from the nebulizer to the scattering chamber is determined by the volume of the drift tube and gas flow rate. The geometry of the drift tube and the location of its heating elements play a significant role in the rate of heat transfer to the aerosol droplets. Consequently, the specific temperature setting for any given separation will vary with the design of the drift tube.

Because the relationship between scattering efficiency and particle diameter varies with the ratio of the particle diameter to the wavelength, the peak area is not a linear function of concentration. Since most nebulizers do not produce a monodisperse aerosol (uniform droplet size) and the concentration of the analyte changes across the peak, the diameter of the particles entering the scattering chamber is not constant. Consequently, the scattering efficiency changes as the peak is eluted. The expected relationship between peak area and concentration is a power function:

$$\text{area} = aC^b \tag{1}$$

where a and b are empirically determined constants. This power function can be transformed by plotting it as a log-log linear curve:

$$\log(\text{area}) = \log(a) + b \log(\text{concentration}) \tag{2}$$

Consequently, demonstrating that the ELS detector follows a log-log linear calibration curve over a limited dynamic range is the appropriate response function qualification test.

Figure 5 is a schematic diagram of a fluorescence detector. Fluorescence detectors differ from typical absorbance detectors in several critical aspects. First, there are two monochrometers that require verification as wavelength-selection devices: one for excitation and a second for emission wavelength selection. Because the source intensity makes a first-order contribution to the observed fluorescence, the apparent excitation spectrum will vary with the emission spectrum of the source (e.g., solutions of quinine sulfate have two excitation maxima at approximately 250 and 350 nm). When a deuterium lamp is used, the peak at 250 nm is much larger than the peak at 350 nm. However, when a xenon lamp is used, the 350-nm peak is larger. Benchtop spectrofluorometers provide a mechanism to normalize the excitation and emission spectra and generate "corrected spectra," which are independent of the emission spectrum of the source and the response spectrum of the photomultiplier tube. Fluorescence detectors for liquid chromatography do not usually provide for such correction because they have been optimized for sensitivity.

Typically, the spectral bandpass (slit width) of monchrometers is 16 nm or larger in fluorescence detectors. Such large slit widths tend to distort spectra with narrow asymmetric peaks. The typical slit width of a tunable absorbance detector

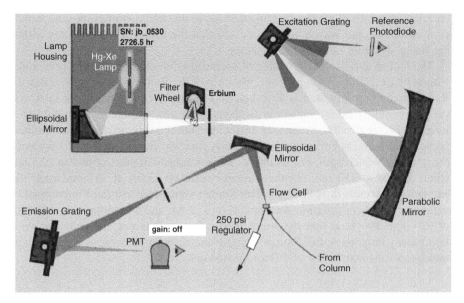

FIGURE 5 Fluorescence detector with mercury-doped xenon lamp.

is 4 to 6 nm and of a diode array detector is 1 to 2 nm. The result of dependency on a source lamp and the relatively large slit widths is that wavelength accuracy is difficult to demonstrate by collecting the spectra of reference compounds.

Fluorescence is an inherently nonlinear phenomenon, as illustrated by the relationship between fluorescence intensity and concentration:

$$F = k\Phi P_0[1 - \exp(-\varepsilon b C)] \tag{3}$$

where k is a constant based on flow cell geometry and optical design, Φ the quantum efficiency, P_0 the source intensity at λ_{ex}, and ε the molar absorbtivity at λ_{ex}. Equation (3) simplifies to

$$F = k\Phi P_0 \varepsilon b C \tag{4}$$

by expanding the term in brackets in a Taylor series and assuming that the higher-order terms can be ignored for small values of $\varepsilon b C$ (when $\varepsilon b C < 0.05$, the error is $< 2.5\%$). It should be noted that the product $\varepsilon b C$ is the absorbance of the sample at the excitation wavelength (Beer's law).

In addition to the approximation for low concentrations, fluorescence can also display nonlinearity associated with the *inner filter effect*, in which the source intensity P_0 is attenuated by the sample in portions of the flow cell that are not imaged onto the emission monochrometer's entrance slit and by self-absorbtion of the emitted photons when the excitation and emission spectra of the analyte overlap. Consequently, verifying the linearity of a fluorescence detector for liquid chromatography will require careful attention to the selection of probe compound, concentrations, and wavelengths to ensure that the test protocol operates within the linear dynamic range of the detector.

The signal-to-noise performance of spectrofluorometers is commonly demonstrated by measuring the signal-to-noise ratio of the Raman scattering of water. In Raman scattering a photon is absorbed and excites a vibrational mode in which the polarizability of the molecule is changed. A second photon is emitted at a longer wavelength. The frequency difference between the exciting and emitted photons, in hertz, is constant. Consequently, the wavelength shift will vary with the exciting wavelength. For example, water produces Raman scatter at 273 nm when excited at 250 nm, but the Raman scattering occurs at 397 nm when the excitation wavelength is 350 nm. The Raman signal-to-noise ratio is commonly measured with an excitation wavelength of 350 nm and an emission wavelength of 397 nm. Provided that the water in the flow cell is not contaminated with fluorescent impurities, the Raman signal-to-noise ratio of water is a convenient measure of the sensitivity of a fluorescence detector.

2 QUALIFICATION OF DIFFERENTIAL REFRACTIVE INDEX DETECTORS

The principal performance specifications to be verified in the qualification of refractive index detectors for HPLC are gain linearity, linearity, and system precision. Because it is not practical to operate the differential refractive index detector as a benchtop refractometer, the distinction between the operational qualification of discrete modules within the chromatograph and the performance qualification of the chromatograph is difficult to maintain. It is appropriate to use the same chromatographic conditions to complete the characterization of the other critical aspects of system performance, such as flow rate linearity and accuracy, injector linearity and accuracy, and column heater precision. Because refractive index detectors respond to the change in refractive index of the mobile phase with gradient elution and with pump-proportioned (i.e., dial-a-mix) isocratic mobile phases and are used with premixed mobile phases, only the flow rate accuracy and linearity of the solvent delivery system must be verified in a chromatograph using RI detection.

There is an ASTM (American Society for Testing and Materials) standard practice document, E 1303-95[1], that describes the characterization of refractive index detectors for liquid chromatography. The protocol describes the measurement of noise, drift, flow rate sensitivity, linear dynamic range, dynamic range, and calibration (in RIU) of refractive index detectors for liquid chromatography. The protocol uses air-equilibrated water as the carrier liquid (no column is used in E 1303-95) and solutions of glycerin in air-equilibrated water (8.72 mg/L to 43.6 g/L) for the dynamic range and calibration. The procedures are appropriate for comparison of refractive index detectors and characterization of the detectors, but they are rather time consuming and not chromatography friendly. The procedures described in this chapter focus on less invasive testing, which is more focused on chromatographic usage.

The selection of a suitable probe compound should reflect the following considerations. First, the chromatography should be fast and simple. The probe compound should form solutions for which the change in refractive index is linear over the range of concentrations chosen. Finally, specialized columns which are either expensive or require unusual maintenance should be avoided. The refractive index of aqueous solutions of sucrose are linear with concentration, but the ion-exchange columns commonly used for their analysis are highly specialized. Additionally, sucrose standards are unlikely to have a good shelf life and would require preparation shortly before the qualification tests. With those considerations in mind, a simple reversed-phase analysis of solutions of caffeine with either a methanol–water or acetonitrile–water mobile phase is more generically appropriate for qualification of refractive index detectors.

Figure 6 shows an overlay of six caffeine chromatograms that demonstrates system precision. The peak area precision for the six injections is expressed with a relative standard deviation of 0.14%. The retention time has a relative standard deviation (RSD) of 0.057% with a standard deviation of the retention time of 0.09 second with a 2.600-minute retention time. The RSD for retention time and peak area should be less than or equal to 1%, while the RSD for peak height should be less than or equal to 1.5%. The mobile phase is 75 : 25 premixed water–methanol (wt/wt), while the sample is caffeine dissolved in 77 : 23 water–methanol (wt/wt). The peak appearing at 1.0 minute is the result of the difference in refractive index of the mobile phase and sample diluent which elutes at the column's void volume. It is a negative-going peak because Figure 2 shows that the refractive index of a 23 wt% solution of methanol in water is less than a 25 wt% solution of methanol in water.

While refractive index can be reported in SI units (RIU) and the dynamic range of most differential RI detectors is less than 100,000 (the ratio of the maximum measurable change in RI/minimum measurable change in RI), which is less than the dynamic range of an 18-bit digital-to-analog converter (DAC), many RI detectors retain a sensitivity or range selection to preserve backward compatibility with established methods. Figure 7 shows a set of four injections of the caffeine sample used in Figure 6 recorded with four different sensitivity settings. The linearity of the sensitivity or gain settings is illustrated in Table 1.

The small variations in area and height sensitivity (area/sensitivity setting and height/sensitivity setting) as measured by RSDs of 0.67 and 0.48%, respectively, clearly demonstrate that the detector is scaling its output linearly with the sensitivity setting. Verification of the linearity of the sensitivity, or gain, settings ensures that methods run at different scale factors will produce consistent results, and methods that change the scale setting within the chromatogram to enhance the detection of small peaks will also generate accurate results. The area/sensitivity and height/sensitivity RSDs should be less than or equal to 4.5%.

The linearity of the detector response to varying concentrations or amounts of analyte injected on a column must also be verified. This is best done by

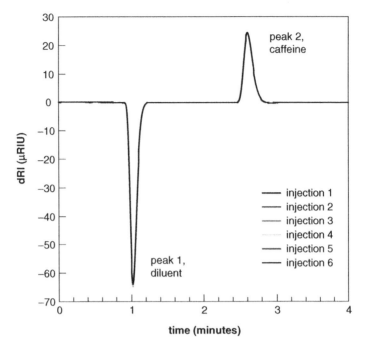

FIGURE 6 System precision for refractive index detection.

TABLE 1 Detector Sensitivity Linearity

Sensitivity	Retention Time	Area	Area/Sensitivity	Height (mV)	Height/Sensitivity
16	2.603	696,870	43,554	78,141	4883.8
32	2.602	1,393,721	43,554	157,562	4923.8
64	2.602	2,787,250	43,551	312,211	4878.3
128	2.603	5,500,439	42,972	623,537	4871.4
Mean	2.603		43,408		4889.3
Std. dev.	0.001		290		23.5
% RSD	0.026		0.67		0.48

injecting a series of solutions with known concentrations and holding the injection volume constant. Manually filling the flow cell with solutions of known refractive index is possible, but the presence of heat exchangers, which are necessary to control the temperature of the detector flow cell and column effluent, makes that task unreliable and tedious. Figure 8 shows a series of injections of caffeine at varying concentrations and several linearity metrics. Figure 9 shows the resulting calibration curves for both peak height and peak area. Both plots show r^2 values greater than 0.999, and the response factors for area and height have relative standard deviations of 3.7 and 3.9%, respectively. The r^2 values should be greater

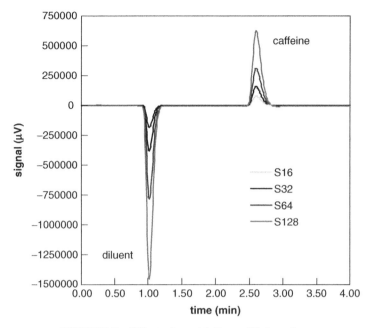

FIGURE 7 Effect of sensitivity on RI detection.

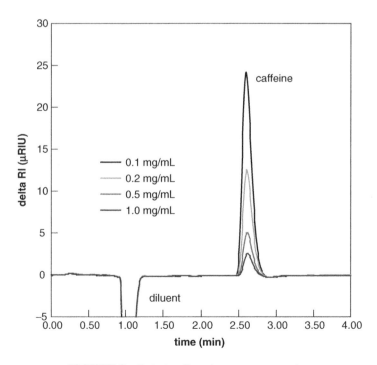

FIGURE 8 Detector linearity vs. concentration.

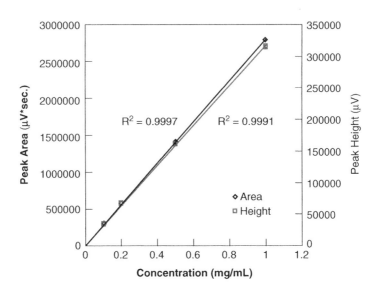

FIGURE 9 Detector calibration curves for caffeine.

than 0.999 for peak area and greater than 0.99 for peak height. The response factors for peak area should have an RSD value below 4.5%, while the response factors for peak height should have an RSD value below 5%.

An additional system-level test that confirms detector linearity can be performed by injecting different volumes of the same solution into the chromatograph. The result is a test of the injector's linearity which includes a de facto test of the detector's linearity. An example is shown in Figure 10. The r^2 value for the linear calibration curve is greater than 0.9999, and the RSD of the response factors (peak area divided by injection volume) is 1.096%. This clearly indicates that the injector *and* the detector operate linearly over the range of injection volumes and ΔRI. It must be noted that an injector linearity test is not a substitute for an independent detector linearity test. The injector linearity calibration curve should have an r^2 value greater than 0.999 provided that the injection volumes are within the sample manager's linear dynamic range.

3 QUALIFICATION OF FLUORESCENCE DETECTORS

The performance parameters of a fluorescence detector that should be verified during qualification include (1) emission wavelength accuracy, (2) excitation wavelength accuracy, (3) detector linearity, (4) photomultiplier gain linearity, and (5) Raman signal-to-noise ratio. System performance parameters that should also be verified include system precision, injector linearity, and carryover. Fluorescence detectors should be characterized using a well-behaved natively fluorescent compound which is not sensitive to oxygen quenching, such as anthracene. While

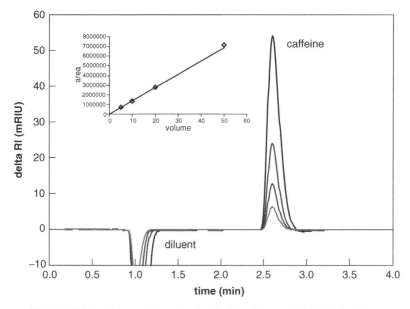

FIGURE 10   Injector linearity for 5-, 10-, 20-, and 50-μL injections.

fluorescence detection is frequently used for the detection of compounds that have been derivatized (both pre- and postcolumn derivatives are used for detection), the results will depend highly on the degree of control of the reaction chemistry.

Because fluorescence detection is used primarily for trace analysis, fluorescence detectors are usually optimized for optical throughput rather than resolution. Typical slit widths found in the monochromters used in absorbance detectors are between 1 and 5 nm (full width at half height); fluorescence detectors commonly have slit widths between 16 and 32 nm. A consequence of the wider slit widths is that the apparent λ_{max} of excitation and emission spectra may shift when the peak in the spectrum is not symmetrical. Figure 11 shows the effect on anthracene absorbance spectra of increasing the slit width from 1.2 nm to 8.4 nm. Note that λ_{max} has shifted by 5 nm with increasing bandpass.

Although there are two monochromters in a fluorescence detector, there is generally only one photodetector, usually a photomultiplier tube (PMT), that monitors the light emitted by the luminescent solutes. Some fluorescence detectors include reference photodiodes in the optical path to reduce the contribution of short-term fluctuations in lamp intensity to the fluorescence signal. In most cases, the excitation beam is trapped in an optical baffle after exiting the flow cell in order to reduce stray light. In fluorescence detectors the light that is detected passes through both monochromters before it is detected by the PMT. This couples the tolerances with respect to wavelength accuracy, increasing them by least $\sqrt{2}$ times the tolerance of the individual monochromters.

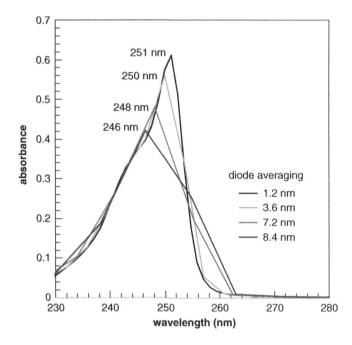

FIGURE 11   Effect of resolution on the apparent λ_{max} value of anthracene.

The excitation spectra observed are sensitive to the choice of excitation lamp. Equation (2) can be modified to highlight that sensitivity:

$$F = k\Phi P_0(\lambda_{ex})\varepsilon(\lambda_{ex})bC \tag{5}$$

Equation (5) highlights the fact that both the source intensity, P_0, and absorbtivity, ε, are functions of the excitation wavelength, λ_{ex}. When excitation spectra are collected, both dependencies will be observed, especially in wavelength regions that are highly structured. The mercury xenon lamp has several peaks from 300 to 400 nm, while the xenon and deuterium lamps have smooth emission spectra for the same wavelength range. The result is that the apparent values of λ_{max} can shift by as much as 4 nm, and the apparent sensitivity is dependent on the choice of source. This combination of shifts in λ_{max} with the slit width of the monochrometers and sensitivity to the selection of source makes the verification of wavelength accuracy difficult to trace to a reference chemical standard.

There are several approaches to verifying the accuracy of the two monochrometers. For instruments in which the flow cell can be removed from the optics bench and ambient light can be safely imaged onto the emission monochrometer's entrance slit, it is possible to take advantage of the ubiquitous presence of mercury in fluorescent lighting fixtures. The mercury emission line at 436 nm can be scanned to verify the wavelength accuracy of the emission monochrometer.

This is usually done at the lowest voltage applied to the photomultiplier tube and may require placing a diffuser over the entrance slit of the monochrometer as well as defeating any safety interlocks, which are designed to limit the exposure of the photomultiplier tube to intense light with high voltage applied to the dynodes.

Some fluorescence detectors include rare earth [such as erbium(III) perchlorate]–doped optical filters as devices which can be inserted into the optical path to verify the accuracy of the excitation and emission monochrometers. Tolerances for such diagnostics are matched to the nominal specifications of the detector, which are typically ±2 nm. This is an excellent tool for routine verification of wavelength accuracy and should be performed with the use of appropriate instrument diagnostics on a schedule prescribed in the laboratory SOPs.

The phenomena of Raleigh and Raman scattering provide a very simple way to verify wavelength accuracy. The emission spectrum for water collected with an excitation wavelength of 350 nm has peaks corresponding to the first- and second-order reflections of the elastic or Raleigh scattered light at 350 and 700 nm as well as smaller peaks for the Raman scattered light at 397 and 794 nm. Fluroescence detectors usually prevent measuring the first-order Raleigh scattering to ensure that the photomultiplier tube is not damaged by exposure to high light levels with high voltage applied. With λ_{ex} set at 350 nm, the second-order reflection will be observed at 700 nm in the emission spectrum. It is much smaller than the first-order Raleigh scattering because the gratings are blazed and there is no likelihood of damage to the PMT. The wavelength tolerance will be doubled because this is the second-order reflection. If the Raman peak for water is measured at the same time, the relationship between the excitation wavelength, usually 350 nm, and the emission wavelength, 397 nm, can also be used to confirm the accuracy of both monochrometers. In measurement of the Raman peak, the excitation wavelength is set at 350 nm, and the emission spectrum is scanned from 360 to 460 nm. Peak height, baseline noise, and λ_{max} can all be measured from the resulting emission spectrum. Some contemporary fluorescence detectors include an automated diagnostic test to measure the Raman signal-to-noise ratio. The λ_{max} value observed in the Raman signal-to-noise ratio test can be used to confirm the wavelength accuracy.

Figure 12 shows a system precision test using anthracene as the probe compound with an excitation wavelength of 252 nm and an emission wavelength of 404 nm. There are six injections, which have an area precision of 0.53% RSD with a retention time precision of 0.09% RSD. The RSDs are expected to be less than 1% for both retention time and peak area. The selection of concentrations that are appropriate to determining detector linearity with respect to concentration requires attention to the peak absorbances observed at the excitation wavelength. Equation (2) contains the term εbC, which is the absorbance, and Eq. (2) is valid only when the absorbance is less than 0.05. Figure 13 shows a pair of calibration curves for anthracene collected with an absorbance detector in series with a fluorescence detector. The absorbance detector was set to the excitation wavelength of 252 nm. The maximum peak absorbance was about 0.45, which exceeded the

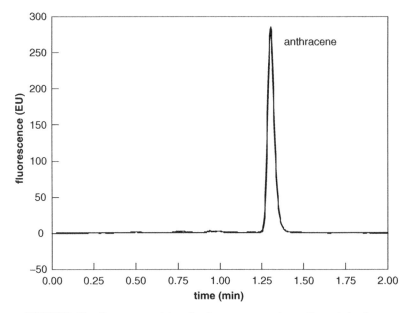

FIGURE 12   System precision for fluorescence: six replicate injections.

conditions necessary for Eq. (2) to apply. The calibration curve for anthracene fluorescence is clearly nonlinear, while the absorbance chromatograms remain linear with respect to concentration. When the mass of anthracene injected on a column is reduced, the peak absorbance falls within the <50 mAU limit associated with a 5% deviation from linearity [Eq. (1)], and the calibration curves for anthracene fluorescence become linear, as demonstrated in Figure 14. The peak absorbances at 254 nm are less than 0.05 mAU. The RSD of the response factors (sensitivity) are 1.5 and 2.0% for area and height, respectively. The RSDs of the response factors should be less than 4.5 and 5.0% for peak area and peak height, respectively. Figures 13 and 14 clearly demonstrate that it is critical to ensure that the fluorescence detector is not operated outside its linear dynamic range.

Fluorescence detectors usually have settable gain values which determine the photomultiplier tube voltage. When samples have components distributed over a wide dynamic range, it is necessary to adjust the detector gain to ensure that the PMT does not saturate when the more concentrated solutes elute from the column. As a result, it is necessary to verify the linearity of the detector's gain settings. This is done by injecting a constant amount of analyte on the column and varying the detector gain for each injection. The ratio of peak area divided by gain should be a constant. Figure 15 shows an example of this measurement. The RSDs of the area divided by gain and the height divided by gain are 1.1 and 0.8%, respectively. This detector's gain settings are highly linear. The RSDs for the gain linearity of peak area and peak height should both be less than 4.5%.

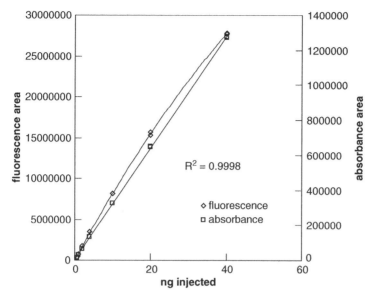

FIGURE 13 Nonlinear fluorescence calibration curve for anthracene.

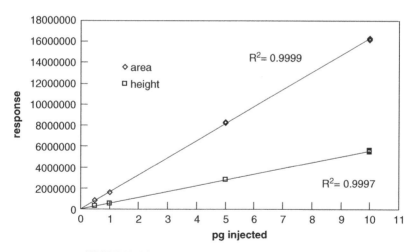

FIGURE 14 Anthracene fluorescence linearity.

Because fluorescence detectors are commonly used for high-sensitivity analyses, the carryover from the sample manager should be verified as a critical system parameter. The sample manager should be operated with its most efficient tools for reduction of carryover. Some sample managers have an injection mode with needle wash by immersion of the sample needle in a "wash vial"; other sample managers feature multiple needle wash solvents or durations of the needle wash

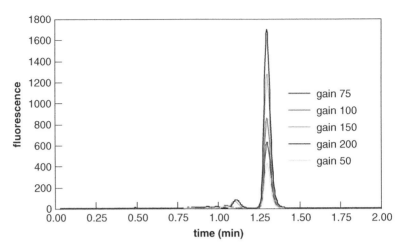

FIGURE 15   Gain linearity for fluorescence detector.

system. Regardless of the specific mechanism for carryover reduction, the needle wash solvent should be chosen to ensure that the carryover probe compound is removed from the injector mechanism and that the needle wash system is fully operational. Because fluorescence can be nonlinear when a wide dynamic range of concentrations is employed, and the usual carryover specifications are at least a 1000-fold reduction (typically, a 10,000- to 20,000-fold reduction), the most reliable sequence for measuring carryover is a series of injections which includes the following steps.

1. Blank injection(s) to verify that no peaks are present.
2. Calibration injections with a dilute standard.
3. One or more challenge injections at a concentration no less than 1000 times greater than the calibration injections. This chromatogram is recorded but is not processed because it is intended to exceed the linear dynamic range of the detector and to ensure that carryover can be measured with a good signal-to-noise ratio.
4. Post-challenge blank injection(s) which are quantitated against the calibration injections. If multiple injections are performed, the response should decrease with each succeeding injection.

Figure 16 shows the chromatograms associated with a carryover measurement. When the post blank chromatograms are processed, the peak areas correspond to less than 0.001% carryover of the challenge sample. In this example pure acetonitrile was used as the needle wash solvent and the mobile phase was a 80 : 20 acetonitrile–water mixture.

General chromatographic performance specifications such as compositional precision and accuracy, flow rate accuracy, injector linearity and accuracy, and

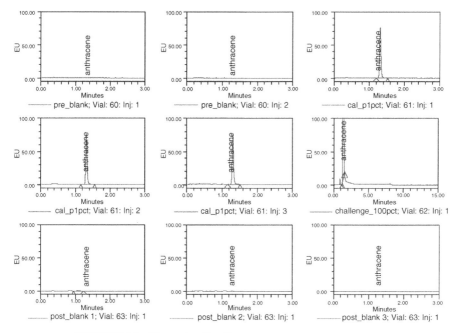

FIGURE 16 Example of anthracene carryover measurement.

column heater performance will require careful planning when verified for a liquid chromatograph equipped with a fluorescence detector. Because fluorescence detectors are used primarily for high-sensitivity assays, it is important to avoid introducing potential sources of contamination into the chromatograph while qualifying it for its intended use. For example, compositional accuracy and precision are commonly measured by applying a series of step gradients between a solvent and the same solvent that contains a marker compound such as acetone or propylparaben. For chromatographs with fluorescence detectors, neither of these commonly used markers will be acceptable since they do not fluoresce. If anthracene were used as the marker compound, there is a real risk that the instrument may be contaminated.

An alternative approach is highly desirable. Figure 17 illustrates a more holistic approach. In this example, a pump-proportional mobile phase of 80 : 20 acetonitrile–water is provided using four different valve-pair combinations in a quaternary low-pressure gradient pump. A total of 12 injections were made in four sets of three injections. The standard deviation of the retention time of anthracene was 0.15 second (RSD = 0.19%) with an area precision of 0.61%. Correlation of the retention time variation with the compositional precision and accuracy specifications of the solvent manager requires knowledge of how retention time varies with composition at the nominal composition. A small molecule (molecular weight <300 Da) eluting at a retention factor between 2 and 3 will show a change of about 3% in retention time for a 0.5% change in mobile-phase

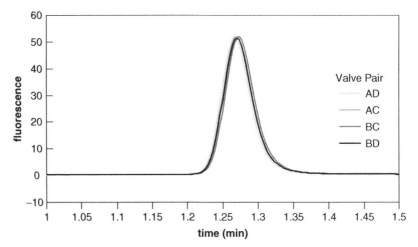

FIGURE 17 Compositional precision with fluorescence detection. Isocratic separation of anthracene with 80 : 20 acetonitrile–water mobile phase generated by quarternary gradient pump.

composition. Consequently, an RSD of less than 3% for retention times should be observed for the valve-pair combinations. Since low-pressure gradient pumps create desired compositions by modulating the duty cycle of metering valves (typically, solenoid driven), this procedure ensures that all four of the valves function correctly and the chromatograph has not been contaminated with a residual of anthracene as a solvent delivery marker.

4   QUALIFICATION OF EVAPORATIVE LIGHT-SCATTERING DETECTORS

ELS detection is a three-step process of nebulization followed by evaporation (desolvation) and finally, detection by measuring the intensity of light scattered by the particles of solute that remain after the droplets are desolvated. The singular advantage of ELS detection is that its selectivity is based on the volatility of the analytes. Nonvolatile analytes are detected, whereas analytes that have a high vapor pressure do not produce particles that scatter light efficiently. This is not related to the presence or absence of functional groups that absorb light (chromophores) or undergo oxidation or reduction (electrophores). It cannot be considered a *universal* detector, but it does respond to many analytes, such as lipids and carbohydrates that do not have chromophores. Additionally, the response factors for ELS detection tend to be less diverse than those observed for absorbance detection with the result that ELS chromatograms tend to reflect the mass balance of the chromatogram more accurately than absorbance chromatograms. This assumes, of course, that none of the sample components are volatile under the detection conditions and that the densities of the dried particles emerging from the desolvation chamber are similar.

Qualification of an ELS detector should demonstrate that the detector is capable of generating precise data, that the gain electronics are linear, and that the response function is consistent with ELS behavior. The effect of varying gas flow rates and desolvation temperatures should be characterized during method development, but they have complex and nonlinear effects on ELS detection, which vary dramatically with the composition of the mobile phase and the flow rate as well as the particulars of the design of the nebulizer and desolvation chamber. Consequently, verifying temperature calibration or gas flow rates is of very limited value and should not be included in ELS qualification.

In most ELS detectors, polychromatic light is used, which eliminates the need to verify wavelength calibration. Because the dynamic range of ELS detection is limited (two orders of magnitude are typical) and smaller than the carryover performance of contemporary auto samplers ($<0.1\%$), an ELS detector is simply not sensitive enough to require a carryover measurement.

Consequently, it is appropriate to limit the qualification tests for an ELS detector to those attributes that are likely to be universally part of the validation and suitability requirements of all chromatography with ELS detection, including system precision, gain linearity, and dynamic response. Acetaminophen and caffeine are excellent choices for the qualification of ELS detectors because they are not volatile under a wide range of analysis conditions, have simple reversed-phase chromatography, are stable, and are available in high purity.

Figure 18 shows an example of a measurement of system precision for an ELS detector in series with an absorbance detector. The RSD of ELS peak area

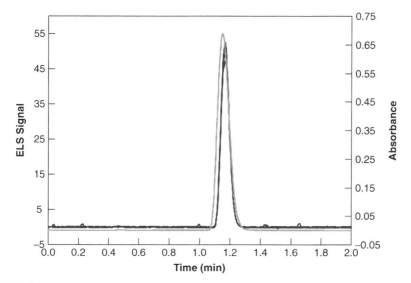

FIGURE 18 System precision for ELS detector (red) with absorbance detector (green) connected in series.

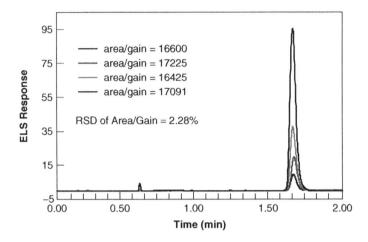

FIGURE 19   Gain linearity for ELS detector.

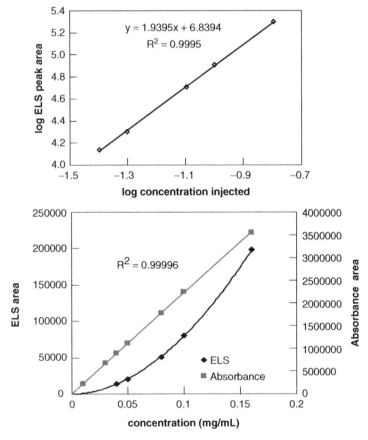

FIGURE 20   Calibration responses for ELS and absorbance detectors.

is 1.37%, while the absorbance detector has an RSD of 0.09%. The retention time RSDs are 0.13 and 0.04% for the ELS and absorbance detectors, respectively. The increase in RSDs for the ELS reflects the lower signal-to-noise ratio and the variability of the peak response that is typical of all nebulization-based detectors. Since the detectors are in series, it is clear that the ELS detector is the primary contributor to the RSD of its peak area. System precision for absorbance detection is generally $\leq 1\%$; with ELS detection it is generally less than 5%.

Figure 19 shows a gain linearity plot for an ELS detector. A constant amount of acetaminophen was injected; and the gains were set at 50, 100, 200, and 500 for the injections. The small RSD value of the peak area divided by the gain setting indicates that the photomultiplier tube's response has been linearized, ensuring that the dynamic range of the detector can be matched to the concentrations of peaks within and between samples. The RSD of the gain linearity is also expected to be less than 5%.

Figure 20 shows a dynamic response plot (log area vs. log amount) for an ELS detector with caffeine as the analyte. It also includes a plot of the area responses for both the ELS and an absorbance detector that was in series with the ELS detector. While the log-log linear plot shows $r^2 > 0.999$, the simple calibration plot makes it clear that the ELS response is inherently nonlinear, which is indicated by the slope value of 1.94 in the log-log linear plot. The nonlinear character of ELS detection suggests that the r^2 value for the log-log linear plot should be greater than 0.99.

The remaining attributes of a chromatograph that uses an ELS detector—autosampler injection volume linearity, carryover, column heater performance, and so on, are best measured with an absorbance detector that has a wide linear dynamic range.

REFERENCE

1. Practice for Refractive Index Detectors Used in Liquid Chromatography. *Annual Book of ASTM Standards*, Vol. 14.02, E 1511-93. ASTM, West Conshohocken, PA, 1999.

12

INSTRUMENT QUALIFICATION AND PERFORMANCE VERIFICATION FOR PARTICLE SIZE INSTRUMENTS

ALAN F. RAWLE

Malvern Instruments Inc.

1 INTRODUCTION

Roderick Jones, now retired from Malvern Instruments in England, once said that Pharmageddon is like Armageddon, but this time, the U.S. Food and Drug Administration (FDA) is involved. The qualification verification of particle size instrumentation presents challenges, and thus corresponding enjoyment, when those challenges are met. In this chapter we aim to explain the differences and subtleties that particle sizing techniques give in comparison to other measurement regimes with which the laboratory manager or analyst may be familiar. Out of the possible 300 or 400 particle sizing techniques, there are only a few that are used routinely within the modern pharmaceutical environment (see, e.g., BS 8471:2007, *Guide to Particle Size Methods*, BSI, Chiswick, UK). Some techniques have withstood the test of time for almost 5000 years (e.g., screening), whereas others are relatively new in technological terms (e.g., light-scattering techniques). We need to consider all of these in this chapter. But before beginning this particle size measurement journey, we should formulate the question that we would want to answer at the destination:

Practical Approaches to Method Validation and Essential Instrument Qualification,
Edited by Chung Chow Chan, Herman Lam, and Xue Ming Zhang
Copyright © 2010 John Wiley & Sons, Inc.

- What is the objective of the measurements?
- Why are we undertaking the measurements?

Careful thought and formulation of these questions will often provide guidelines to the analytical method that should be selected and the route to solving the real issue. We must avoid just providing a number on a spreadsheet or a single point on a statistical quality control chart. Heywood [1], in the closing speech at the third major international particle size congress in 1966 (predecessors were in 1947 and 1954), expressed this succinctly: "However, it must be realised that particle size analysis is not an objective in itself but is a means to an end, the end being the correlation of powder properties with some process of manufacture, usage or preparation." Note the use of the word *powder* in contrast to other formulation types that we may encounter: suspensions, emulsions, or aerosols as well as tablets. We also need to consider these alternative drug delivery systems in terms of the instrumental verification process.

Powders and suspensions may contain agglomeration in the finished formulation and we need to ask the question: Is the bulk size in the finished formulation desired, or is the primary (dispersed) particle size required? The only way that we can distinguish between primary particles and agglomerates is by means of energy input and quantification of the (apparent or real) size changes that occur on such application. Only the user of the result can decide which result is appropriate to the question or objective formulated.

The form of a bulk powder controls such properties as flowability, viscosity, filter blockage, tendency to agglomerate ("balling"), and dusting tendency. The primary particle size may relate to properties (typically, surface properties) such as the dissolution rate described by the Noyes–Whitney equation [2], chemical reactivity, moisture, and gas absorption. It can be noted at this stage that the energies found in most particle size "dispersion" units or techniques are considerably higher than can be found in normal conventional plant processing conditions. Thus, if we have a problem related to the inability of a powder to flow, such as blockage in a silo, it is ridiculous to try to correlate this to a primary dispersed size achieved by either high (Δ) pressures in a dry unit or sonication to stability in a wet particle size measurement.

The preferred approach in, say, a tableting example, may be to produce powders of different size distributions and perform the appropriate tests to determine at what sizes or size ranges failure would result. These could be dissolution or other tests. The ideal situation is to have a link between the particle size specification and the efficacy predicted in the patient or any detrimental effect on subsequent manufacturing steps. The difficulties of this are seen when absorption is considered (example from www.rxlist.com/cgi/generic/ranit_cp.htm):

Pharmacokinetics:

Absorption: Zantac is 50% absorbed after oral administration, compared to an intravenous (IV) injection with mean peak levels of 440 to 545 ng/mL occurring 2 to 3 hours after a 150 mg dose. The syrup and EFFER dose formulations are bioequivalent to the tablets. Absorption is not significantly impaired by the administration of food or antacids. Propantheline slightly delays and increases peak blood levels of Zantac, probably by delaying gastric emptying and transit time. In one study, simultaneous administration of high-potency antacid (150 mmol) in fasting subjects has been reported to decrease the absorption of Zantac.

Now, particle size or particle size distribution has an impact on virtually every important property associated with solid-form design. In an article in the *Journal of Pharmaceutical Innovation* [3], particle size distribution is stated as being of importance in relation to quality attributes and processing behavior for the following parameters: impact flow, blending, wetting, drying, mechanical, dissolution, and stability. No other property appears to exert as much influence as does the particle size distribution.

The careful reader will probably note several themes repeating through this chapter. This repetition should be a guide to the reader of the importance of these topics, and more care and attention should be paid to these areas.

2   SETTING THE SCENE

In the draft document Guidance for Industry: Analytical Procedures and Methods Validation; Chemistry, Manufacturing, and Controls Documentation [4], the FDA describes a number of particle sizing methods. To quote: "Particle size evaluation can include characteristics of size, morphology, surface, and characterization of drug substances and drug products."

2.1   Particle Size Methods

Types of particle size methods include, but are not limited to:

1. Nonfractionation methods that evaluate an entire population of particles
    - Microscopy (optical, electron)
    - Light scattering (dynamic, photon correlation, laser diffraction)
    - Electrozone sensing
    - Photozone sensing
2. Fractionation methods that use physical techniques to separate particles on the basis of size
    - Sieving
    - Cascade impactor
    - Sedimentation
    - Size-exclusion chromatography

These are techniques that will be discussed in more detail in relation to verification, but at this stage it is important to realize a number of basic considerations relating to particle size analysis in general. It could be considered a reasonable expectation that all the techniques listed above will provide exactly the same answer for any material presented to them. Unfortunately, this is not the reality.

We will not labor the point, as there are plenty of articles on the subject, but the basic cause of differences between particle sizing techniques and other metrology applications revolves around real-world particles being three-dimensional but needing a single one-dimensional descriptor—thus, there is an inherent shape dependence, and particle sizing techniques measure different aspects of a particle. For further information on the basic philosophical issue, see an article by Rawle [5]. There are also many ways of expressing the statistics—not all obvious (contrast number and volume) and many (correct, but different) ways of gathering the data—different methods provide different properties of the particle. This needs to be dealt with by using different instruments that deal with particle size as well as shape distributions and is summed up in the *U.S. Pharmacopeia's* (USP's) General Test ⟨776⟩ as follows [6]: "For irregularly shaped particles, characterization of particle size must include information on particle shape." In general terms, we will need more than one "sizing" technique to characterize our systems. Visualization is essential. There are statistical implications of the measurement of small numbers of particles that need understanding. We will deal in slightly more detail with these issues when we consider method validation.

The second challenge revolves around the third word in the phrase "particle size distribution." The implication and emphasis in the word *distribution* are that both the quantity, the y (percent: more correctly, density distribution)-axis, as well as the x [size parameter (number, surface, or volume, based in the simplest or equivalent forms)]-axis, need specification and also that statistics will play the role in defining the situation [7]. This double-jeopardy type of challenge is combined with the need for statistical understanding of the methods and numbers they generate, for which most end users are ill-equipped. For example, there is confusion among users and even within the regulatory authorities of the difference between particle counting/particle counters (number-based analyses) and a particle size distribution analyzer (usually, volume or mass based). Again, we endeavor not to belabor the points, but without setting these basic building blocks in place, the entire structure that we are building is in danger of collapse.

3   PARTICLE COUNTING TECHNIQUES

Particle counting is the basis of a number of techniques, which merit further discussion in the context of the pharmaceutical industry. These include:

- Microscopy (manual or electron)
- Image analysis
- Light obscuration or shadowing techniques for air or liquid applications
- Electrozone sensing

The use of counting should be considered if our requirement is hinted at by any of the following points, known as the "3C's":

1. Concentration
2. Contamination
3. Cleanliness

The number of size classes in these techniques can be very limited (often, corresponding to regulatory specifications), and typically, the number of particles measured is very small (at least in comparison to an ensemble technique). In addition, the dynamic range of a counting instrument tends to be very limited. Often, a light obscuration (or "shadowing") counter is used for particle counting purposes. Using either a laser or white light source, the typical range of such a counter used in the pharmaceutical industry is quite limited (e.g., 1 to 100 μm for liquid, about 0.3 to 10 μm for air). It need not do more than this, as (for example) USP calls for measurement at 10 and 25 μm for parenterals (USP General Test $\langle 788 \rangle$) and 0.5 and 5 μm only for air counting (Federal Standard 209D/E). The larger the size of the particle to be counted, the lower the absolute count or concentration can be. Typically, a 1- to 100-μm liquid counter cannot deal with more than about 50,000 particles/cm^3 of fluid, well below the 1 g of SiO$_2$ particles that would be no trouble to an ensemble technique such as laser diffraction.

A particle counter operating on the obscuration or shadowing principle is a secondary form of measurement. A reference standard is delivered through the measurement zone and the response adjusted until the signal sits in the required or "correct" channel. Note that it is quite difficult for standard latex material larger than a micrometer or so to be nebulized so that it can pass through the measurement zone of an air counter. Furthermore, although the x-axis (size) can be calibrated quite easily, it is another matter to confirm that the absolute count (or concentration) is correct. In summary, in informal terms, a particle counter will tend to answer the question: How many?

4   PARTICLE SIZE ANALYSIS AND DISTRIBUTION

In contrast to a particle counter, a particle size analyzer is "interested" in the answer to the question: How much? For example:

- How much (volume or weight) material is greater than x μm?
- How much (volume or weight) material is less than y μm?
- How much material is between two size bands?
- What are the mill efficiency calculations (Tromp curves)?
- What is the average size? How has this changed on a quality control chart?
- How much is the value of my material?

Note that these questions imply that something is needed to describe or deal with the quantity or y-axis that we discussed earlier. In these situations we already have a powder, emulsion, suspension, or spray, and we are concerned with obtaining a particle size distribution. There is a subtle difference between "How many?" and "How much?" This is sometimes not always appreciated by the average purchaser of an instrument. Then the question "How many?" is answered either by some form of imaging or a light obscuration counter (counting single particles; number distribution), and the question "How much?" by a laser particle size analyzer (ensemble method; volume or mass distribution) in a modern laboratory.

5 INSTRUMENT QUALIFICATION FOR PARTICLE SIZE

They are a variety of instruments that may be used for particle size characterization. The qualification will depend on the basic technique that the instrument uses in the measurement. Techniques will deliver results in either a continuous (e.g., light scattering) or a discontinuous, sometimes called *discrete* (e.g., screens) format. Techniques may rely on the separation and quantification of particles (e.g., screens, sedimentation, fractionation) or will be able to deal with all the particles on a normalized basis (e.g., light-scattering techniques). We will define, in general terms, those qualification protocols that will need examination for any sizing technique.

Two definitions issued by the FDA set the scene (italic added):

- "Methods validation is the process of demonstrating that analytical procedures are *suitable for their intended use*" [4].
- "Establishing documentary evidence which provides a high degree of assurance that a specific process will consistently produce a product meeting its predetermined specification and quality attributes" (FDA definition, 1997).

We need to define terms that are in regular use within the particle sizing field that may have different apparent meanings in other technologies. These (particle sizing) definitions then allow the variety of variables that contribute to a total error balance to be assigned to the contributing causes of sample heterogeneity, changes of the sample during the measurement and instrumental variable. In this way, the instrument can be validated correctly [especially with respect to product qualification (PQ)] as opposed to sample-dependent variation. It is important that this PQ is actually related to a fit-for-purpose specification, as overspecification is both costly and unnecessary in many instances.

5.1 General Definitions and Terminology

In particle sizing we define the relevant terms as follows:

Accuracy. Is the measurement correct? This verification is accomplished by comparison with an appropriate certified reference material. The variation or stated tolerances in the standard and any predefined manufacturing tolerances need to be considered in this analysis.

Precision. How close are repeated, consecutive measurements? It is obvious that a single measurement is perfectly precise and has no meaning or value! The precision acts as a guide to the inherent stability of the overall system: instrument and sample. We note that in the case of a sample where dispersion is the objective, precision will provide guidelines to the attainment (or otherwise) of stability. It is usual to state precision in the form of *relative standard deviation* (RSD), also known as the *coefficient of variation* (CV), and defined as $100\sigma/mean$, where σ is the standard deviation.

Repeatability. This is a measurement of precision where the same group of particles is measured multiple times. It therefore applies only to a measurement under wet conditions, where the same group of particles is present throughout the analysis. In the pharmaceutical environment, this is termed *intra-assay precision*.

Reproducibility. This term refers to repeated measurements on other than the same group or set of particles. For example, multiple measurements of a dry system involve taking separate subsamples or different groups of particles. This is thus subject to the homogeneity or otherwise of the sample and is not an inherent instrumental variable (although, of course, it enters into the entire precision account). We can also talk of reproducibility in terms of different operators, different laboratories, different times of day, and so on.

Robustness. This is an examination of any parameter that exerts or may exert an influence on the end result. For example, in light scattering we may examine the effect of optical properties on the end result, with the objective of knowing how much accuracy or variation in these constants is required for analysis. Other commonly utilized parameters within laser diffraction that would need examination in dry powder measurement would include the ΔP (differential pressure across the venturi, which governs the energy imparted to the particulate system).

The last three tenets of quality control are commonly referred to as the "3R's." The expression the "3M's" (machine, method, and material) is also used. All of these variables can be measured experimentally and appropriate predefined specifications can (and should!) be formulated for the material(s) in question. We can divide up the total variation in a set of experiments among the following:

- Instrument
- Changes in the material during the experiment itself
- Sampling or homogeneity/heterogeneity of the material in question. This can apply to dry powders, slurries, or aerosol precursors. An aerosol initiation is a dynamic event—once the aerosol or metered dose inhaler or similar is activated, we are again dealing with a heterogeneous and dynamic system.

We can see immediately then that the factors that control the *total* precision of any experiment can be broken down as follows:

- Repeatability
  - Instrument
  - Changes in the sample during measurement
- Reproducibility
  - Instrument
  - Changes in the sample during measurement
  - Sample heterogeneity (i.e., sampling regime)

The instrumental variable is the smallest (yes!) and the easiest to deal with. Verification is with a known (certified or prespecified) material with predefined pass–fail criteria. Ideally, this reference material allows traceability back to an international unit, such as the meter.

Changes in the material during measurement are not usually considered in other metrologic techniques, but they may be an objective of particle size analysis (i.e., dispersion, deagglomeration) or they may be undesired events (i.e., dissolution, attrition/milling, settling). Such changes can never be seen with a single measurement, and thus multiple consecutive measurements are a prerequisite of any method development.

Last, and not least, is the sample heterogeneity that makes one subsample different from another subsample from the same bulk lot. In particle science this is termed the "breakfast cereal" problem. If material larger than around 75 or 100 μm is present in the sample, this is actually the largest variable in the total precision balance. We consider next each variable in turn.

Sampling   Modern instrumentation is very sensitive and can also deal with small samples. However, the heterogeneity of a particle size distribution has statistical implications: in particular, with respect to the minimum mass of material needed to satisfy any particular required standard error or specification.

The key factors in determining the minimum mass of sample are:

- *The standard error required by the user*. To specify a smaller or "better" requirement will obviously require that more sample needs to be taken for the measurement. Indeed, as there is a square relation between the standard error and the number of particles, then to modify the standard error required from, say, 10% to 1% will require 100 times the number of particles. The maxim here that needs answering is: "fit for purpose"? This requirement needs to be specified by the user prior to any measurements being taken.
- *The width of the distribution*. Obviously, if every particle were the same, only one particle would be representative of the entire population. The wider the distribution, the more sample that will need to be taken to satisfy the standard error requirements for the tails of that distribution.

- *The top-end size of the distribution*. This is because larger particles make up a corresponding larger proportion of the mass (as the mass of a particle is proportional to d^3, where d is the diameter of the particle).
- *The density of the particles*. As the standard error (S.E.) is related to the *number* of particles (S.E. $\propto 1/n^{0.5}$) in any required size band, the denser the particulate system, the greater the mass required.

More details may be found in the International Standards Organization's ISO 14488 [8] and also in Malvern application note MRK 456-01 [9].

A few brief pertinent examples will illustrate the importance of sampling in any metrologic application. If the mean of a distribution is required to 10% S.E., only 100 random particles need measuring. Potential sampling issues arise when either the required S.E. is lower than this nominal value or when a distribution is required, the key factor being what top end of the distribution is required or specified: x_{90}, x_{95}, or x_{99}. Imagine that we have a silica material (density ~ 2.5 g/cm^3) and wish to specify the 99% point (x_{99}) with a standard error of 1%. What is the minimum mass needed when the top end of the distribution is 1, 10, 100, and 1000 μm?

Now the S.E. is inversely proportional to the root n, where n is the number of particles. Thus, for a standard error of 1% ($= 0.01$), it is trivial to show that the required number of particles is 10,000. To calculate the total mass required, we have only to calculate the mass of 10,000 particles of the x_{99} size and then remembering that this point makes 1% of the total mass of the system, multiply our mass of 10,000 particles by 100.

As an illustration, taking the x_{99} point of the silica to be 100 μm, the minimum mass required is

$$(100 \times 10,000)\frac{\pi}{6}(100 \times 10^{-4})^3(2.5) \sim 1.3 \text{ g}$$

Now this mass is around the sample size (1 to 2 g) typical of many metrologic techniques. Therefore, if there are particles in a system larger than around, say, 75 μm, sampling becomes a (indeed, *the*) key issue in defining the possible precision of the measurement. As the minimum mass scales as the cube of the particle diameter, a simple Excel spreadsheet is easily constructed, as in Table 1.

Now if the x_{90} (not x_{99}) is required to a S.E. of 1%, the minimum mass in the last column is reduced by a factor of 10 (the x_{90} point representing 10% of the total mass of the system). These statistical features of the tails of distributions are not often evaluated by end users of particle size measurement equipment. It was necessary in ISO 13320:2009 (the laser diffraction standard) to employ a form of words indicating that the use of an x_{100} value is "specifically forbidden." It should be understood that these requirements are not particle sizing specific but originate solely from statistical considerations no different in reality from the sampling of red and white balls from a hat, an example probably familiar to those undertaking a tenth-grade mathematics class. Two more examples follow to illustrate different issues.

TABLE 1   Minimum Mass of Powder Required for a 1% Standard Error at the x_{99} Point[a]

x (μm)	x(cm)	x^3	π/6	ρ (g/cm^3)	Mass of one Particle (g)	Mass of 10,000 Particles	Minimum Mass (g)
1	0.0001	1E−12	0.525266	2.5	1.31317E−12	1.31317E−08	1.31E−06
10	0.001	0.000000001	0.525266	2.5	1.31317E−09	1.31317E−05	1.31E−03
100	0.01	0.000001	0.525266	2.5	1.31317E−06	0.01313165	1.31E+00
1000	0.1	0.001	0.525266	2.5	0.001313165	13.13165	1.31E+03

[a]The x_{99} point of the distribution is x μm and the density of the material is 2.5 g/cm^3.

- A system with Gaussian normal distribution has a mode at 400 μm with an x_{99} value of 750 μm. The particle density is 3.5 g/cm^3. What is the minimum mass needed to specify this x_{99} to 5% standard error? If only 0.3 g is sampled, what is the best standard error that can be achieved?
- The minimum number of particles for 5% S.E. is easily found as 20^2, or 400. This minimum mass = $(400 \times 100)(\pi/6)(750 \times 10^{-4})^3.3.5 \sim 31$ g. If 0.3 g is actually taken, this represents roughly 1/100 of the minimum mass indicated above and thus would contain 4 particles only at the x_{99} point. The S.E. is then $1/(4^{0.5})$ or 1/2 (50%!). It can be seen that the mean of the distribution is irrelevant in this calculation; rather, the top end of the distribution will define the minimum mass required. One would expect in this scenario that repeated samplings of 0.3 g (reproducibility; discussed later) would not take exactly 4 particles each time. The x_{99} point would therefore vary significantly as, for example, if 2 or 5 particles were present in each of the repeated samplings. The heterogeneity of the distribution in the top-end tail is the real problem here.

The "needle-in-a-haystack" problem is one that is frequently encountered and is usually couched around some apparently rational statement such as "I wish to detect a small amount of agglomeration in my system." This statement is made with the assumption that the particle size distribution is continuous and that <u>any</u> sample is expected to show this tiny amount of material in the tails (ppm or ppb) and assumes that the measurement technique has the appropriate sensitivity and resolution. The reality is that this agglomerate represents the single particle in a million or billion (the needle in the haystack) and that on repeated samplings we would not expect to find this particle except in an exceptional case. This is the nature of statistical quality control in the classic sense.

Let's take, for example, the removal of a number of balls in a hat (say, 10) looking for the one red ball in the 1 million total balls in the system. Obviously, the answer to the question "Do we see the red ball?" would be answered in the negative in virtually every case. Even when the amount of agglomeration is significant, the top end of the distribution can make the sample amounts frightening. The following is a real example of a customer with around a believed 1%

TABLE 2   Minimum Mass of Material Needed to Specify the x_{99} Point to a Standard Error of 1% [a]

D (μm)	Diameter (cm)	Radius (cm)	Density (g/cm^3)	Weight in Top Size Fraction[b] (g)	Total Weight[c] (g) (= Last Column × 100)
1	0.0001	0.00005	1.5	7.9×10^{-9}	0.00000079
10	0.001	0.0005	1.5	7.9×10^{-6}	0.00079
100	0.01	0.005	1.5	0.0079	0.788
200	0.02	0.01	1.5	0.063	6.3
500	0.05	0.025	1.5	0.99	98.5
1,000	0.1	0.05	1.5	7.9	788
1,500	0.15	0.075	1.5	26.60	2,660
2,000	0.2	0.1	1.5	63.1	6,305.2
10,000	1	0.5	1.5	7881	788,148

[a]The x_{99} point is at 2000 μm and the density of material is 1.5 g/cm^3.
[b]This is the weight of 10,000 particles, assuming spheres.
[c]This is where 1% of the particles are in the top size band.

of agglomeration at 2 mm (2000 μm) in an aqueous pharmaceutical slurry (20% solids by mass). Assuming a particle density of the solid of 1.5 g/cm^3, we can calculate the minimum mass of sample to have a 99% confidence limit in seeing this agglomeration (Table 2).

The mass calculated is around 6.3 kg, representing over 30 L of 20% slurry. We can consider this in another way—as the minimum amount of material that is needed to show homogeneity at the 99% confidence level—each 30 L or so will be statistically equivalent (99% confidence limits) or homogeneous. Now, even if we are sampling 100 cm^3 at a time (300 samples), we expect many negative tests and only an occasional positive test. A single test or even 10 repeated samplings could not state with certainty that 1% of the universe of material was at 2000 μm. The entire 30 L would need evaluating. In this case, if this problem material is present or occurs in storage, it is probably best not to test for it (we cannot inspect quality into a batch; rather, it must be built in), but simply to filter or remove before use or reprocessing. Thus, any potential problem arising from the presence of the agglomerate will not occur. We probably would have to do this anyway even if we were certain that the material was present.

The above is the essence of the statistical argument (embodied in ISO 13320:2009) that the x_{100} value is a meaningless specification for a continuous distribution (it is not meaningless for a discrete distribution. For example, it is easy to say that 100% of a material passes a 2-mm screen and thus that the x_{100} point is 2 mm). Now the x_{100} represents the largest particle in the system. Assuming that we have "captured" this particle, then, as an example: 1 g of SiO$_2$ density 2.5 g/cm^3 contains around 760×10^6 particles if the particle size is 10 μm on average. The largest particle therefore represents 1 particle in 760 million. This is the 99.999999% (six 9's after the decimal point) point. No technique has the ability to be able to resolve to such a level. Indeed, with a

Gaussian distribution this point would represent more than 7 standard deviations from the mean (6σ is 3.4 parts in 1 billion).

We can then ask:

- Do we actually have this largest particle in our subsample? Clearly, this is unlikely unless we take the entire lot from which we are sampling.
- If not, have we got the second largest?
- If we do have the largest particle, can we actually measure it in our system, and is it statistically valid?

Try considering these questions and this analysis applied to a glacial moraine. The corollary to this is that techniques evaluating only a tiny amount of sample (e.g., electron microscopy) must be suspect on the quantitative level when particle size *distribution* is considered. In terms of measurement, then, the laboratory can only be as good as the sample submitted, so taking the primary sample is of major importance. The division of the primary sample into subsample aliquots (which must be measured in their entirety) is covered in standard texts (e.g., [10], Chap. 1) and the Malvern application note (MRK 456-01) mentioned earlier. The two key points to note, as a summary, from these texts are:

- The only route to bringing the sample-to-sample variation down to <1% RSD is with a spinning riffler.
- Scoop sampling leads to an RSD of 1σ slightly above 5%. Thus, 3σ is on the order of 15 to 20%. This sets the best-case scenario for a specification where systematic sample division is not employed.

ISO 13320:2009 states: "For dry dispersion the complete fractional sample should be used for the measurement." It is the stages of primary sampling and subsequent division ("the laboratory can only be as good as the sample submitted") that should reduce this heterogeneity variable to a minimum. Specifications must deal adequately with the inherent sample heterogeneity, and, of course, particle size standards (see Section 5.2) should have statistics stating the expected bottle-to-bottle variability. The objective of instrument verification should be to quantify this variable and not measure the variation within a sample, standard or otherwise. Modern instruments are very sensitive and are easily capable of quantifying the variation within a sample, and this can be misinterpreted as an instrument variation by an unknowledgeable user. The interesting point here is that the statistical variation within a powder can easily be predicted as we have detailed earlier, and thus meaningful and realistic specifications can and should be set.

5.2   Standards

In this section we explore the written and material standards available in the particle size field and comment on their use and suitability.

General Standards We normally consider two forms of standards: reference material standards and (paper) methodology or method standards. Both types arise from national (prestigious) bodies. An excellent review article [11] on reference materials has been written by Jolyon Mitchell. The main source of reference materials is organizations such as the National Institute of Standards and Technology (NIST) in the United States, AIST and APPIE in Japan, BCR in the European Community, or NPL in the UK and a very small number of manufacturers, such as Duke Scientific (United States; now part of ThermoFisher), Polysciences (United States), and Whitehouse Scientific (UK).

This section deals with standards in the form of a "recipe" (method) and/or a philosophical or scientific (methodological) approach to verification in the particle size area as well as standard materials. The differences need to be appreciated and all are needed—verification cannot proceed without a (standard/certified) material and a protocol for its use. There is a clear difference between a method and a methodology. A *method* is akin to a recipe and a *methodology* (the science of methods) is more akin to a philosophy. A method may state "Stand on your head while taking the measurement" and needs to be followed rigidly. The methodology will explain the rationale behind the method or general technique and thus will explain the reasons for standing on one's head and when it may or may not be important to do so. The methodology or practice guide usually has sufficient background to enable a reasonable method to be developed and may require certain materials to validate the technique or instrument in question. The term *standard operating procedure* (SOP) is frequently encountered within the pharmaceutical and other industries and is a formalization of the method or measurement protocol. Paper standards arise mainly from committees within the ISO and ASTM (American Society of Testing and Materials). In particular, we cite ISO TC24 (Particle Sizing Methods Including Sieving), ISO TC229 (Nanotechnologies), ASTM E29 (Particle and Spray Characterization), and ASTM E56 (Nanotechnology).

Reference Materials The nature and philosophy of particle sizing techniques means that a large range of reference material standards exists, but they may not always be transferable between techniques. We must distinguish between materials used for calibration of techniques and those used for verification. Particle size techniques that involve calibration procedures (i.e., altering the response of an instrument to give the correct or certified value) include light obscuration and electrozone sensing counters. Other particle sizing techniques tend to verify the performance of the equipment (i.e., using a reference standard to confirm that the instrument is within the combined tolerances of the standard and that of the manufacturing stage). These include light-scattering techniques, which are termed "first principles" methods in that there is nothing to adjust (e.g., a potentiometer) in order to bring an instrument into agreement with the reference standard. Rather, an intervention needs to occur (e.g., cleaning the system, adjustment of optics) and some (documented) corrective action initiated, followed by a new test to confirm that the intervention has succeeded. *Testing into compliance* (taking the first correct result after a consecutive stream of failures) is an action not permitted by the FDA, and there are a large number of court cases in relation to this (*USA* v. *Barr Laboratories*, 1993, is one defining case).

The ideal particle size reference material is one that challenges both the size and the quantity axes of the appropriate distribution display (frequency or histogram; cumulative or discrete). The correct display is the cumulative frequency distribution. To avoid ambiguity in the size parameter (and to allow traceability to the appropriate SI unit—the meter), the ideal standard should be spherical. The vertical (or quantity) axis should be challenged by having a range of sizes; that is, the distribution should be polydisperse. Such spherical polydisperse standards are available through:

- *Organizations such as the National Institute of Standards and Technology (NIST).* Examples of polydisperse spherical standards include NBS 1003c and NBS 1004a.
- *Whitehouse Scientific.* Seven standards, covering the size range 0.1 to 2000 µm; produced as single-shot vials from 25 mg to 10 g. These standards were developed to support the official Community Bureau of Reference (BCR) particle metrology standards.
- *Duke Scientific* (now part of ThermoFisher). There are, for example, ranges of polydisperse glass microspheres (Sales Bulletin 124A) from 1 to 230 µm and polystyrene–divinylbenzene copolymer standards of 1 to 50 and 100 to 500 µm (Sales Bulletin 123A).

Obviously, other manufacturers exist, and the list above cannot be regarded as comprehensive in any way. See more details in the Appendix (which again is not comprehensive).

Other nonspherical, polydisperse standards exist in the marketplace, but the certified values will only be applicable to a particular technique, and the values, almost certainly, not transferable to any other particle sizing technique. A good example is the BCR standards (crushed quartz), made under the auspices of the European Economic Community in Brussels. These have been certified by sedimentation and sieve techniques. The smallest (BCR 66), nominally 0.5 to 2.5 µm, certified by sedimentation, cannot be expected to deliver the certified results using a light-scattering technique. This is because, at these sizes, there is a competitive vector to gravitational settling (namely, Brownian motion), and this tends to give a sedimentation result smaller than the laser diffraction–generated volume distribution. So this standard is reasonable enough for verifying sedimentation apparatus working on gravitational settling (as all devices should work in the same manner). This standard could be used to verify the *precision* of a laser diffraction apparatus, but the accuracy against the certificate valued is clearly different. This could be catered for with a different set of values for laser diffraction as is the case with NBS 1003c (available from NIST), now certified on the basis of sieves, electrozone sensing, and laser diffraction.

We need to be clear on which standards are applicable to which techniques, as some standards may be too narrow for certain techniques. *Monodisperse* spherical standards (e.g., polystyrene latex), actually being easier to manufacture than polydisperse variants, are very common (see the organizations and manufacturers named above). Such standards are used in the calibration of electrozone sensing and light obscuration (shadowing) counters. They fix the signal for calibration of the x- or size axis. Narrow standards are also used in the verification of image analysis systems that can deal with such distributions.

Ensemble techniques such as sieves or light scattering usually need wider distributions to verify performance. In the case of laser diffraction, it is important to verify over a wide range of (angles and thus) scattering detectors. Thus, the international standard for laser diffraction (ISO 13320-1, November 1999) recommends a width of distribution exceeding 1:10 to verify the performance of such instrumentation. This paper standard states that, ideally, a verification material should be spherical and polydisperse as well as certified. Such a material would be a primary standard and allow traceability back to the meter. Once an instrument has been verified in this manner, it should then be possible to use a precharacterized (with appropriate pass–fail criteria) performance qualification (PQ) material that is irregular in shape to confirm the working performance of the instrument—and some manufacturers do so. This would clearly be a secondary verification, as there would be no traceability of such a PQ standard back to the (international) meter. Setting the pass–fail criteria for such materials would be problematic as well, with no other means of verification, such as microscopy. The instrument manufacturer would be in the commanding position of being able to set criteria based on its own particular setup, without the ability to perform any independent verification. Such a situation would be different with polydisperse, spherical standards, which should prove verifiable with other techniques (subject, of course, to factors such as minimum numbers of particles in line with the appropriate point of the cumulative distribution).

For historical reasons we mention another "material" device used to verify the optical equipment used in spray and aerosol characterization. Generation of a standard verifiable aerosol is obviously a task fraught with difficulty (although such devices as the Sinclair–La Mer nebulizer are in existence). Nebulizing material larger than 1 or 2 μm from a suspension is also very difficult; the defunct BS 5295 counter specification was difficult to verify at the 10- and 25-μm levels. Thus, the reticle was designed to verify the hardware and software performance of optical devices designed for aerosol or spray measurement. This device forms part of an ASTM specification (E 1458-92) and consists of chromium "particles" (characterized by microscopy) on a glass plate manufactured and analogous to the photomask used in the electronics industry. Unfortunately, a combination of factors prevented this device from having universal acceptance other than in certain spray and dry powder applications:

• Fundamentally, it does not test any dispersion unit that would be part of an instrument.

- The fixed nature of the spots indicates that a standing wave was set up and the results that were calculated, which were based on the spot distribution alone, need modification to cater to this phenomenon.
- Fusion of some spots was apparent from microscopy evaluation, and thus a round-robin exercise became necessary to obtain agreement as to the certified mean size and distribution.
- There were no spots in the smallest regions. In particular, the sub-5-μm and sub-1-μm sizes were either underrepresented or nonexistent. It is the region 0.5 to 5 μm that is crucial for lung entry, and thus to be missing data in this area was not ideal.

Formal (Paper) Standards   The last 25 years has seen a dramatic increase in both the numbers of standards available and also the quality thereof within the particle sizing community. The main proponents in this arena are the International Standards Organisation (ISO) and ASTM (originally, the American Society for Testing and Materials). The published standards of most use are probably the "Practice guides" (in ASTM parlance) and, more simply, methodologies—the science of methods—to be found within the appropriate ISO standards.

At the time of writing this chapter (January 2008), a resolution has just been approved to change the name of the ISO TC24 Committee from Sieves, Sieving and Other Sizing Methods to Particle Characterization Including Sieving. For a number of years, the sieving side of the TC24 Committee took only a maintenance role, based on the success of the ISO 3310-x group of screening standards, and one subcommittee (TC24/SC4, Particle Characterization) became the only really active part of the grouping. The new nomenclature should therefore reflect the strength of the SC4 committee in developing (around 48 at last count) standards within its 17 working groups. The working groups presently in existence are:

WG1	Representation of Analysis Data
WG2	Sedimentation, Classification
WG3	Pore Size Distribution, Porosity
WG5	Electrical Sensing Methods
WG6	Laser Diffraction Methods (do not anticipate meeting)
WG7	Dynamic Light Scattering
WG8	Image Analysis Methods
WG9	Single-Particle Light Interaction Methods
WG10	Small-Angle X-Ray Scattering Method
WG12	Sample Preparation and Reference Materials
WG12	Differential Electrical Mobility Analysis for Aerosol Particles
WG14	Acoustic Methods
WG15	Particle Characterization by Focused Beam Techniques
WG16	Characterization of Particle Dispersion in Liquids
WG17	Methods for Zeta Potential Determination

One should also note the recent formation of ISO TC229 (Nanotechnologies) and the fact that particle size standards exist within other technical committees of the ISO, including those relevant to paints and coatings, ceramics, metal powders, and other industrial groupings. A similar situation exists within ASTM, with the following committees having most significance:

E29   Particle and Spray Characterization (formed in 1969) and originator of the highly successful ASTM E 11-04 specification for wire-cloth sieves

E56   Nanotechnology (formed in 2005 ahead of the corresponding ISO grouping)

Again, other committees (e.g., C01 Cement, C28 Advanced Ceramics) have made extensive use of particle sizing techniques within their published standards. As standards are a "moving feast," the appropriate Web sites should be consulted to ensure that the latest standard or draft is appropriate for the purpose envisaged.

In terms of standards relevant to this chapter, a very nice overview is provided within ASTM E 1919-03, Standard Guide for Worldwide Published Standards Relating to Particle and Spray Characterization, published by the E29 committee cited above.

In particular, relating to methodologies we deal with in this chapter, the following represent a short list of well-used (and abused!) standards published by ISO TC24/SC4:

ISO 9276-1:1998 Representation of Results of Particle Size Analysis, Part 1: *Graphical Representation*

ISO 9276-2:2001 Representation of Results of Particle Size Analysis, Part 2: *Calculation of Average Particle Sizes/Diameters and Moments from Particle Size Distributions*

ISO 13317-1:2001 Determination of Particle Size Distribution by Gravitational Liquid Sedimentation Methods, Part 1: *General Principles and Guidelines*

ISO 13318-1:2001 Determination of Particle Size Distribution by Centrifugal Liquid Sedimentation Methods, Part 1: *General Principles and Guidelines*

ISO 13319:2007 Determination of Particle Size Distributions: Electrical Sensing Zone Method

ISO 13320-1:1999 Particle Size Analysis: Laser Diffraction Methods, Part 1: *General Principles*

ISO 13321:1996 Particle Size Analysis: Photon Correlation Spectroscopy

ISO 13322-1:2004 Particle Size Analysis: Image Analysis Methods, Part 1: *Static Image Analysis Methods*

ISO 13322-2:2006   Particle Size Analysis: Image Analysis Methods, Part 2: *Dynamic Image Analysis Methods*

ISO 14488:2007   Particulate Materials: Sampling and Sample Splitting for the Determination of Particulate Properties

ISO 14887:2000   Sample Preparation: Dispersing Procedures for Powders in Liquids

ISO 21501-2:2007   Determination of Particle Size Distribution: Single-Particle Light Interaction Methods, Part 2: *Light Scattering Liquid-borne Particle Counter*

We note that the more useful of paper standards provide knowledge relating to the technical background and understanding of the technique in question rather than detailed recipes for measurement of particular materials or entities. Other standards are also well utilized within the pharmaceutical industry. In particular, the U.S., European, British, and Japanese are those most frequently encountered, and the reader of this document may be much more familiar with these standards in a general way than is the present writer. However, there are instances where the writer has special expertise in these areas, especially where the standards have been based on ISO standards; for example, USP General Test ⟨429⟩ (for laser diffraction) closely follows the earlier ISO 13320-1 (November 1999) standard.

6   QUALIFICATION OF INSTRUMENTS USED IN PARTICLE SIZING

In general terms, verification or qualification of any instrument will look at accuracy and precision of that technique or instrument.

We look at broad definitions, in practical terms, following the "4Q's" system:

Specification qualification (SQ): statement of the requirements of a product or system from the end user's or customer's perspective, which defines what is required and practical, measurable performance criteria.

Installation qualification (IQ): documentation of the installation procedures for the system and the verification that the equipment is supplied and delivered as per the requirements specification and is installed correctly and provided with the necessary services for use. This is normally carried out on site by a trained (and documented) service engineer from the supplier. The components of IQ that need dealing with include:

• Whose responsibility is the IQ? If it is deemed to be the supplier's, has this been factored into the purchase price?

• Does unpacking of the instrument by the customer invalidate the warranty? Must this be done by the manufacturer's representative?

• If the manufacturer carries out the IQ (and then usually the operational qualification), is the representative or engineer qualified or certified to do so?

- Does the supplier's packing list correspond with that ordered?
- Do the components, instruments, accessories, consumables, and so on, listed on the packing list correspond to those within the box or boxes?
- Is there external or water damage to the boxes? This may involve an insurance claim.
- Are certificates of compliance and other testing (e.g., quality control) protocols in place? (These may have been prespecified or supplied by the manufacturer as a matter of course.)
- Record the serial numbers on the instruments, accessories, and other components. Note the software version and check the compatibility against the computer (if the computer is supplied by the customer).
- Ensure that the place where the instrument is to be located is ready. Responsible manufacturers supply a preinstallation checklist that can include provision of services such as deionized water, oil, and water-free air, electricity of the correct phase and voltage, vibration-free tables, fume or extract hoods, and so on.
- Assemble and power-up the instrument.
- Wait the recommended (warm-up) time and begin the OQ.
- Record any discrepancies in the above and discuss the remedial steps, if appropriate, with the manufacturer's representative. Note that "if it is not written, it is a rumor."

The main points that we concern ourselves with in this chapter relate to the following:

Operational qualification (OQ): confirmation that the equipment is performing to the predefined manufacturer's specifications. That is, the instrument is the same as when it left the factory. It will need demonstrating that the instrument performs over its stated range (within the appropriate specifications) and that the system is complete and working on the customer's site. This qualification occurs once the instrument has been installed successfully on-site. The customer and supplier will undertake a detailed sign-off procedure for this stage and the IQ stage described previously. This follows the assembly and power-up of the instrument and will involve the calibration or verification stages that we have been discussing.

Performance qualification (PQ): the verification or confirmation that the instrument is performing correctly on the customer's own materials. In theory, this is the sole responsibility of the customer, but responsible suppliers assist with the appropriate sensible and meaningful tests and tolerances for this stage. This stage is performed as often as dictated by the policies of the appropriate departments within the customer's facility. This is where raw data need to be saved (and in light-scattering and other indirect methods, raw data are thus different from the first result) and backed up. Predefined specifications need to be in place and appropriate SOPs formulated and justified on the 3R's principle. The material(s) need to be tested against these predefined protocols and pass–fail recorded together with appropriate remedial action. The entire document set above should be available for the FDA auditor and forms part of the validation exercise and paper trail (now the electronic trail). It is clearly important to get it right the first time, as the cost of making a change at the testing point has been stated to be 50 times that of making a change at the requirement or specification stage. This is where a well-designed SOP is of paramount importance.

At this stage we deal exclusively with the IQ/OQ aspects of instrument qualification; indeed, this is paraphrasing the chapter title! We deal with the major particle sizing techniques, stressing the similarities and differences.

6.1 Sieves

These are not really instruments in the classical sense, but automation (e.g., the Ro-Tap) may be present. Sieves are normally calibrated by image analysis methods and can be proved in practice by the use of an appropriate certified sieve standard. In many instances (but not desirable) such a standard could be recovered. An excellent description is presented in Terence Allen's book [10]. PQ can be performed with an appropriate material of the customer's (see above). A particularly easy-to-understand and helpful animation appears on the Whitehouse Scientific Web site (www.whitehousescientific.com/sieve.htm). ISO 3310 and ASTM E 11-04 can be consulted as to the permitted tolerances on screens. From the latter reference, we see, for example, that for screens below #325 (45 μm), the "permissible variation of average opening from the standard sieve designation" is ±3 μm, that the "opening dimension exceeded by not more than 5% of the openings" is 57 μm for a 45-μm screen, and finally, that the "maximum individual opening" for the 45-μm screen is 66 μm. What this means, in practice, is that screen calibration should be performed on a much more regular basis than one commonly sees in industry. Section A1 in E 11-04 provides a test method for specification confirmation ("Test Methods for Checking Wire Cloth and Testing Sieves to Determine Whether They Conform to Specification"). The main advantage of a screen in addition to the apparent simplicity of measurement is that the material is actually separated into different-sized fractions.

It must be remembered that screen separate on the basis of shape and size (as well as density for a mixture) and that a true weight distribution is not produced by a screening process. This is easily seen by considering cylinders of fixed diameter and varying length, all of which could pass a screen with an aperture slightly larger than the cylinder's diameter.

6.2 Sedimentation

The system can be proved with the appropriate standard, of which a reasonable number are in existence, ranging from garnet through quartz (e.g., the BCR standards) and materials such as calcium carbonate. In certain cases the optimum dispersion conditions will need defining. Density can be determined by means of a gas pycnometer or specific gravity bottle (if the material wets with the appropriate fluid). Viscosity can be determined in a number of ways, including the simple Ostwald viscometer.

6.3 Light Scattering

We will spend most time and effort within this area, as this type of instrumentation is now extremely common within the pharmaceutical industry and probably provides the most challenge to users. The functions that require verification or qualification within light-scattering equipment are the following:

- *Hardware.* Has the optics been set up correctly, and is the system clean?
- *Software.* Has the calculation been carried out correctly? In light scattering generally, a light intensity vs. angle or other function is deconvoluted to provide a particle size distribution. This can be proven separately by means of proving known inputs (with correspondingly known outputs) and checking that the answer generated is as expected. Clearly, knowledge that the vendor is qualified (e.g., TickIT) in the appropriate manner for software writing is important.
- *Dispersion unit.* In many instances there is a need to disperse agglomerates to primary particle size. Furthermore, large or dense material may need more agitation to keep it in suspension and circulated in front of the laser beam—contrast a peristaltic pump and a centrifugal pump. The performance of any wet or dry accessory will need verification. In some instances, external dispersion may take place, although this is not desirable.

All of the above can be verified simultaneously by means of the appropriate certified reference standard(s). A narrow (monodisperse) latex standard or sets of latex standards can be used to verify the x-axis in a laser diffraction instrument, and only a small number may be needed to verify the entire angular detector set. However, such standards do not challenge the vertical (or quantity) axis and do not confirm the correct assignation of any proportion within a distribution. Further, such standards do not confirm the correct working of a sample dispersion unit. For example, a polydisperse or dense material could conceivably undergo some form of flow segregation (flow fractionation) if the agitation in a wet dispersion is not adequate. Just pumping the small material in front of the laser beam with gentle agitation is one way of getting a small result but is clearly not representative of the bulk material! Thus, a combination of latex and polydisperse samples will be used in practice to verify all aspects of an instrument's performance.

ISO 13320-1 has spawned USP ⟨429⟩ and EP 2.9.31. We base our discussion around the recommendations in these documents. The verification of a laser diffraction instrument normally involves the confirmation of the optics, hardware, and software calculations by means of the appropriate latex standards. As a monodisperse latex scatters over a wide angular range, despite its narrow distribution, only a small number of latex standards are needed to confirm the entire detector set and collimating or focusing optics in an instrument. The (wet or dry) dispersion unit is then verified by use of an appropriate polydisperse material (spherical, polydisperse over a range of 1 : 10; ISO 13320-1). Primary verification of the instrument with traceability back to the meter can take place with an appropriate NIST standard, and this will check all aspects of an instrument's performance. This is not usually undertaken in practice except perhaps once in the manufacturing stage. Subsequent use of traceable latex and the designated manufacturer's OQ standard will be sufficient to prove that the instrument is in the same state as when it left the manufacturer's site. Any deviation from the predefined specifications of these materials must be examined carefully. This could arise from some issue in the optics (seen by one or more of the latex samples), which could be as simple as dirty cell windows or could be a missing detector element. At this stage, the (observation of the) deviation will be noted formally in writing. Possible causes will be examined and an intervention suggested, discussed, and performed. The results of that intervention will again be noted formally in writing within the documentation provided by the manufacturer. This process will continue until a documented change has brought the instrument back into the predefined specification. This is different from testing into compliance, where the first correct result could be taken. The fact that the instrument is within specifications would be demonstrated in practice by repeated consecutive measurements of the standard—in theory, at least. This is, of course, unacceptable.

ISO 13320:2009 permits a variation of 2% in x_{50} and 4% in x_{10} and x_{90}, with a doubling of these parameters allowed under 10 μm. Thus, this standard assumes that we have dealt with the sampling issues or, alternatively, that the width of the distribution is narrower than 1 decade, the recommended width for a verification standard. USP ⟨429⟩, which is based loosely on ISO 13320-1, allows 10% on x_{50} and 15% on x_{10} and x_{90}, again with a doubling under 10 μm. Instruments are capable of values at least an order of magnitude better than these stated ISO 13320-1 values, so these values are practical and sample-specific, reflecting the inherent homogeneity (or otherwise) and dispersibility of the material.

Entering into the tails of a distribution is a journey fraught with statistical dangers. For example, specifying the x_{98} based on the ISO numbers quoted above and based on a Gaussian distribution, we can easily calculate reasonable "permitted" deviations. We would expect to specify the x_{98} with a (permitted) tolerance of more than $(2.05/1.28) \times 5\%$; that is, a minimum of 8% standard deviation should be permitted on this parameter. (The calculation is based on the fact that the x_{90} point represents 1.28 standard deviations from the mean and the x_{98} point is 2.05 standard deviations.) Thus, $3\sigma \sim 25\%$ will set a reasonable specification. It is easy to see that a ±25% tolerance is 50%! What if the material is under 10 μm? Then ±50% is suggested! Note that this (apparently large) variation may not have any bearing on the efficacy of the material. Note, too, that, statistically, 3 in 1000 measurements *would be expected* to lie outside a $\pm 3\sigma$ tolerance level.

Last, and by no means least, ISO 13320:2009 states: "Quotation of an x_{100} value by laser diffraction is specifically deprecated by this standard."

6.4 Photon Correlation Spectroscopy

ISO 13321 quotes the use of a (submicron) latex standard to verify the workings of an instrument based on the photon correlation spectroscopy (PCS) technique. Only the z-average size and polydispersity index (PI) should be stated. A derived distribution is algorithm dependent, so conversion from intensity (proportional to volume squared) to volume is only possible with a relatively narrow distribution but can indicate whether or not a higher peak in the size spectrum is present in significant mass. Conversion to a number is never permissible.

The reference material standard specified in ISO 13321 is 100 nm with three consecutive measurements having a 5% RSD on the mean as verifying the performance, obviously with the mean falling within the values certified. With a single monodisperse standard, obviously the quantity axis is not verified, and this fits in with the ill-conditioned nature of the PCS technique. The other comment is that the latex materials used are often certified on the basis of electron microscopy (number distribution), and even a slight polydispersity of latex materials (typically, <3% on the mean) can provide a smaller certified value than the PCS-derived z-average (intensity mean $\propto$ volume squared). Often, a latex will need either rapid measurement (if diluted in deionized water) or dilution in millimolar KCl (for example) to suppress the electrical double layer. If the double layer is not suppressed, the apparent size of the particle will appear larger. Latex materials for PCS are typically used (and available) in the range 20 nm to 1 μm or so, although some vendors do specify equipment to as high as 10 μm. A polydisperse submicrometer standard quoted as 0.1 to 1 μm is available from Whitehouse Scientific, but as this is of a higher-density material (soda-lime glass $\sim$2.5 g/cm^3) than the latex materials (polystyrene–divinylbenzene nominally 0.99 to 1.02 g/cm^3), settling in water means that it is not suitable as a PCS standard. With laser diffraction the x_{50} of this standard is around 850 nm (0.85 μm), with some material up and over 2 μm or so.

A 100-nm latex standard from NIST (SRM 1693) has been extensively characterized. The first batches of this material have agglomerated slightly over time, which means that care needs to be exercised in their use. For some techniques this agglomeration will not be a problem. In 2008, NIST released extensively characterized gold colloids. These 10-, 30-, and 60-nm (nominal) materials (RM 8012) probably represent the greatest characterization work carried out on gold sols—and this is saying something, given that Gustav Mie's classic paper [12] was concerned with the colors of gold sols in relation to their size.

The 10-nm Au sol standard mentioned above is the first sub-20-nm reference standard material available. Below this size, nature and molecules provide the only proof of performance. Here materials of known sizes based on molecular modeling can be measured by PCS and the results compared. A concentrated sucrose solution was measured by PCS to confirm the 0.6-nm lower limit of one instrument [13].

6.5 Particle Counting Techniques

Here a latex standard will be used to adjust the response of an instrument in order that it reads correctly. Air counters are normally specified in accordance with Federal Standard 209-D, where, for example, the air under a Class 100 hood will contain no more than 100 particles of size 0.5 μm and above. Only this size class and 5 μm need to be measured. The metric version (209E) of this standard has simply not caught on. The 209E standard was officially withdrawn by the General Services Administration of the U.S. government in November 2001. The replacements nominated were ISO 14644-15 for air cleanliness classification and ISO 14644-27 for proving continued compliance of a clean room with ISO 14644-1. These standards use 0.1 μm as the reference size rather than 0.5 μm. The Class 100 hood still hasn't died out, though.

To calibrate an air counter, the standard latex will be nebulized and passed through the counter in an isokinetic flow manner and the peak position adjusted until the stated value is in line with the certificate. BS 5295 was used for some time by at least one major pharmaceutical company, but there are practical problems in nebulizing 10-μm and, especially, 25-μm latex beads which this standard calls for. Obviously, for a liquid counter dealing with particles in suspension, 25 μm is not a problem. Thus, air counters typical have quoted ranges of, say, 0.1 to 10 μm or so, and we note that the numbers of particles counted are relatively low.

Condensation nucleus counters (CNCs) do not provide a size distribution but simply count the numbers of particles by condensing a liquid onto the particle and counting in the prescribed manner. This brings the lower limit of air counters below 0.1 μm but does not provide a particle size distribution. Liquid counters generally are used in the pharmaceutical industry for small- and large-volume parenteral materials (USP ⟨788⟩ is a typical standard in this area). Liquid counters also find usage in many other industrial areas, such as hydraulic fluids, and a wealth of CHARN, ISO, and military standards exist. Liquid counters again provide a limited particle size and (low) concentration range, in line with the needs of the pharmacopeias and counters in general. USP ⟨788⟩ calls for measurements at 10 and 25 μm and the equivalent BP at 2 and 5 μm, so that a typical sensor in a liquid counter will be specified to probably less than 50,000 particles/mL and a size range of 1 to 100 μm.

There are some concentration standards available to confirm the resolution and quantity axis response. It is not clear whether these could be applicable to an air counter, as preferential nebulization of one component seems possible, if not probable. Care needs to be taken with such standards, as any evaporation will obviously alter the actual concentration.

6.6 Electrozone Sensing

Again, polystyrene latex standards are used to calibrate the instrument in the formal sense of that word. The response of the instrument is adjusted to meet a predefined or certified value. Any subsequent measurement then compares (interpolates) the signal given by the material against the calibration signals stored in the instrument A mass balance can also be used to verify the instrument.

7 METHOD DEVELOPMENT

Method development is actually the crux of the PQ step. This stage bridges the vendor's instruments (which are qualified at the factory) to the needs of the final customer and the objective of fitness for the purpose of the material. There are two appropriate questions that need answering before method development is embarked upon:

- Is a primary or bulk size needed? The answer to this question will dictate the energy to be employed in any dispersion or sample preparation stage.
- What is the acceptable quality level? Put another way, what specification is needed to meet a "fit for purpose" or "just good enough" criterion?

It is a cliché, but quality cannot be inspected into a material and, indeed, repeated inspections until a "pass" is obtained, or blending of failed batches, is not something the FDA will tolerate. Quality needs to be built into a product and the PAT (process analytical technologies) and QbD (quality by design) initiatives stress this obvious point. It may be seductive to feel that a single pass–fail on a spatula scoop of material is adequate for control purposes, and therefore the challenges of constant monitoring of a production line for particle size have been resisted in the past. Handling the large amounts of data obtained, say every second, in online measurements may be a computer memory headache, but worse still is the interpretation of the sets acquired. Particle size distribution is based on statistical considerations, and these do not fit easily into the digital pass–fail scenario. We need more gray scale, or varying degrees of pass or fail. If we look hard enough, we will find—be it parts per trillion of contaminant or a single "boulder" in the production line. The question that should be asked but is difficult to answer (and harder to prove to the FDA!) would be: Is this level of contamination or "boulder" detrimental to the next stage (which may be the insult to the patient)? What is often the case is that a specification is rigidly imposed that simply cannot be met, due to the statistical heterogeneity of the ingredients in the formulation. The particle size of aspirin could probably vary as much as 400% without having much impact on a patient. Although such a specification would frighten the average quality assurance manager, this is clearly a fit-for-purpose specification and one that is justifiable. The natural urge of a manager is to tighten this specification to, say, first 40% and then 4%. The latter will certainly end in tears and repeated failures and, most important, with a huge impact on the cost of the material. It will not help (or hinder) the efficacy of the treatment.

We deal next with the design of experiment considerations in any method development, assuming that we are aware of the fit-for-purpose criterion and an adequate predefined specification that meets the needs of the end use of the material.

7.1   Design of Experiment or Development of a Robust Standard Operating Procedure

A single result has absolute precision and therefore no statistical merit. Thus, the move toward a standard operating procedure (SOP) should be based around solid statistical quality control. In the author's opinion, an SOP that does not generate a mean and standard deviation based on a minimum of 10 experiments (be they consecutive measurements or repeated subsamplings from a lot) has little or no statistical merit. Quite why the number "3" is used extensively for numbers of measurements in the pharmaceutical industry is a complete mystery to the writer. It is quite clear that two of the values generated will be close and one will be different, or an *outlier*. Such is the nature of the 1 standard deviation from a Gaussian distribution (68% of measurements are expected to lie within $\pm 1\sigma$ of a Gaussian distribution and thus one in three measurements lies approximately outside $\pm 1\sigma$).

To return more directly to the subject, the three prime objectives of any quality control metrology are stability, stability, and stability (the mathematical expression is *precision*). In R&D we may be looking at kinetics and changes, but these generally need to be avoided in an analytical laboratory—or at least, fully understood and specified appropriately.

In particle size distribution analysis, the design of experiment revolves around inputting controlled energy into the system (be it a powder, suspension, or emulsion) and evaluating the effect of that energy on the system in terms of the parameters derived. This step allows the quantification of agglomerates and, possibly, the friability/fragility of the system (see Figure 1).

It is easiest to illustrate these points by describing method development for both dry and wet analysis in laser diffraction. The generalities of these descriptions can be used for any other instrumental technique.

Two parameters that we defined earlier dictate what should be examined:

1. Any changes that take place in the material during the time of the experiment itself. The "experiment" could mean long-term storage or accelerated lifetime studies. Often, though, in practical terms, it means the repeatability of measurement for a single sample over a (relatively short in comparison to the age of the earth) period of time with controlled energy input. This is the intra-assay precision or repeatability.

2. The heterogeneity (others prefer to use the term *homogeneity*, but this is a moot point in a particle size distribution) of the material defined by

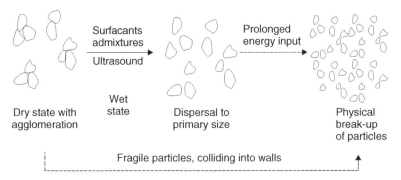

FIGURE 1 Generic diagram indicating the effect of increasing energy on a particulate system. The stages of agglomeration, dispersion, and attrition are noted.

repeated subsamplings and measurement as above under predefined conditions so that one-on-one comparisons can be made. The RSDs from repeated samplings should be used to define a workable specification based on the attributes of the material and its use, not on some whims of a nonstatistically qualified engineer or manager. Incidentally, George Klinzing has stated that the average American college student of chemical engineering spends around 20 minutes in four years dealing with particle size and distribution. This is probably why such considerations are alien and need to be trained into users of particle size equipment.

So, in brief, the design of the experiment is simple. We look at the effect of energy on a single (or first) sample and define a stable or required portion of the size–energy plot. Then repeat that process on subsamples from the bulk lot. Experimental statistics will then be calculable relating to the experimental parameters and the batch variability. It is these real experimental values that should be used in the formation of a realistic specification. Trying to make a universal specification applicable or taken from another technique is not sensible. When all this is welded together in an appropriate and sensible specification, the end result is a material that can be made by production and will pass its quality attributes in line with the end use of the material.

In laser diffraction, method development follows the guidelines of ISO 13320-1, in particular Section 6.2.3:

- Ensure that the equipment has formal traceablility by verification with NIST-traceable standards appropriate to the specific material.
- Ensure that a representative sample of the bulk material is taken; this may necessitate the use of a spinning riffler if any material above 75 μm is present.
- *Dry:* pressure–size titration followed by wet to decide on an appropriate pressure (ISO 13320-1, Section 6.2.3.2) or to reject dry if significant attrition is occurring (as with many organics).

- *Wet:* consecutive measurements before, during, and after sonication in order to obtain a plateau of stability. For organics the formation of a stable dispersion will follow the steps of wetting, separation, and stabilization (if the latter is needed). This may require some work with solvents and perhaps surfactants. Such liquids and admixtures are only vehicles in which to circulate the powder (or slurry/suspension, etc.).
- Export to a program such as Excel to calculate the means and RSDs of the measurements.
- SOP formulated around the stable dispersed region for primary size or against the input of minimum energy for the bulk size (possibly only one measurement may be possible for the latter if dispersion is occurring in the wet).
- Formulation of a reasonable, fit-for-purpose specification based around the statistics of the experimental evidence and the predefined requirement specification for the product.
- Constant feedback and review.

The key features of any method must be:

- Repeatability
- Reproducibility
- Robustness

These 3R's were defined earlier and are explored further in detail by Rawle [14], to which the reader is referred.

As ISO 13320-1 indicates, a wet–dry comparison is extremely useful in method development. If the wet and dry results are identical, this indicates that the material is in the same state of dispersion (probably full) in the method. If different, the changes in the material need to be understood and the appropriate measurement route selected in line with the quality control attributes of repeatability and reproducibility. The form of wording in the standard is: "It is necessary to check that comminution of the particles does not occur and conversely that a good dispersion has been achieved. This is usually done by direct comparison of dry dispersion with a liquid one: ideally, the results should be the same." Wet–dry comparisons can also be used in order to verify a refractive index for a material or, at least, to reject unsuitable values [15]. As an illustration, the following particle size plots and comparisons involved in method development for an organic compound should be easy to understand without further comment.

Dry Powder Analysis   In the laser diffraction field, the mnemonic for the key term *pressure–size titration* is PST (pronounced "pssst"). Our (theoretical) expectation of the effect of pressure on (apparent or measured) size is shown in Figure 2. The practical reality is shown in Figure 3, although when care is taken at differential pressures (ΔPs) close to 0 bar, it will be seen that the true form of

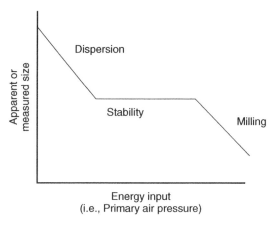

FIGURE 2 Theoretical plot of a dry powder pressure–size titration in laser diffraction.

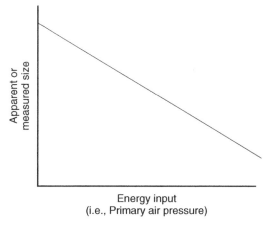

FIGURE 3 Real and practical plot of a dry powder pressure–size titration in laser diffraction.

the plot is exponential, indicating the obeying of Rittinger's law of comminution. Indeed, the construction of a dry powder disperser in diffraction is analogous to that of an air jet mill in a micronizing process.

The reason for this expectation difference is obvious in hindsight and arises from two possibilities:

1. Dispersion and milling are parallel rather than sequential processes. We do not get a stepwise sequence of dispersion followed by milling. Agglomerates can mill as well as single particles.
2. Velocity biasing—the particles in the vacuum collection device are stationary, but the particles are traveling up to tens of meters per second (up

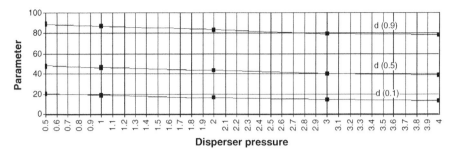

FIGURE 4 Dry measurement in laser diffraction: pressure–size titration according to ISO 13320-1 [14]. The plot displays the apparent or measured size of the key parameters of x_{10}, x_{50}, and x_{90} for 10 measurements at each differential pressure.

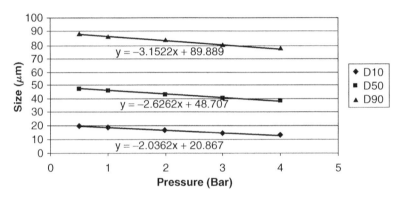

FIGURE 5 Dry measurement in laser diffraction. This is an initial check on method. Note that the magnitude of the slopes is in the order $x_{90} > x_{50} > x_{10}$. The magnitude of the slopes (around -2 to—3 μm/bar increased pressure here) indicates little or no attrition, but this needs confirmation by comparison with a wet measurement.

to a theoretical 300 m/s, the speed of sound) in the venturi. Thus, the particles must be slowing down through the measurement zone. Due to inertial effects the larger particles will slow less than the smaller particles. Thus, there can be a tendency to overestimate the fine fraction unless the dispersion mechanism is designed carefully. Only with a monodisperse system can we expect to see a flat pressure–size titration—the slope of the pressure–size plot reflecting a combination of the polydispersity of the material and the friability/fragility. The greater the slope, the greater the width of the distribution (and we have to carry the larger particles through in an air current) and/or a more friable nature of the material (e.g., needles or plates are more easily broken than spherical material). If we recirculate the particles, this effect is obviously eliminated, but this is usually impossible in a dry method.

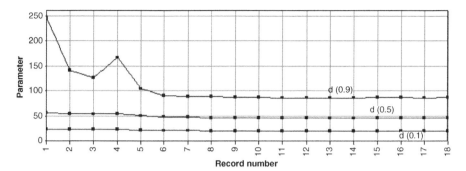

FIGURE 6 What differential pressure should we measure in Figures 4 and 5? This figure illustrates a wet measurement of the same material with sonication (dispersion) to stability: consecutive measurements before, during, and after the application of ultrasound energy. The key parameters are plotted as before for the dry measurements.

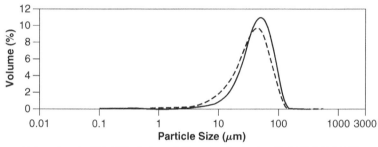

— Malvern Powder With 50% u/s then off, Thursday, September 21, 2000 5:38:21 PM
-- "Malvern Powder" 4Bar, Thursday, September 21, 2000 2:56:05 PM

FIGURE 7 Wet–dry comparison according to ISO 13320-1. 4-bar differential pressure, ΔP, compared with dispersed wet result. Note the lack of overlap.

It is thus obvious, that reproducibility is examined only in a dry measurement, and repeated measurements again give a measure of homogeneity or otherwise of the material. Practical measurements are indicated and described in Figures 4 to 9.

Wet Analysis   The mnemonic for the key term *before, during, and after sonication* is BDAS (pronounced "bee-dass"). The formalization of the methodology falls in line with the needs of wetting, separation, and stabilization required for a stable dispersion. In this case the expectation aligns with practice. The general form of the plot looks slightly more complicated than that of dry plots (Figure 10). Examination of the apparent (measured) size–energy plot indicates a plateau or stable region, and if dispersed results are required, we may design an SOP around this stable region. We also note that the ratio of the "immediate" to dispersed result is some measure of the agglomeration in the system. There is

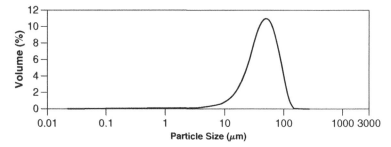

— Malvern Powder With 50% u/s then off, Thursday, September 21, 2000 5:38:21 PM
-- "Malvern Powder" 0.5Bar, Thursday, September 21, 2000 3:04:18 PM

FIGURE 8   Wet–dry comparison according to ISO 13320-1. 0.5 bar differential pressure, ΔP, compared with dispersed wet result. Note the excellent overlap, indicating that the material is in the same state of dispersion and that the optimum ΔP is 0.5 bar.

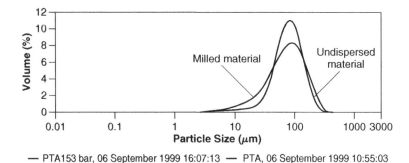

— PTA153 bar, 06 September 1999 16:07:13 — PTA, 06 September 1999 10:55:03

FIGURE 9   Comparison of wet and dry dispersion in laser diffraction when attrition is occurring. Note that we can "make" the median, x_{50}, results agree, but the dry distribution is wider—some particles are being milled while others are not fully dispersed.

only a single "first" or immediate measurement—this is the measurement (nominally at time = 0) that reflects the bulk size of the material (agglomerates and all) "as is." We also note that we can get repeated measurements on the same group of particles–the definition of repeatability. This stable region may approach the practical limitations of the instrument if the material is stable (e.g., a standard latex or glass beads) with RSDs expected in the 0.2% or below region for 10 or more consecutive measurements. As the inherent sampling variation is expected to be more than an order of magnitude above this value, the true effect of energy can be studied on the sample without the complications of sample variation. This is the reason why in situ sonication is preferred to external sonication: Sampling from a slurry is notoriously difficult and we may have reagglomeration once the sample from the slurry is diluted in, say, deionized water.

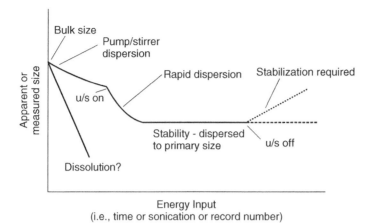

FIGURE 10 Theoretical and practical plots for ultrasonic dispersion of a material in wet suspension.

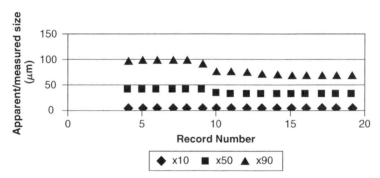

FIGURE 11 Dispersion of a sparingly soluble material (calcium sulfate, gypsum) in water.

Last, a word on measurement in saturated solution. Don't! There is no need to make a saturated solution—in fact, it is dangerous to do so (seeding and crystallization effects). However, it is seductive to believe that preferential dissolution of the small material will bias a result, but swamping the dispersant liquid with particles prevents this, especially when we consider a mass distribution. Indeed, a little dissolution works well for dispersing sparingly soluble materials, and there is little change in refractive index in this situation. Note, too, that to obtain a 1% change in diameter requires a 2% change in specific surface area. It is preferable to make up a saturated solution in situ by constantly adding solid and working at higher-than-normal obscurations [obscuration = (1 − transmission)] to minimize solubility and reflect powder size. The dissolution of solid bridged material is a feature of stress corrosion milling. An example of the measurement of calcium sulfate (gypsum) is shown in Figures 11 and 12 and Table 3.

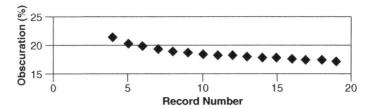

FIGURE 12 Dispersion of a sparingly soluble material (calcium sulfate, gypsum) in water, indicating the changes in obscuration ($= 1 -$ transmission) over time (equally spaced record numbers).

TABLE 3 Repeatability of Measurement for a Sparingly Soluble Material in Water (Calcium Sulfate, Gypsum)

	Calcium Sulfate After Use			
Record Number	$\times 10$	$\times 50$	$\times 90$	Obsc.
15	7.723	30.536	71.427	17.85
16	7.734	30.485	71.435	17.67
17	7.748	30.409	71.327	17.49
18	7.757	30.38	71.488	17.33
19	7.787	30.414	71.595	17.21
Mean	7.75	30.44	71.45	17.51
SD	0.02	0.06	0.10	0.26
% Variation	0.32	0.21	0.14	1.47

Despite a significant change in obscuration, the results are stable and water is considerably less expensive in use than a solvent such as 2-propanol. We also note that in a material with high solubility in water, the refractive index of the saturated solution is likely to be similar or identical to the particle itself. Thus, the relative refractive index will tend to unity, there will be less or no optical contrast of the particle and the medium in which it resides, and thus the scattering will be negligible or zero (the latter where the optical parameters are index-matched).

Two generalizations may be useful at this point:

- If we have material larger than around 75 μm, representative sampling must be accomplished to lower any sample-to-sample variations.
- If we have material smaller than 10 μm present in the sample, control of the dispersion stage becomes paramount.

7.2 Dispersion of Nano- and Micromaterial

Generally speaking, a stable dispersion is made from a powder with the steps of wetting, separation, and stabilization. These steps will be familiar to those in the ceramics and paints/coatings industry. If the fundamental particle size is small (typically, less than 100 nm), a phenomenon known as *solid bridging* occurs arising from the solid–solid diffusion of small particles caused by thermal energy. Breakage of this solid bridge is difficult, if not impossible, and is usually only achievable by a process called *stress corrosion milling*, where a change of pH or chemistry can cause the dissolution of this solid bond. Thus, truly nano material (below 100 nm) is achieved more easily with a bottom-up (examples are decomposition of nickel carbonyl to form nano-Ni and chemical reduction of gold salts in solution to produce gold colloids) rather than a top-down (milling/comminution) process.

In pedantic terms, the only dispersing agent is energy. In the dry powder state, achievable energy input limits analysis down to 0.5 to 1 μm at best. This is because the shear forces (ΔP across the venturi and particle collisions with each other and with walls) that can be input are inadequate to separate material very much smaller than this size. In the wet state, materials can be made in colloidal form and also stabilized (sterically or with adequate charge/zeta potential). If a top-down or comminution process is employed, typically a stabilizer (an ionic stabilizer such as phosphate for inorganic oxides; an anionic surfactant for other systems) will need employing in order that fractured surfaces are immediately covered immediately and are not accessible to recombination. If this is not carried out correctly, the rate of agglomeration equals the rate of fracture and breakage, limiting the bottom-end size to 1 μm or so (similar to dry). With suitable stabilizers and sufficient energy (and time!), dispersion to 100 nm or even 50 nm can be achieved, but this certainly cannot be considered routine.

On the other hand, if the fundamental particle size is relatively large (e.g., >20 μm), the inertial forces dominate over the attractive van der Waals forces and separation is easily effected by relatively mild shear forces (Silverson mixer in the coatings industry, a pump-stirrer in a wet dispersion unit for light scattering). Indeed, it is the relationship between the inertial and van der Waals forces that dictates if irreversible aggregation will occur in any system. The dividing line begins below 5 μm or so and is very important in the submicrometer region. There is plenty of literature dealing with this topic, and detailed discussion is outside the scope of this chapter.

7.3 Robustness

The robustness of an analytical procedure is a measure of its capacity to remain unaffected by small but deliberate variations in method parameters and provides an indication of its reliability during normal use. Note that, ideally, robustness should be explored during the development of the assay method. By far the most efficient way to do this is through the use of a designed experiment [16].

Intermediate precision (e.g., another operator/laboratory). This is often included within the reproducibility section described above, but it is convenient to place this within the robustness section, as method transfer between sites and instruments requires a robust method and an understanding of the statistical variants that affect the end result. Understanding of these key parameters will allow tear-free transfer of methods between sites in different laboratories or countries.

Pump-stirrer. This is the mechanism by which a wet suspension is kept from settling and should normally be set close to the allowable maximum as the default. Large or dense material is obviously more difficult to keep in suspension (this also applies to air). Use of surfactants should be kept to an absolute minimum to allow wetting only as, clearly, foaming can be generated under high shear conditions.

Concentration/obscuration. Large particles scatter effectively, but the obscuration is low. Small particles scatter poorly but the obscuration is high. The defining factor in a good measurement is an adequate signal-to-noise ratio. The level of background and generated signal can often be accessed directly. For particles under 1 μm or so, the concentration dependence should be explored, as the possibilities of multiple scattering (which reduces the apparent size of the material) could occur.

Sonication time in wet and measured size. Sonication should be continued until dispersion stability is achieved—if, of course, the primary particle size is desired. Use of $x\%$ power for y seconds is not an adequate means of describing the energy input. This is because the actual energy input by an ultrasound probe is dependent on a large number of conditions, including tip geometry, size and shape of the vessel within which the liquid is contained, type, volume and volatility of liquid, solids content, air or oxygen content of the liquid, and duty cycle of the ultrasonic probe, among others. These parameters can never be fixed and a typical ultrasound probe in an ultrasonic cleaning bath may lose up to 10% of its power within 12 months. Manufacturers may increase the power supplied to the ultrasonic transducer to compensate for this loss. Ultrasonic probes may need repolishing on a routine maintenance basis. The calibration of an ultrasound probe or the measurement of the power supplied is fraught with difficulties and can be attempted, but it is simpler to work with the maxim "dispersion to stability."

ΔP *in dry against measured size.* In dry powder dispersion, the ΔP defines the shear forces placed on the particulate system. The flow rate is also linked here, as higher ΔP's will correspond to higher flow rates.

Refractive index (RI) values especially imaginary for a small material (<25 μm). A robustness matrix for the imaginary is easily set up once the real part of the RI is known. Changes to the RI and the effect on the derived result can be explored by the 3S's (small, sensible, and systematic changes to the parameters). The form (shape) of the curve should be noted and the results should lie within expectation. Apparent multiple peaks need investigation in a system known to contain only one material.

Time! (*Stability*). Multiple consecutive measurements will indicate if a material is dispersing, dissolving, or even reacting in a wet system. Consecutive measurements in a dry system indicate the heterogeneity of the material or otherwise.

8   VERIFICATION: PARTICLE SIZE DISTRIBUTION CHECKLIST

- How is the primary sample taken?
- How does it get to the laboratory?
- How will we/do we take a laboratory sample?
- How much do we take?
- What is our required standard error?
- Can we achieve it or must we alter the specification or the sample division route?
- Have we examined the material under the microscope and obtained representative images? Note that the form of the material and/or polymorphs will govern key absorption and stability properties.
- Have we measured the real part of the refractive index for the material? Or do we know this from literature: CRC/Google/Luxpop, Cargille immersion fluids/Becke Lines, solutions extrapolated to 100% solids [17], ChemSketch (molar refractivities), manufacturers' Web sites (e.g., www.malvern.com) and literature such as the *Handbook of Optics*?
- Have we measured the real density of the material with a gas pycnometer?
- Have we obtained or do we have knowledge of the solubility of the material in water and other solvents (e.g., hexane or heptane)? Would measurement in these solvents be indicative of what may happen in application of the material (e.g., in the stomach)?
- How much sample do we have? Do we know an approximate top end of the distribution (e.g., is it a sieve-cut material)? If we have 20 mg only, we'll need to measure it all or we may want to forget about particle sizing this time around.
- What is the anticipated delivery or dosage form? Tablet? Nasal spray? Parenteral? The intended delivery form will dictate the type of distribution and numbers we seek.
- What about stability? Will this require a long shelf life? Tropical usage? Humidity?
- What's going to happen next? More processing (e.g., micronizing or formulation or blending or tabletting)
- What are the consequences (economic, political) of getting it wrong?

So, shouldn't we verify our systems on a regular basis?

An excellent AMC technical brief (No. 28, September 2007), issued by the Royal Society of Chemistry (RSC) in the UK, is titled "Using Analytical

Instruments and Systems in a Regulated Environment." Within this document there is a section called "A Framework for Analytical Data Integrity" attributed to C. Burgess, *Valid Analytical Methods and Procedures* (London: RSC, 2000). This section contains eight points of such importance that they will be reproduced in full:

- The instrumentation used has been qualified and calibrated.
- The method selected is based upon sound scientific principles and has been shown to be robust and reliable for the type of test material.
- The laboratory sample is representative and sufficiently close to homogeneous.
- A person who is both competent and adequately trained has carried out the analysis.
- The integrity of the calculation used to arrive at the result and its uncertainty is correct and statistically sound.
- Internal quality control is carried out in every analysis run. (A "run" is the period within which repeatability conditions prevail.)
- Proficiency testing is undertaken whenever practicable.
- Independent audits and assessments of the whole analytical system are carried out at intervals.

Next we look at how the foregoing principles can be applied in the verification and validation of particle size instrumentation.

9   COMMON PROBLEMS AND SOLUTIONS

Sample related issues (sampling and dispersion) need dealing with first, as these are often the largest variables in the machine, method, and material (3M's) approach. We assume that the sample has been taken correctly, that the sample has not changed over time, and that a valid calibration or verification (OQ) standard is employed. In this section we deal with possible causes and remedies of OQ failure rather than PQ failure where the sampling and dispersion issues need confirming prior to instrument evaluation.

9.1   Sieves

Problems here normally revolve around lack of calibration (this tends to happen very frequently) and obvious damage to screens (soldering, distortion, or breakage of holes). Small agglomerated and sticky materials present obvious problems in gaining reproducibility. The obvious tolerances in sieve aperture need factoring into any measurement, and the smaller the sieve aperture, the more practical the difficulties.

9.2 Sedimentation

We assume that basic theoretical concepts such as a system being too concentrated (particle–particle interactions; not free settling) and wall effects have been taken care of. Nonapplicability of the Stokes equation or working in a regime where it is not applicable is clearly not desirable!

For certain materials (porous, flocs) the correct input for the density presents some concerns. Selection of the type of verification standard needs care: It must exhibit attenuation to x-ray beams, and it needs to be of sufficient density difference with respect to the medium. Generally, the latter two requirements mean that the polystyrene standards have limited applicability. Standards such as almandine garnets are more common—density around 4300 kg/m^3 and high attenuation, due to the iron and aluminum content. Temperature fluctuation is to be avoided, as this generates thermal convection currents and local areas of different viscosity. Obviously, these need to be avoided and a suitable equilibration time set. Below a certain size, Brownian motion competes with gravitational settling, and standards are problematical in this area. However, the quartz standard BCR 66 has been used in this area. Water quality may be an issue if surfactants have been used for prior determinations.

Certain verification samples may need appropriate dispersion to break up agglomeration The extinction signal may need correction for absorption and scattering by the dispersing media, so baseline x-ray extinction data need collection akin to laser diffraction, and an inadequate baseline could be used to correct the analysis data. Last but not least, the temperature-measuring device within the instrument must be calibrated properly.

9.3 Light Scattering

We deal with laser diffraction (typically, 0.1 to 3000 μm) and photon correlation spectroscopy (typically, <1000 nm).

Laser Diffraction The most common cause of (rare) failure at the OQ stage is poor or fluctuating background. The former can be caused by dirty optics (usually, the windows through which a wet sample passes). Fluctuating background can be caused by inadequate temperature equilibration, especially if a volatile dispersant liquid is employed. Misalignment or gross movement of optics (usually, after transportation issues) is encountered occasionally. Although theoretical possible, missing or faulty detectors may also give rise to a failure at the OQ stage, but the author has not seen such an instance in the last decade or so. A poor connection between detector and computer is again an occurrence occasionally seen in the distant past. Again, size standards should be carefully chosen to have a spread in the distribution of 1:10 or greater if verification of the entire system, including the dispersion unit, is required (as it usually is).

For the smaller standards (<1 μm), water quality can again be important, and the dispersion of agglomerates will need to be accomplished successfully. Agglomeration can occur in small standards over time or if excessive time is used for the measurement protocols. The quality of most deionized water is based simply on a conductivity measurement, and this does not deal with small quantities of organic material that can alter the zeta potential significantly. Small standards also pose another potential problem, in that high quantities of surfactant can induce foaming and measurement of bubbles (with obvious detriment to the end result). Similarly, large standards need adequate agitation to keep them in suspension.

Subsampling of single-shot materials (to save money) is encountered too frequently (once is too much!). This phenomenon can also happen inadvertently with plastic vials of the standard where electrostatic charge can conspire to keep a small amount of material in the vial. Carefully removing this "stuck" material with a flush from a wash bottle will cure this issue. Standard functioning of the wet or dry accessory is obviously vital, and gross errors here will give a failure. Examples are again rare but include features such as a failed pump or stirrer on a wet accessory or a blocked or partially blocked venturi on a dry accessory. The latter has been observed in powder coatings plants with materials of low T_g (glass transition temperature) which fuse onto the inside of the venturi, thus altering the dispersion conditions (more aggressive) and inducing a OQ failure when a polydisperse standard is run.

PCS   The use of standard materials applicable to the technique is vital. Standards that are too large or that have settled or are sedimenting are obviously undesirable. The standard should be in free suspension. Standards below 1 μm that exhibit polydispersity are difficult to find. The IRMM 304 SiO_2 standard may be one material, although it does not come with distribution data for PCS (z-average and polydispersity information is given, though). In PCS, failure at the OQ stage is normally through optical misalignment (often caused by poor transport; some manufacturers fit temperature and shock sensor badges both visibly and invisibly to diagnose such issues) and is easily cured with a small (service engineer) "tweak." A more common issue is that of inadequate temperature equilibration and subsequent generation of convection currents in the cuvette. Leaving the sample to sit for a minimum of 120 seconds on the Peltier or other temperature control device normally cures this issue.

9.4   Electrozone Sensing

The use of blocked, damaged, or inadequate (wrong size range) orifices is clearly not desirable. Incorrect or out-of-date calibration data leads to inadequate sizing of the latex or other standard, and this is encountered routinely. A standard that has agglomerated may be usable after dispersion, but is usually to be avoided. The absolute mass balance method needs to be employed if accuracy is paramount, and this is an excellent confirmation that particles are not being missed in the system.

10 CONCLUSIONS

So, where are we now? Verification of particle size distribution should be a logical process, but without understanding of the principles involved, it has the probability of turning into some form of witchcraft. The link between IQ/OQ and PQ is one, in particular, that needs understanding, as it is certain that specification for PQ needs to be wider than those for the OQ stage. Hopefully, this chapter will guide the end user to sensible questions to ask the instrument vendor at each stage in the verification life cycle.

Acknowledgments

The author is extremely grateful and thanks Dennis Ward of GSK wholeheartedly for reviewing the manuscript at very short notice and making some excellent suggestions. Any mistakes or debatable opinion are the sole responsibility of the author—I can't say that I haven't been told!

Last, but not least, I would like to extend my thanks to my colleagues and friends in the Malvern worldwide family, who have made my life so interesting and bearable over nearly the last two decades or so. Without them I would probably have been doing something completely different.

REFERENCES

1. H. Heywood. Concluding remarks. In *Proceedings of the First Particle Size Analysis Conference*, 1966, pp. 355–359.
2. A. A. Noyes and W. R. Whitney. The rate of solution of solid substances in their own solutions. *J. Am. Chem. Soc.*, 9:930–934, 1897.
3. A. J. Hlinak, K. Kuriyan, K. R. Morris, G. V. Reklaitis, and P. K. Basu. Understanding critical material properties for solid dosage form design. *J. Pharm. Innov.*, 1:12–17, 2006.
4. *Guidance for Industry: Analytical Procedures and Methods Validation; Chemistry, Manufacturing, and Controls Documentation*. FDA, Washington, DC, 2000. Available at www.fda.gov/CDER/GUIDANCE/2396dft.pdf.
5. A. F. Rawle. Basic principles of particle size analysis. 1993. Available at www.malvern.com/malvern/kbase.nsf/allbyno/KB000021/file/Basic_principles_of_particle _size_analysis_MRK034-low_res.pdf.
6. *U.S. Pharmacopeia*, 34, General Test ⟨776⟩. USP, Rockville, MD, 2000, pp. 1965–1967.
7. K. Somner. 40 years of presentation particle size distributions—yet still incorrect? *Part. Part. Syst. Char.*, 18:22–25, 2001.
8. *Particulate Materials: Sampling and Sample Splitting for the Determination of Particulate Properties*. ISO 14488: 2007. ISO, Geneva, Switzerland, 2007.
9. A. F. Rawle. *Sampling for Particle Size Analysis*. Application Note MRK456-01. Malvern Instruments, Malvern, UK, 2003. Available at www.malvern.com.

10. T. Allen. *Powder Sampling and Particle Size Determination*. Elsevier Science and Technology, New York, 2003.

11. J. Mitchell. Particle standards: their development and application. *KONA Powder Part.*, 18:1–18, 2000.

12. G. Mie. Beitrage zur Optik trüber Medien, speziell kolloidaler Metallösungun. *Ann. Phys.*, 4(25):377–455, 1908.

13. M. Kaszuba, D. McKnight, M. Connah, F. McNeil-Watson, and U. Nobbmann. Measuring subnanometre sizes using dynamic light scattering. *J. Nanopart. Res.*, 2007; DOI 10.1007/s12051-007-9317-4.

14. A. F. Rawle. Attrition, dispersion and sampling effects in dry and wet particle size analysis using laser diffraction. Paper 0208 2000. Presented at the 14th International Congress of Chemical and Process Engineering "CHISA 2000," Praha, Czech Republic, Aug. 27–31, 2000.

15. A. F. Rawle. Refractive index verification via comparative wet and dry laser diffraction measurements. Presented at the World Congress of Particle Technology—5, Orlando, FL, 2006.

16. M. M. W. B. Hendricks and J. H. De Boer, Eds. *Robustness of Analytical Chemical Methods and Pharmaceutical Technological Products*. Data Handling in Science and Technology, Vol. 19. Elsevier Science, New York, 1996.

17. H. Saveyn, D. Mermuys, O. Thas, and P. van der Meeren. Determination of the refractive index of water-dispersible granules for use in laser diffraction experiments. *Part. Part. Syst. Char.*, 19(6):426–432, 2001. First presented at Partec 2001 (Nürnberg) Proceeding Session, Mar. 28, pp. 10:15–10:40, Particle Characterization I Internal Number 212. See also Malvern Application Note MRK 529, 1999.

Appendix: Selection of Standard/Certified Reference Materials for Particle Sizing

Designation[a]	Type	Size Range (μm)	Spherical	Polydisperse[b]	Suitable or Certified for (Technique)[c]	Supplier	Source and comments
SRM 1003c	Glass beads	20–45 (635–325#)	Y	Y	Screens, electrozone sensing, laser diffraction	NIST	ts.nist.gov/measurementservices/referencematerials/index.cfm
SRM 1004b	Glass beads	40–150 (270–120#)	Y	Y	Screens, electrozone sensing, laser diffraction	NIST	srmors.nist.gov/tables/view_table.cfm?table=301-1.htm
SRM 1017b	Glass beads	100–400 (140–45#)	Y	Y	Screens, electrozone sensing, laser diffraction	NIST	
SRM 1018b	Glass beads	220–750 (60–25#)	Y	Y	Screens, electrozone sensing, laser diffraction	NIST	
SRM 1019b	Glass beads	750–2450 (20–10#)	Y	Y	Screens	NIST	
SRM 1021	Glass beads	2–12	Y	Y	LLS, electrosensing zone, sedimentation	NIST	
RM 8010	Sand	A (30–100#)	N	Y	Screens	NIST	ASTM C 429, Standard Test Method for Sieve Analysis of Raw Materials for Glass Manufacture
		C (70–200#)	N	Y	Screens	NIST	
		D (100–325#)	N	Y	Screens	NIST	
SRM 659	Silicon nitride	0.2–10	N	Y	Sedimentation	NIST	
SRM 1978	Zirconium oxide	0.33–2.19	N	Y	Sedimentation	NIST	
SRM 1982	Zirconium oxide	10–150	N	Y	SEM, laser diffraction, sieving	NIST	
SRM 1984	Tungsten carbide/cobalt	9–30	N	Y	SEM, laser diffraction	NIST	Acicular
SRM 1985	Tungsten carbide/cobalt	18–55	N	Y	SEM, laser diffraction	NIST	Spheroidal
SRM 1690	Polystyrene spheres	1	Y	N	TEM, light scattering, EZS, disc centrifuge + others	NIST	
SRM 1691	Polystyrene spheres	0.3	Y	N	Electron microscopy, DLS, centrifuge	NIST	Developed with ASTM
SRM 1692	Polystyrene spheres	3	Y	N	Microscopy	NIST	Developed with ASTM
SRM 1961	Polystyrene spheres	30	Y	N	Counters, microscopy	NIST	Developed with ASTM
SRM 1963a	Polystyrene spheres	0.1	Y	N	Electron microscopy, DLS, DMA	NIST	Developed with ASTM
SRM 1964	Polystyrene spheres	0.06	Y	N	Electron microscopy, DLS, DMA	NIST	Developed with ASTM
SRM 1965	Polystyrene spheres (on slide)	10	Y	N	Manual microscopy	NIST	Developed with ASTM
RM 8011	Gold colloid	10 nm	N	N	AFM, electron microscopy, DMA, DLS, SAXS	NIST	www.thermo.com/com/cda/product/detail/0,1055,10137986,00.html
RM 8012	Gold colloid	30 nm	N	N	AFM, electron microscopy, DMA, DLS, SAXS	NIST	
RM 8013	Gold colloid	60 nm	N	N	AFM, electron microscopy, DMA, DLS, SAXS	NIST	
BCR 66	Quartz	0.35–3.5	N	Y	Gravity sedimentation	IRMM	
BCR 67	Quartz	2.4–32	N	Y	Gravity sedimentation	IRMM	
BCR 68	Quartz	160–630	N	Y	Sieving	IRMM	
BCR 69	Quartz	14–90	N	Y	Gravity sedimentation	IRMM	
BCR 70	Quartz	1.2–20	N	Y	Gravity sedimentation	IRMM	
BCR 130	Quartz	50–220	N	Y	Sieving	IRMM	
IRMM-304	Colloidal silica	Nominal 40 nm	N	?	DLS, disk centrifuge (SAXS)	IRMM	www.irmm.jrc.be/html/reference_materials_catalogue/catalogue/attachments/IRMM-304_product.pdf
MS0009	Glass microspheres	7.78–10.44	Y	N	Image analysis, microscopy (not suitable for LLS)	Whitehouse Scientific	www.whitehousescientific.com/
MS0040	Glass microspheres	38.9–41.4	Y	N	Image analysis, microscopy (not suitable for LLS)	Whitehouse Scientific	A large number of others of this type are available.

(Continued overleaf)

Appendix: (Continued)

Designation[a]	Type	Size Range (μm)	Spherical	Polydisperse[b]	Suitable or Certified for (Technique)[c]	Supplier	Source and comments
MS0589	Glass microspheres	572–615	Y	N	Image analysis, microscopy (not suitable for LLS)	Whitehouse Scientific	A large number of others of this type are available.
PS180	Glass	0.1–10	Y	Y	Centrifuge, SEM, LLS	Whitehouse Scientific	
PS190	Glass	1–10	Y	Y	Andreasen pipette, ESZ	Whitehouse Scientific	
PS200	Glass	3–30	Y	Y	Andreasen pipette, ESZ, microscopy	Whitehouse Scientific	
PS211	Glass	10–100	Y	Y	Andreasen pipette, ESZ, microscopy	Whitehouse Scientific	
PS222	Glass	50–350	Y	Y	Sieve, microscopy	Whitehouse Scientific	
PS232	Glass	150–650	Y	Y	Sieve, microscopy	Whitehouse Scientific	
PS240	Glass	500–2000	Y	Y	Microscopy	Whitehouse Scientific	
XX015	Clear glass	50–250	Y	Y	Image analysis	Whitehouse Scientific	
XX025	Clear glass	170–710	Y	Y	Image analysis	Whitehouse Scientific	
XX030	Clear glass	500–2000	Y	Y	Image analysis	Whitehouse Scientific	
XX035	Clear glass	1400–1500	Y	Y	Image analysis	Whitehouse Scientific	A large number of others of this type are available.
GP0042	Glass microspheres	38–45	Y	N	Sieve	Whitehouse Scientific	
GP0463	Glass microspheres	425–500	Y	N	Sieve	Whitehouse Scientific	
GP3775	Glass microspheres	3350–4000	Y	N	Sieve	Whitehouse Scientific	
BM0083	Basalt microspheres	75–90	Y	N	Sieve, LLS?	Whitehouse Scientific	Opaque
BM0550	Basalt microspheres	500–600	Y	N	Sieve, LLS?	Whitehouse Scientific	A large number of others of this type are available.
BM2200	Basalt microspheres	2000–2400	Y	N	Sieve, LLS?	Whitehouse Scientific	
4009A	Polystyrene microspheres	1	Y	N	Microscopy, EZS	ThermoFisher	http://www.thermo.com/com/cda/product/detail/ l..10136016,00.html
4209A	Polystyrene microspheres	9	Y	N	Microscopy, EZS	(was Duke Scientific)	A large number of others of this type are available.
4240A	Polystyrene microspheres	40	Y	N	Microscopy, EZS	ThermoFisher	
4316A	Polystyrene microspheres	160	Y	N	Microscopy, EZS	(was Duke Scientific)	A large number of others of this type are available.
4320A	Dry polystyrene spheres	200	Y	N	Microscopy, EZS	ThermoFisher	
4350A	Dry polystyrene spheres	500	Y	N	Microscopy, EZS	(was Duke Scientific)	
4400A	Dry polystyrene spheres	1000	Y	N	Microscopy, EZS	ThermoFisher	
3020A	Polystyrene microspheres	20 nm	Y	N	DLS, AFM, electron microscopy	(was Duke Scientific)	A large number of others of this type are available.
3100A	Polystyrene microspheres	100 nm	Y	N	DLS, AFM, electron microscopy	ThermoFisher	
3900A	Polystyrene microspheres	900 nm	Y	N	DLS, AFM, electron microscopy	(was Duke Scientific)	
8100	Silica microspheres	1	Y	N	TEM	(was Duke Scientific)	www.dukescientific.com/www.dukescientific.com/pdfs/ sales/Bull18A%20Silica.pdf
64004-15	Polystyrene microspheres	40 nm	Y	N	Disk centrifuge	Polysciences	www.polysciences.com/Catalog/40/categoryId_4/
64019-15	Polystyrene microspheres	500 nm	Y	N	Disk centrifuge	Polysciences	A large number of others of this type are available.
64028-15	Polystyrene microspheres	950 nm	Y	N	Disk centrifuge	Polysciences	
64030-15	Polystyrene microspheres	1	Y	N	Disk centrifuge	Polysciences	
64120-15	Polystyrene microspheres	9	Y	N	Disk centrifuge	Polysciences	
64130-15	Polystyrene microspheres	10	Y	N	Single-particle optical sensing (SPOS)	Polysciences	A large number of others of this type are available.
64190-15	Polystyrene microspheres	50	Y	N	Single-particle optical sensing (SPOS)	Polysciences	
64235-15	Polystyrene microspheres	175	Y	N	Single-particle optical sensing (SPOS)	Polysciences	
NT02N	Polystyrene microspheres	40 nm	Y	N	Electron microscopy?	Bangs Laboratories	www.bangslabs.com
NT22N	Polystyrene microspheres	5	Y	N	Electron microscopy?	Bangs Laboratories	A large number of others of this type are available.
NT40N	Polystyrene microspheres	175	Y	N	Electron microscopy?	Bangs Laboratories	

[a] This list should not be considered exhaustive, and listing of a material does not mean that it it is suitable for any given application. Note that a large number of standards exist for techniques such as BET surface area, which is related to (but not the same as) particle size distribution determination.

[b] Polydisperse implies a width of distribution greater than 1.5:1 or so.

[c] Suitability for a particular technique or instrument should be discussed with the supplier and the appropriate documentation/certification studied.

13

METHOD VALIDATION, QUALIFICATION, AND PERFORMANCE VERIFICATION FOR TOTAL ORGANIC CARBON ANALYZERS

José E. Martínez-Rosa

JEM Consulting Services Inc.

1 INTRODUCTION

Total organic carbon (TOC) is the quantity of carbon attached to an organic compound and is frequently used as a surrogate to water quality in several industries and fields, such as the pharmaceutical industry, microelectronics, potable water, and wastewater management. The classic TOC analysis measures both the total carbon (TC) and the inorganic carbon (IC) concentration in an aliquot of water. Subtracting the IC from the TC produces the amount of TOC. A different variant of TOC analysis involves removing the IC portion first and then measuring the residual carbon. The latter method involves purging the sample with carbon-free air or nitrogen and then acidifying it prior to the sample measurement. This method is called nonpurgeable organic carbon (NPOC). For further details, see Figure 1.

Practical Approaches to Method Validation and Essential Instrument Qualification, Edited by Chung Chow Chan, Herman Lam, and Xue Ming Zhang Copyright © 2010 John Wiley & Sons, Inc.

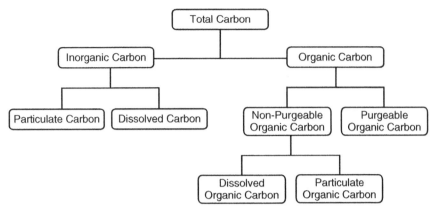

FIGURE 1   Relationship of total carbon.

1.1 Definitions

Dissolved organic carbon (DOC): organic carbon (OC) left behind in a sample after filtering the sample using a 0.4- to 0.7-μm filter, a 0.45-μm filter being the most common [1]

Inorganic carbon (IC): carbon drawn from nonliving sources [e.g., carbonate (CO_3^{2-}), bicarbonate (HCO_3^-), and dissolved carbon dioxide (CO_2)]

Nonpurgeable organic carbon (NPOC): organic carbon remaining in a sample after purging the sample with carbon-free air or nitrogen

Organic carbon (OC): carbon derived from living organisms and their metabolic activities

Purgeable (volatile) organic carbon (POC): organic carbon that has been removed from a sample with carbon-free air or nitrogen

Suspended inorganic carbon or *particulate inorganic carbon*: the inorganic carbon in particulate form

Suspended organic carbon or *particulate organic carbon*: the carbon in particulate form that is retained after filtration with a 0.4- to 0.7-μm filter

Total carbon (TC): all the carbon present in a sample, counting both inorganic and organic carbon

1.2 Potable Water

OC is present in water as a complex mixture of molecules such as carbohydrates, amino acids, hydrocarbons, fatty acids, and phenolics. Other components are natural macromolecules and colloids, sewage and industrial particulates, soil organic matter, microorganisms, and plant material. Organic carbon can also occur as traces of lubricants, fuels, fertilizers, and pesticides. The TOC in water is a useful indication of the degree of pollution, particularly when concentrations can be compared upstream and downstream of potential sources of pollution.

In surface waters, TOC concentrations are in general less than 10 ppm, and in groundwater less than 2 ppm, unless the water receives wastes or has natural organic materials [2].

There is a relationship among the concentration of by-products of drinking water disinfection and trihalomethanes (THMs). The higher the concentration of TOC, the higher the concentration of THMs would be. Therefore, several environmental agencies in the world are recommending <4 ppm for source water and <2 ppm for treated water [3].

1.3 Pharmaceutical, Biologics, and Biotechnology Industries

In purified water (PW) systems OC is introduced into water by microorganisms, either dead or alive. Other sources of OC are the purification and distribution systems and even the environmental air. The pharmaceutical, biologics, and biotechnology industries use water purified to different levels according to the intended use [PW, highly purified water (HPW), and water for injection (WFI)]. It is well known that there is an association between microbial contamination and the amount of TOC present in purified water systems. The TOC may come from planktonic and/or sessile bacteria. In addition, endotoxins contribute to the concentration of TOC in pharmaceutical water distribution systems. In fact, in HPW and WFI systems, TOC is an indirect indicator of endotoxin contamination. The *United States Pharmacopeia* (USP), *European Pharmacopoeia* (EP), and *Japanese Pharmacopoeia* (JP) include TOC as a compendial test for PW and WFI. Additionally, the EP requires HPW to be tested for TOC. Consequently, TOC is a process control tool in the pharmaceutical, biologics, and biotechnology industries to assure the effectiveness of the water purification and distribution systems.

Regulatory agencies such as the U.S. Food and Drug Administration (FDA) and the EMEA allow the use of TOC as a means of assessing the cleanliness of manufacturing equipment. To make certain that there is no cross-contamination between different drug products manufactured using the same pieces of equipment, various cleaning procedures are performed. After cleaning, TOC samples are taken and analyzed. Afterward, TOC test results are compared to preestablished acceptance levels of TOC. If the TOC measurement is at or below the acceptance criterion, the piece of equipment is considered clean. This approach is very useful in clean-in-place (CIP) applications.

1.4 Microelectronics and Semiconductors

Semiconductors and microelectronics manufacturers require ultrapure water (UPW) because particulate, chemical, and biological contamination has been shown to be a major cause of manufacturing problems and process failures [4–6]. In particular, bacteria contain metals and other substances that are harmful to integrated circuits. The levels of nutrients in UPW are extremely low. Actually, UPW is an oligotrophic environment. But despite this fact, some bacteria can grow to a wearisome concentration [7]. Several studies have demonstrated that most high-purity water systems have biofilms [6,8–10]. The primary source of planktonic bacteria isolated from these systems is detached microorganisms from biofilms. Therefore, TOC can be used to monitor the development of biofilms in UPW systems.

1.5   Wastewater Management

TOC measurements are been used as a way of determining contamination levels of wastewater. The TOC determination yields more reliable and reproducible data than chemical oxygen demand (COD) and biological oxygen demand (BOD) analyses. The need for rapid determination of TOC levels in wastewater has led to introduction of TOC instruments into treatment plant laboratories. The TOC values will generally be less than COD values, because a number of organic compounds may not be oxidized in the TOC analysis. Usual values of TOC for domestic wastewater may vary from 100 to 300 ppm [11].

2   TOC METHODOLOGIES

There are two ways of measuring TOC. The first, generally called the TC–IC method, is a two-stage analysis. As the name implies, it measures both IC and TC separately; but the IC is measured in an acidified aliquot of the sample. The IC concentration in a sample is found by lowering its pH below 2. This releases the IC from the sample as CO_2. Next, carbon-free air or nitrogen carrier gas sparges through the sample. By this means the IC gas is carried through the instrument, removing unnecessary water vapor and potential interfering substances before reaching the instrument's detector.

The TC analysis involves no acidification, only oxidation of the carbon, which releases the CO_2 out of the sample. Then carbon-free air or nitrogen carrier gas sparges through the sample, which is also sent through the instrument, removing unwanted water vapor and interfering substances before reaching the instrument's detector. The TOC value is obtained by subtracting the IC value from the TC value: TOC = TC − IC.

The second method measures TOC in the sample directly by acidifying the sample to a pH below 2 to release the inorganic carbon as CO_2. Yet this IC gas-stripping step also removes some of the organic molecules, which are recaptured, oxidized to CO_2, and measured as purgeable organic carbon (POC). The remaining nonpurgeable organic carbon present in the sample is oxidized to produce CO_2, which is carried to the detector and quantitated as nonpurgeable organic carbon (NPOC). In this method, TOC is the total of POC and NPOC. That is, TOC = POC + NPOC. In PW, HPW, WFI, and UPW the quantity of POC is insignificant and can be discounted. Therefore, in these types of water, NPOC is equivalent to TOC. Regarding of the methodology used, a TOC analysis may be divided into three main steps: acidification, oxidation, and detection and quantization.

2.1 Acidification

CO_2 is soluble in water, in which it interchanges spontaneously between CO_2 and H_2CO_3 (carbonic acid). The concentrations of CO_2 and H_2CO_3, and of HCO_3^- (bicarbonate) and CO_3^{2-} (carbonate), are determined by the pH. At pH >6.5, the concentration of HCO_3^- is >50%, and at very alkaline conditions (pH >10.4) CO_3^{2-} is the prevalent anion (>50%). However, as the pH drops below 6.5, the concentration of HCO_3^- also drops and the equilibrium of the reaction is moved toward the production of CO_2 and H_2O. When the pH is ≤ 2, all HCO_3^- is converted to CO_2.

The chemical reaction depicting this process is the following:

$$\text{low pH} \xleftarrow{\hspace{2cm}} CO_2 + H_2O \leftrightarrow H^+ + HCO_3^- \leftrightarrow 2H^+ + CO_3^{2-} \xrightarrow{\hspace{2cm}} \text{high pH}$$

The exclusion and venting of IC and POC from the sample by acidification and gas stripping occurs as follows:

$$HCO_3^-, CO_3^{2-}, CO_2, POC \xrightarrow{\text{carrier gas, acid}} CO_2 + POC$$

2.2 Oxidation

The second step is the carbon oxidation to CO_2 and other gases in the outstanding sample. Current TOC analyzers perform this oxidation step in one of several ways:

- Photooxidation
- Photochemical oxidation
- Thermochemical oxidation
- High-temperature catalytic combustion oxidation

Photooxidation (Ultraviolet Light) In this oxidation method, ultraviolet (UV) light, <190 nm in wavelength, oxidizes the carbon contained by the sample to produce CO_2. The means of UV-promoted oxidation relies on negatively charged hydroxyl (HO^-) radicals generated from the irradiation of water with UV energy. With sufficient exposure, all dissolved organics can be oxidized to yield CO_2. The UV oxidation method presents the most reliable and low-maintenance method of testing TOC in high-purity waters.

A variant of the UV oxidation method is the UV partial oxidation (UV partial) method. As the name indicates, the UV partial method does not perform a complete oxidation of the organic carbon in the sample. Instruments based on this methodology measures the incoming conductivity in a stream of water, which then goes through a UV oxidation reactor to oxidize the organic molecules present and measures the conductivity change after oxidation. The change in conductivity correlates to the concentration of organic carbon in the water sample [12].

UV-Chemical (Persulfate) Oxidation   Like the photooxidation method, UV light is the oxidizer, but the oxidation potential of the reaction is increased by the addition of a chemical oxidizer, typically a persulfate reagent such as ammonium persulfate. The mechanisms of the reactions are controlled by the formation of radicals and the excitation and oxidation of the organic molecules.

Formation of radicals:

$$S_2O_8^{2-} \xrightarrow{h\nu} 2SO_4^-$$

$$H_2O \xrightarrow{h\nu} H^+ + OH^-$$

$$SO_4^- + H_2O \rightarrow SO_4^{2-} + OH^- + H^+$$

Excitation of organic molecules: $R \xrightarrow{h\nu} R^*$ Oxidation of organic molecules: $R^* + SO_4^- + OH^- \rightarrow nCO_2 + \cdots$

Thermochemical Oxidation   The sample is mixed with a quantity of persulfate solution and heated to a high temperature. After a predefined reaction time, the resulting CO_2 is gas-stripped for detection. This oxidation technique is more energetic than UV alone.

High-Temperature Catalytic Combustion (HTCC) Oxidation   With the combustion technique, the sample is injected into a furnace at a temperature $\geq 680°C$, usually in the presence of a catalyst, with a stream of hydrocarbon-free compressed air or oxygen to oxidize the organic molecules. Both dissolved and particulate organics are completely oxidized to CO_2 under these settings. HTCC instruments employ a variety of diverse catalysts, including cobalt oxide, cupric oxide, titanium dioxide, or platinum based. In HTCC instruments the oxidation temperatures may fluctuate from 680 to $1000°C$, depending on the application.

HTCC oxidation methods are useful in those situations where difficult-to-oxidize compounds and particulates are present because they provide complete oxidation of organics under these circumstances. The major downside of HTCC instruments is its unsteady baseline levels, which limits its effectiveness for high-purity water analysis. This is caused by the accretion of nonvolatile residues inside the high-temperature reactor. These residues produce false and variable TOC background levels to the analysis, yielding erratic results [13]. HTCC instruments in general exhibit low sensitivity and require high maintenance.

2.3   Detection and Quantification

Conductivity and nondispersive infrared (NDIR) are the two common detection methods used in current TOC instruments (Table 1).

Conductivity   There are two types of conductivity detectors (sensors), direct conductometric (DC) and membrane conductometric (MC), also known as gas

TABLE 1 Official TOC Testing Methods

Oxidation Method	Detection Technology	Analytical Range (mg/L of C)	Official Method
HTCC	Thermal conductivity detector	5000–1,000,000	AOAC 955.07
	Conductometric	10,000–1,000,000	ASTM D 4139
	NDIR	0.004–25,000	Standard Method 5310C; EPA 415.1, 9060A; ASTM D 2579; ISO 8245; AOAC 973.47, JP XV
UV-persulfate	NDIR	0.002–10,000	Standard Method 5310C; EPA 415.2, 9060A; ASTM D 2579; ISO 8245; AOAC 973.47, USP ⟨643⟩, EP Method 2.2.44, JP XV
	Direct conductometric, membrane conductometric	0.0005–50	Standard Method 5310C, USP ⟨643⟩, EP Method 2.2.44, JP XV
Heated persulfate	NDIR, direct conductometric, membrane conductometric	0.002–1000	Standard Method 5310C; EPA 415.2, 9060A; ASTM D 2579; ISO 8245; AOAC 973.47, USP ⟨643⟩, EP Method 2.2.44, JP XV
UV	Direct conductometric, membrane conductometric, NDIR	0.0005–2.0	USP ⟨643⟩, EP Method 2.2.44, JP XV

permeation. DC offers simple, low-cost ways of measuring CO_2. This method uses no carrier gas and has excellent sensitivity, but has a limited analytical range. MC detector uses the same technology as a direct-conductometric detector but incorporates a semipermeable membrane that limits the diffusion of potential interfering substances.

Both DC and MC measure the sample conductivity before and after oxidization. The difference in conductivity is attributed to the TOC contents of the sample. During the sample oxidization step, CO_2 and other gases are formed. The dissolved CO_2 forms carbonic acid, which changes the sample conductivity as the organic molecules are oxidized.

DC instruments take for granted that only CO_2 is present in the oxidized sample. Provided that this is true, then, the TOC measurement is accurate. Nonetheless, depending on the chemical groups present in the sample and their resulting oxidized species, they may produce a positive or a negative analytical interference. Several of the interfering chemical species include acid gases, iodine, organic acids, hypochlorous acid, halogenated organics, nitrite ions, sulfide ions, and other ions with dissolved gas phases. In addition, minute changes in pH and temperature may add variability to the TOC analysis in DC instruments.

MC methods have been proved to be simple, sensitive, rapid, and easy to operate. The gas permeation analysis used for the determination of TIC with membrane separation was first reported by Carlson [14]. The method consisted of the transfer of CO_2 by diffusion via silicone-rubber hollow fibers into a flowing stream of deionized water, followed by electrical conductivity detection. Instruments based on the MC technology are resistant to interference substances. But they have their own specific situations, such as true selectivity, clogging, and microbial growth. MC instruments have been improved by integrating the use of hydrophobic gas permeation membranes to allow a more "selective" transfer of the dissolved CO_2 [15].

Nondispersive Infrared (NDIR) The most common form of detection found in TOC instruments is the NDIR detector. It measures the infrared light (IR) absorbed by CO_2 as it passes through a flow-through IR absorption cell. Interferences from other IR-absorbing gases are minimized by use of a highly wavelength-specific detector [16].

As the carrier gas sweeps the derived CO_2 through the NDIR detector and IR energy is absorbed at the absorption frequency of CO_2, about 4.26 mm, the NDIR detector generates a nonlinear signal that is proportional to the instantaneous concentration of CO_2 in the carrier gas. A second measurement that is obtained from a reference cell, which is filled with a nonabsorbing gas such as N_2, is also taken and the differential result correlates to the CO_2 concentration in the sample detector at that moment. As the gas continues to flow into and out of the detector cell, the sum of the measurements results in a peak that is integrated. The area obtained is then compared to stored calibration data, and the sample concentration is calculated [17]. Commonly, sample and reference cells are located side by side.

The principal advantage of using NDIR is that it measures the CO_2 generated by oxidation of the organic carbon directly and specifically. NDIR detection of the CO_2 gas formed in the oxidation process has no interferences. NDIR instruments can operate at the low TOC levels required for meeting requirements of high-purity water, but they require much more effort from users to keep reagent blanks under control. Unfortunately, it is difficult to produce CO_2-free reagent blanks. Just the act of processing a sample can contaminate it with environmental CO_2. These limitations are exacerbated in NDIR technology because it measures CO_2 directly rather than carbonic acid. Field experience has demonstrated that both DC and MC methods produce better TOC results with less calibration, maintenance, and support than the NDIR analyzers at the TOC levels found in high-purity water systems.

New Advances in NDIR Technology A new advance of NDIR technology is static pressurized concentration (SPC). SPC detection technology allows OC to be oxidized and the resulting CO_2 swept and pressurized inside the NDIR detector. The valve situated at the detector's outlet is closed, preventing the escape of the CO_2 from the detector. With the valve closed and all the CO_2 enclosed inside the NDIR detector, a single measurement is performed to measure the amount of CO_2. Consequently, there is an increase in sensitivity and precision since all oxidized OC is measured in one measure, compared to flow-through technology, where variation is introduced in measuring flowing CO_2 over time [18,19].

2.4   Off-line vs. Online TOC Technologies

There are several advantages in using online TOC instruments versus TOC laboratory (off-line) instruments [20].

- *Minimization of samples*. This is self-explanatory.
- *Negligible contamination*. Elevated TOC levels can be observed in grab samples due to contamination by organic compounds from the sample bottles or during collection.
- *Real-time profiling*. Continuous online determinations of TOC enable users to profile changes in their water systems and minimize the risk of missing excursions and out-of-specification results.
- *Decreased laboratory labor*. Costs associated with collecting and analyzing samples for TOC are reduced.
- *No glassware required*. In off-line TOC testing, vials that have not been precleaned can contribute significant background TOC. Certified precleaned vials must be used. If no alternative is available, the vials should be rinsed thoroughly with low-TOC water before use to minimize contamination.
- *Less variability*. Online TOC analyses have less variability than off-line analyses. When a sample is analyzed off-line, there are several variables that are added to the analysis, such as sampling procedure, sample handling, glassware, instrument variability, and analyst technique. On the other hand, online analyzers only account for the instrument variability [20].

3   PARAMETERS FOR METHOD VALIDATION, QUALIFICATION, AND VERIFICATION

In general, TOC analyzers should be qualified as laboratory instruments regardless of where they are used (e.g., desktop analyzers or online analyzers). Performance qualification (PQ) for online instruments should be performed in parallel with desktop instruments if the latter are available. Equivalency between off-line and online instruments must not be pursued since online measurements of TOC are usually lower than those of desktop units. Although the behavior should be the same in online and off-line test results, the online test results are about one-third to one-half the values of the off-line results. As explained above, off-line testing

TABLE 2   **Sources of Contamination and Variability in TOC Analyses**

Off-line TOC Test	Online TOC Test
Collection of water sample: prone to CO_2 contamination from environmental sources such as sample container, organic fumes, and breathing from the person collecting the sample *Sample handling and storage:* prone to bacterial contamination *Sample analysis:* contamination from glassware and reagents, instrument variability, and analyst technique	*Sample analysis:* contamination from reagents (if used) and instrument variability

involves the execution of several independent steps with their own variability, while online testing has only instrument variability and the contamination from reagents (when used). For further details, see Table 2.

3.1   Relevant Analytical Parameters

Four analytical parameters are extremely important to consider in TOC analyzes: recovery, blank contribution, detection limits, and particulates with organics. For high-purity water systems, particulates with organics are not an issue.

Recovery   The technical literature is equally divided as to whether it is better to oxidize dissolved organic compounds by persulfate digestion or combustion. Only when testing difficult particulates such as crystallized cellulose does the combustion method show a significant advantage. With high levels of organic particulates or TOC, it is clearly the best method. The latest UV and UV–persulfate TOC analyzers achieve much higher recoveries than older-generation models because of higher persulfate concentrations and/or advances in UV lamp design. So both methods are perfectly suited for routine tests of high-purity water samples.

Blank Contribution   In most cases, this analytical parameter is important only for low-level TOC measurements such as in potable water, HPW, PW, and WFI. Blank contribution is carbon associated with the TOC analyzer itself. Some sources include persulfate, dilution water, and the catalysts used in combustion-based TOC analyzers. The amount and stability of a blank has a major effect on an analyzer's detection limits and reproducibility. Blank contribution to sample analysis by UV, heated persulfate, and UV–persulfate methods is well known to be smaller than that of the combustion method.

Detection Limits   TOC testing of high-purity water samples requires a very low instrument background. Analyzers that use UV, UV–persulfate, or heated-persulfate oxidations typically show low background interference. High blank contribution is the primary disadvantage of combustion methods. The UV, heated-persulfate, and UV–persulfate methodologies can easily quantify TOC below USP ⟨643⟩ requirements (limit of detection = 50 ppb).

Particulates with Organics   One of the most difficult water samples to analyze is that with particulates, especially if these particulates contain organic molecules embedded within. The diverse nature of particulates can cause results with poor precision and accuracy [21]. In extreme situations particulates can cause the clogging of lines and valves in contact with the sample. This problem is corrected by filtration of the sample. This step traps particulates on the filter surface and lets DOC go through. But many countries in the European community, worried about organics in particulates, do not allow filtration of the sample. In these countries, ISO method 8245-13 and EN method 1484-13 are in place to validate an instrument's ability to measure samples with particulates [22].

3.2   Relevant Test Scripts

Specificity: generally refers to a method that produces a response for a specific analyte only. However, the terms *selectivity* and *specificity* are often used interchangeably. Selectivity refers to a method that provides responses for several substances that may or may not be distinguished from each other. Applies to release testing: as system suitability, monitoring, and cleaning validation.

Accuracy: refers to the degree that test results produced by the method and the true value agree. Applies to monitoring and cleaning validation.

Precision: has two components: repeatability and intermediate precision (ruggedness). Repeatability is the variance experienced by an analyst on a specified instrument. Repeatability does not discriminate between variance from the instrument and variance from the sample preparation process. Intermediate precision (ruggedness) points out variations within a laboratory or different laboratories (i.e., different days, instruments, or analysts). Applies to monitoring and cleaning validation.

Linearity: refers to the method's ability to obtain a response that is proportional to the amount of analyte in the sample. Applies to monitoring and cleaning validation.

Limit of detection: the lowest amount of analyte in a sample that can be detected by the analytical method but not necessarily quantitated as an accurate value. Applies to monitoring and cleaning validation.

Limit of quantization: the lowest amount of analyte in a sample that can be quantitated by the analytical method with the specified precision and accuracy. Applies to monitoring and cleaning validation.

Although, USP ⟨643⟩ has system suitability as the sole requirement for release testing of USP waters (PW and WFI), it is recommended to include accuracy,

TABLE 3   Example of Precision and Accuracy Test Results

Reading	Actual Results[a] (ppb)		Precision Data/Results	Accuracy Data/Results
	R_w	R_s		
1	22.5	517	Average: 519 ppb	Blank TOC (R_w) average: 23 ppb
2	23.0	517		Certified TOC concentration: 500 ppb
3	23.5	518	SD: 2.3 ppb	Expected standard concentration: 523 ppb
4	23.8	521		Average of standard TOC readings: 519 ppb
5	22.1	522		Carbon concentration difference: −4 ppb
Average	23	519	RSD: 0.5%	Accuracy (error value) results: −1%

[a] R_w, instrument response to the reagent water used to prepare the standards; R_s, instrument response to the 500 ppb sucrose standard as C.

precision, and linearity to the qualification activities to have a robust qualification package. Tables 3 and 4 show examples of test results for precision, accuracy, and linearity.

4   QUALIFICATION, VALIDATION, AND VERIFICATION PRACTICES

4.1   Regulatory Requirements for Qualification and Verification

There are no regulatory or normative requirements to qualify laboratory instruments. Also, guidelines from regulatory bodies are not available. However, this void has been filled with guidelines from organizations such as the American Association of Pharmaceutical Scientists (AAPS) and Good Automated Manufacturing Practice (GAMP). For instruments such as TOC analyzers, which have proven technology and a long history in the market, there is no need for a cumbersome and costly qualification [23,24].

4.2   Common Requirements for Validation of Analytical Procedures

- Qualified and calibrated instruments
- Documented methods
- Reliable reference standards
- Qualified analysts
- Sample integrity

TABLE 4 Example of Linearity Test Results

Standard	Measured TOC Average (ppb)	Measured TOC Average Minus Rw (ppb)	Standard Deviations Results (ppb)	Adjusted Standard Concentration (ppb)	Correlation Coefficient Result
Reagent water	23	—	—	—	Correlation coefficient (r^2): 0.9993
Sucrose standard 250 ppb	269	246	1.3	273	
Sucrose standard 500 ppb	519	496	2.3	523	
Sucrose standard 750 ppb	789	763	2.9	773	

4.3 Regulatory Requirements

United States Pharmacopeia ⟨643⟩, European Pharmacopoeia Method 2.2.44, and Japanese Pharmacopeia General Test 2.59   USP test monograph ⟨643⟩ calls for TOC tests to be performed using a calibrated instrument, with periodic demonstration of its suitability [25]. It also requires analytically dependable low limits of quantization for TOC analysis and a true TOC measurement for standardization and validation. It also states that the frequency of recalibration is a function of instrument design, degree of use, and so on. Water having a TOC concentration of not more than 0.10 mg/L must be used. Implemented July 1, 1999, *European Pharmacopoeia* (EP) method 2.2.44 is the equivalent of USP ⟨643⟩ and has the same analytical requirements, except for the following. Water for the reagent blank and preparation of standards must use highly purified water complying with the following specifications [26]:

- *Conductivity:* not greater than 1.0 μS/cm at 25°C
- *TOC:* not greater than 0.1 mg/L

As in the case of the EP, the *Japanese Pharmacopeia* General Test 2.59 (test for total organic carbon) [27] has the same analytical requirements as USP ⟨643⟩. Except for the following:

- The oxidation device should be capable of generating not less than 0.450 mg/L of organic carbon when using a solution of sodium dodecylbenzenesulfonate (theoretical value of total organic carbon in this solution is 0.806 mg/L) as the sample.

- The water used for preparing standard solutions or decomposition aid or for rinsing the instrument should have not more than 0.250 mg/L (250 ppb) of organic carbon.
- The standard for organic carbon is potassium hydrogen phthalate.
- Includes sodium hydrogen carbonate as the standard for inorganic carbon.
- Does not mention system suitability test.

USP ⟨1325⟩, Validation of Compendial Procedures USP ⟨1325⟩, like many others documents in the pharmaceutical and biotechnology industries, is not a regulatory binding document—it is only a guideline. However, it presents best practices and methods that when followed should comply with regulatory agencies' requirements. In addition, it presents information of the utmost importance regarding the validation of compendial analytical methods.

The first paragraph on the first page of USP ⟨1325⟩ states that users of analytical methods described in the *U.S. Pharmacopeia* and the *National Formulery* (NF) are not required to validate the accuracy and reliability of these methods, but merely verify their suitability under actual conditions of use [28]. In addition, it is indicated that according to Section 501 of the Federal Food, Drug, and Cosmetic Act, assays and specifications in monographs of the USP and the NF constitute legal standards.

The quote above means that when analytical methods published by the USP and NF are implemented, the user does not require to validate the accuracy and reliability (precision and reproducibility) of these methods, only their suitability for the intended purpose; and suitability means appropriate to a purpose. Another definition of suitability is fitness for a purpose. Therefore, we have that under USP ⟨1325⟩ the validation of a *compendial analytical method* is to demonstrate that when correctly applied, the method produces results that are fit for the purpose.

Additionally, as deemed by Porter [29], the quote above is consistent with the rationale that the term *validation* should be reserved for the process whereby one determines if a given method is suitable for its intended purpose, that is, should be reserved for the demonstration that the conditions under which the method is to be performed are indeed appropriate for the method [29]. The amount of validation work required for an analytical method such as TOC in pharmaceutical-grade water is reduced. Especially when TOC-⟨643⟩ is not a quantitative assay, it is just a limit test. Yet, when TOC is used in cleaning validation, it becomes a quantitative assay and is subject to more requirements.

In general, to be fit for the purpose intended, the analytical method must meet certain validation characteristics. Typical validation characteristics, which should be considered, are accuracy, precision, specificity (provides an exact result that allows an accurate statement of contents in a sample), limit of detection, limit of quantization, linearity, range, and robustness. Since compendial test requirements vary from extremely demanding analytical determinations to subjective evaluation of attributes, different test procedures require different validation schemes.

USP ⟨1325⟩ includes only the most common categories of tests for which validation data should be required. These categories are as follows:

- *Category I:* analytical procedures for quantization of major components of bulk drug substances or active ingredients (including preservatives) in finished pharmaceutical products
- *Category II:* analytical procedures (i.e., quantitative assays and limit tests) for determination of impurities in bulk drug substances or degradation compounds in finished pharmaceutical products
- *Category III:* analytical procedures for determination of performance characteristics (e.g., dissolution, drug release)
- *Category IV:* identification tests

Table 5 shows the parameters for method validation as presented in USP-31 ⟨1325⟩.

Note: Already established general procedures (e.g., titrimetric determination of water, bacterial endotoxins) should be revalidated to verify their accuracy (and the absence of possible interference) when used for a new product or raw material.

Following the scheme in Table 2, one may deem that the data elements required for validation required for the TOC method in pharmaceutical-grade water are specificity and detection limit. However, the detection limit is not required because USP ⟨643⟩ demands that the TOC analyzer must have a manufacturer's specified limit of detection of 50 ppb or lower. Nevertheless, for HPW systems we may need to evaluate the limit of quantization because these systems usually have a TOC below 10 ppm. What is more, many PW, HPW, and PWI

TABLE 5 Data Elements Required for Validation of Analytical Methods

Analytical Performance Characteristics	Category I	Category II		Category III	Category IV
		Quantitative	Limit Tests		
Accuracy	Yes	Yes	[a]	[a]	No
Precision	Yes	Yes	No	Yes	No
Specificity	Yes	Yes	Yes	[a]	Yes
Detection limit	No	No	Yes	[a]	No
Quantization limit	No	Yes	No	[a]	No
Linearity	Yes	Yes	No	[a]	No
Range	Yes	Yes	[a]	[a]	No

Source: Data from USP-31 ⟨1325⟩.
[a]May be required, depending on the nature of the specific test (means that this characteristic is normally not evaluated).

systems have TOC concentrations below 50 ppm. The specificity is demonstrated by the use of standards called for in USP ⟨643⟩: USP 1,4-Benzoquinone, USP Sucrose RS, and the water blank.

ICH Topic Q2 (R1) Validation of Analytical Procedures: Text and Methodology [30] This guidance document states that the typical validation characteristics that should be considered during the validation on an analytical procedure are accuracy, precision, repeatability, intermediate precision (means ruggedness in USP ⟨1325⟩), specificity, detection limit, quantization limit, linearity, and range. It also indicates that robustness, although not listed, should be considered at an appropriate stage in the development of the analytical procedure.

All the relevant analytical parameters addressed in ICH Q2 (R1) are included in USP ⟨1325⟩. In fact, if you follow USP ⟨1325⟩, you are taking care of the guidelines in ICH Q2 (R1).

4.4 Compliance Background

In general, the pharmaceutical and biotechnology industries are following an overkill approach for the qualification of laboratory instruments (and maybe production equipment as well). There are several compliance and business factors to consider here:

- Analytical instruments are purchased from a vendor and are rarely customized.
- Business and compliance decisions depend on the truthfulness of analytical results.
- Intended use of the equipment and risk involved.
- In off-the-shelf instruments such as TOC analyzers, hardware and software are interconnected from development (but this does not apply for custom developed software).

What is to be avoided:

- *Overkill approach—Avoid risk.* just do all.
- *Hands-off approach.* Do nothing.
  What we need here is a strategy that is:
- *Comprehensive.* Includes hardware and software together because we use them as one.
- *Realistic/logical.* Does not require overdocumentation or for underdocumentation.
- *Coherent.* Is applied plantwide based on a risk analysis.
- *Inclusive.* Can be applied to any instrument.

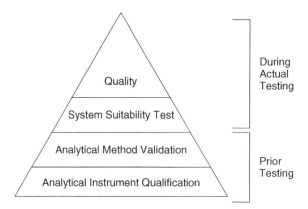

FIGURE 2 Quality hierarchy in AAPS.

4.5 Overview of AAPS Guidance Document

See Figure 2.

1. IQ, OQ, and PQ

   Installation qualification (IQ) is defined as the review and approval of the receipt of an instrument plus any associated user documentation. The IQ also guarantees that the instrument is installed correctly.

   Operational qualification (OQ) is defined as on-site testing and documentation review providing verification that an analytical instrument is functioning as the vendor intended. OQ guarantees correct performance of modules within the instrument.

   Performance qualification (PQ) is defined as on-site testing that verifies reliable and correct operation of the analytical instrument. It also guarantees correct holistic performance and suitability for its intended use.

2. Role of manufacturer

   Provides design qualification (DQ) documents

   Validation of the manufacturing process for the instrument

   Validation of firmware and software

   Provides qualification and validation documents to user

   Provides critical functional test scripts to qualify the instrument at the user's site

   Provides training and support

   Notifies user of defects discovered after delivery

3. Role of user

   Responsible for operation and quality of data

   Assures appropriate training is provided to analysts

Is (are) the best qualified person(s) to design instrument test scripts and requirements or specifications

4. Role of quality assurance

Understanding of the qualification process

Gaining knowledge of instrument's performance by working with users

Reviews qualification process to determine if it meets regulatory requirements and has a valid scientific rationale

5. Software validation

Firmware: validated as a component of the instrument and requires no separate validation activities.

Control, acquisition and processing software: manufacturer validates software following the life-cycle approach and provides validation documents to the user and the user qualifies software as a part of the instrument.

Stand-alone software: validation performed by software developer, based on the FDA's general principles of software validation (2002).

4.6   Overview of GAMP Validation of Laboratory Computerized Systems

- Categories in GAMP-VLCS were derived from the software categories in GAMP 4, Appendix M4.
- The system has an IT viewpoint.
- GAMP-VLCS depends on GAMP 4 for quality practices such as risk management and vendor evaluation.
- Categories are based on risk to data integrity, with secondary prominence on data complexity.
- For custom software it separates the implementation life cycle from the development life cycle.

4.7   Two Strategies: AAPS and GAMP-VLCS

Together they:

- Create categories for analytical instruments based on risks
- Include comprehensive qualification of equipment and software
- Require vendor assessment based on risk analyses
- Use DQ/IQ/OQ/PQ approach for design and testing, as required by a risk analysis
- Are not specific about end-user software (e.g., Excel applications in VBA); imply use of the overkill approach

Relevant items:

- AAPS focuses on hardware.
- GAMP-VLCS leans toward software.
- AAPS is the best choice at this moment. If necessary, GAMP-VLSC can be used as a supplement for AAPS.
- A qualification plan should be based on a risk analysis.

In general, we may say that the fewer qualification activities you include, the more quality control checks you should include during actual test execution. However, quality control checks should be based on the analytical method validation and analytical instrument qualification.

4.8 USP ⟨1058⟩: Analytical Equipment Qualification

USP ⟨1058⟩ is based on the AAPS guidance document. It emphasizes that the thoroughness applied to the qualification process will depend on the complexity and intended use of the instrument. It puts the burden on analytical instrument qualification for the overall process of obtaining reliable data from analytical instruments [31].

4.9 Qualification Strategy

1. The online TOC analyzers and the inline CMs at the PW system will be qualified as laboratory instruments (like desktop TOC analyzers and conductivity meters), although they are installed in the utilities area.
   a. As stated earlier, according to 21 CFR 211.194(a)(2), users of analytical methods described in the USP and NF are not required to validate the accuracy and reliability of these methods, but just verify their suitability under actual conditions of use.
   b. TOC is a USP monograph, and in addition, it is a limit test. Therefore, it is not required to evaluate performance characteristics such as precision, specificity, detection limit, quantization limit, linearity, or range. The only legal requirement is the system suitability test. However, when the TOC values are also intended for a monitoring program, then linearity, limit of quantization, accuracy, and precision should be included in the qualification in order to provide TOC values within a predetermined variation.
2. Performance qualification for online TOC instruments should be performed in parallel with desktop analyzers. Equivalency between off-line and online units will not be pursued; online measurements of TOC (and conductivity) are usually lower than those of desktop units. Although the behavior is usually the same in online and off-line test results, the online results are

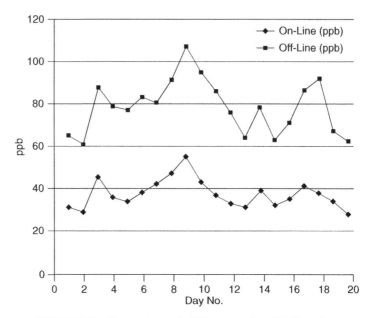

FIGURE 3   Comparison of off-line to online TOC results.

about one-half to one-third the value of the offline results. For additional details, see Table 6 and Figure 3.

PQ activities should take place for 20 to 30 calendar days.

3. Additional sampling points must be installed parallel to the online TOC analyzers. The online TOC analyzers and the new sampling point will be fed by the same water drop. During the PQ stage, PW water samples will be taken from the new sampling points in order to visualize and get a feeling for differences in test results between the two test methods. These

TABLE 6   Example of Off-line vs. Online TOC Results

Day Number	Online (ppb)	Off-line (ppb)	Day Number	Online (ppb)	Off-line (ppb)
1	31	65	11	37	86
2	29	61	12	33	76
3	45	87	13	31	64
4	36	79	14	39	78
5	34	77	15	32	63
6	38	83	16	35	71
7	42	80	17	41	86
8	47	91	18	38	92
9	55	107	19	34	67
10	43	95	20	28	75

new sampling points will be used as a backup procedure in case of the failure of an online TOC analyzer. In that case a sample will be taken and analyzed for TOC at the QC laboratory, as needed.

4. The analytical tests per se will be qualified as limit tests as specified in USP ⟨1325⟩, validation of compendial methods, assay category II (quantitative and limit tests).

5. The intervals for system suitability tests should be established empirically according to approved and separate protocol.

6. Once the online TOC analyzers are qualified, they will replace the off-line TOC tests performed on the sampling and use points.

7. The TOC instrument calibration time interval should be determined.

4.10 Qualification Protocol

Subjects of a Validation Protocol for an Analytical Method
- Statement of purpose and scope
- Responsibilities
- Documented test method
- List of materials and equipment
- Test scripts. Should contain at least the following contents:
  - Make, model, and maker's manual
  - Modifications
  - Installation and operational qualification
  - Maintenance schedules
  - Calibration programs
  - Statistical analysis
  - Acceptance criteria for each performance parameter

Final Summary Report Upon completion of the protocol execution, a final summary report must be written and approved. This final report will consist of a discussion of the results obtained for the various tests contained in the protocol, an evaluation of these results against the corresponding acceptance criteria, discussion of the discrepancies encountered during the protocol execution and their respective resolutions, and a conclusion with regard to the final disposition of the executed protocol.

4.11 Instrument Calibration

TOC instrument calibration time interval is not dealt with in the USP ⟨643⟩ nor in the EP. However, the JP XV 2.59 indicates how to calibrate the instrument. The JP points to calibrating the instrument for OC using a standard solution of potassium hydrogen phthalate and for IC to use a standard solution of either sodium hydrogen carbonate or sodium carbonate decahedra in accordance to instrument's requirements.

USP ⟨643⟩ affirms that a "calibrated instrument" must be used. In real life it means that the instrument's manufacturer must provide the user with a calibration procedure or that the manufacturer should perform the calibration in accordance to their procedure. The calibration time interval must be established by the empirical and theoretical data. The calibration procedure and interval must be included in the qualification plan and protocols. It is strongly recommended that the instrument be calibrated shortly before the qualification activities to avoid any interference due to calibration drift.

4.12 System Suitability Testing

System suitability testing (SST) is an integral part of many analytical procedures. The tests are based on the concept that the equipment, electronics, analytical operations, and samples to be analyzed constitute an integral system that can be evaluated as such. System suitability test parameters to be established for a particular procedure depend on the type of procedure being validated. The most generally applied SST considers the precision of the analysis (i.e., the repeatability); that is, standard deviation must not exceed a predefined value. That is, SST monitors analysis reproducibility, where the standard deviation or relative standard deviation of monitored parameters must not exceed the specified range. The system is then declared suitable only if the response is within given limits.

SST, measured in a response efficiency test, is a requirement of USP-31 ⟨643⟩ and EP 2.2.44 for the determination of TOC. USP-31 ⟨643⟩, Total Organic Carbon, specifies that TOC tests must be performed using a calibrated instrument, and the suitability of the apparatus must be demonstrated periodically. EP Method 2.2.44, Total Organic Carbon in Water for Pharmaceutical Use, is the equivalent of USP-31 ⟨643⟩. The purpose and the value of the SST are based on the premise that TOC measurements are not direct measurements of organic carbon, but they are indirect measurements of other carbon-based chemicals. Since organic carbon appears in various forms in nature and subsequently in water systems, a wide variety of oxidation states and chemical forms are found in water systems. The SST intention is to challenge the instrument by verifying that it responds equally to two types of chemicals (sucrose and 1,4- benzoquinone) that challenge its measurement capability. Sucrose (the standard solution) is theoretically an easy-to-oxidize solution that gives an instrument response at the attribute limit. The analytical technology is qualified by challenging the capability of the instrument using 1,4- benzoquinone (a solution that is theoretically difficult to oxidize) in the system suitability portion of the method. The challenge to the TOC analyzer is to oxidize these two chemicals equally. Because of their quite different chemical structure, both chemicals will challenge the bond-breaking and oxidation capability of the TOC measurement technology. These two chemicals are specified in the USP and EP chapters. USP-31 ⟨643⟩ characterizes the relationship between the instrument's response to these compounds as the instrument's response efficiency:

$$\text{response efficiency} = 100 \left(\frac{r_{ss} - r_w}{r_s - r_w} \right) \tag{1}$$

where r_{ss} = is the instrument response to 500 ppb as C to the 1,4-benzoquinone standard, r_s the instrument response to 500 ppb as C to the sucrose standard, and r_w the instrument response to reagent water used to prepare the standards. The TOC analyzer is suitable if the response efficiency is not less than 85% and not more than 115% of the theoretical response.

USP-30 ⟨643⟩ asserts that the suitability of the apparatus must be demonstrated periodically. Hence, it is of the utmost importance to establish in the validation exercise the time interval for the SST. The best way to find the correct time interval for the SST is by using a matrix. Table 7 shows what a matrix for the SST time interval would look like, and Table 8 presents representative test results. At the end of the first year of using the empirical SST time interval, the data should be evaluated and the frequency adjusted as appropriate.

5 COMMON PROBLEMS AND SOLUTIONS

5.1 Sampling and Handling

Extreme care must be exercised on the preparation of the 500-ppb TOC standards for the SSTs since there are several critical issues to consider. Cleanness of glassware must be maintained thoroughly to decrease contamination. The sucrose standard is sensitive to microbial contamination, which will skew the TOC value. Therefore, standard preparation techniques, storage conditions, and shelf life must be meticulously managed. Additionally, the 1,4-benzoquinone is sensitive to light and must be protected accordingly. Biologically active samples should be analyzed immediately or preserved by freezing.

5.2 Interfering Substances

Halogenated compounds may conduct to a positive bias when the direct conductometric method is used. When they are present, elemental carbon, carbides, cyanides, cyanates, isocyanates, isothiocyanates, and thiocyanates are determined as organic carbon using the methods discussed. In addition, ionic organic substances or organic substances containing nitrogen or sulfur are a cause for concern [32]. Compounds such as nicotinamide, methanol, isopropanol, acetic acid, formic acid, trimethylamine, and chloroform can cause positive or negative bias on TOC analyses, depending on the detection technology used [33].

5.3 Instrumental Background

The quantity and variability of the blank have a major influence on the detection limits and precision. UV, UV–persulfate, and heated-persulfate instrumental

TABLE 7 Matrix for System Suitability Test Verifications

Unit	Day 0	Day 1	Day 3	Day 5	Day 7	Day 14	Day 21	Day 28
A	Calibration	Response efficiency	—	Response efficiency	—	Response efficiency	—	—
B	Calibration	—	Response efficiency	—	Response efficiency	—	Response efficiency	—
C	Calibration	—	—	Response efficiency	—	Response efficiency	—	Response efficiency

TABLE 8 Example of Results for System Suitability Test Verifications

Unit	Day 0	Day 1	Day 3	Day 5	Day 7	Day 14	Day 21	Day 28
A	Passed	95.0%	—	95.7%	—	93.7%	—	—
B	Passed	—	94.8%	—	94.7%	—	105.7%	—
C	Passed	—	—	100.4%	—	91.7%	—	97.2%

backgrounds are smaller in proportion to the combustion analysis. In fact, this is referenced in Standard Method 5310, which states that the HTCC methods accumulate nonvolatile residues in the analyzer, and this accumulating effect on the blank yields a shifting unstable blank; in the UV–persulfate or heated-persulfate oxidation methods, reaction residuals are drained from the analyzer. The latter oxidation methods generally provide better sensitivity for lower-level samples (<1 mg/L) [34].

5.4 Shifting Through Zero

Given that it is impossible to obtain reagent water that is carbon-free, a correction in the calibration curve must be made when measuring samples. Since the standards are prepared with the same water that is used as the reagent water, the same correction must be made for the standards as the reagent water. In this correction, known as *shifting through zero*, the entire curve is moved down proportional to the area counts of the reagent water [35].

Shifting through zero should not be mistaken with forcing the intercept through zero. When a curve is forced through zero, the entire curve does not move down. The curve is "pivoted" to run through the origin, thus changing the slope of the curve. This method does not consider the carbon concentration in the reagent water and will yield false low results.

When a standard is tested as an unknown sample using a curve that is shifted through zero, the concentration of carbon in the reagent water must be subtracted from the result. That is, the same principle that has been applied to the standards in the calibration curve is applied to the unknown. The standard concentration, which has been spiked with an amount of stock solution appropriate to give a concentration of 500 ppb will actually have a concentration of 500 ppb plus the concentration of the reagent water. When the calibration curve is shifted through zero, the reagent water is removed from the calibration. While shifting through zero accounts for the amount of carbon in the reagent water, it does not account for the instrument background [35].

REFERENCES

1. T. Karanfil, I. Erdogan, and M. A. Schlautman. (2003). Selecting filter membranes for Measuring DOC and UV254. *J. Am. Water Works Assoc.*, 95(3):86–100, 2003.

2. B. Wallace, M. Purcell, and J. Furlong. Total organic carbon analysis as a precursor to disinfection byproducts in potable water: oxidation technique considerations. *J. Environ. Monit.*, 4:35–42, 2002.

3. *Disinfectants and Disinfection Byproducts Rule (Stage 2 DBPR) Implementation Guidance*. EPA 816-D-002. EPA, Washington, DC, 1998. Available at www.epa.gov/OGWDW/mdbp/dbp1.html. Accessed April 29, 2010.

4. S. Kim, G. H. Lee, and K. J. Lee. Monitoring and characterization of bacterial contamination in a high-purity water system used for semiconductor manufacturing. *J. Microbiol.*, 38(2):99–104, 2000.

5. R. A. Governal, C. Gerba, and F. Shadman. Characterization of organic impurities in high-purity water systems by AOC and TOC. *Ultrapure Water*, pp. 19–24, Apr. 1993.

6. W. Harned. (1986). Bacteria as particle source in wafer processing equipment. *J. Environ. Sci.*, pp. 32–34, May–June 1986.

7. I. S. Kim, S. Kim, and J. Hwang. Nutritional flexibility of oligotrophic and copiotrophic bacteria isolated from deionized-ultrapure water made by high-purity water manufacturing system in a semiconductor manufacturing company. *J. Microbiol. Biotechnol.*, 7:200–203, 1997.

8. W. F. Harfst. Back to basics: fundamentals in microbiology for high-purity water treatment. *Ultrapure Water*, pp. 33–35, July–Aug. 1992.

9. G. R. Husted, and A. A. Rutkowski. Microbials: response of oligotrophic biofilm bacteria in high-purity water system to stepwise nutrient supplementation. *Ultrapure Water*, pp. 43–50, Sept. 1994.

10. J. Martyak, J. Carmody, and G. R. Husted. Characterizing biofilm growth in deionized ultrapure water piping systems. *Microcontaminations*, 11:39–44, 1993.

11. A. Pacquiao-Sincero, Sr., and G. Alivio-Sincero. *Physical–Chemical Treatment of Water and Wastewater*. CRC Press, Boca Raton, FL, 2003.

12. Determination of total carbon in liquid samples. U.S. patent 43,44,918, issued Aug. 17, 1982. Available at www.patentstorm.us/patents/4344918/fulltext.html. Accessed April 29, 2010.

13. R. Benner and M. A. Storm. Critical evaluation of the analytical blank associated with DOC measurements by high-temperature catalytic oxidation. *Mar. Chem.*, 41:153–160, 1993.

14. R. M. Carlson. Automated separation and conductimetric determination of ammonia and dissolved carbon dioxide. *Anal. Chem.*, 50:1528–1531, 1978.

15. W. Yanlin, and L. Jinming. Determination methods for total carbon concentration in water samples. *Chem. J. Internet*, 6(1):6, 2004. Available at www.chemistrymag.org/cji/2004/061006re.htm. Accessed April 29, 2010.

16. R. M. Emery, E. B. Welch, and R. F. Christman. The TOC analyzer and its application to water research. *J. Water Pollut. Control Fed.*, 43:1834, 1971.

17. *Theory and Operation of NDIR Detectors*. Technical Note TN-169 rev 1. Available at www.raesystems.com/~raedocs/App_Tech_Notes/Tech_Notes/TN-169_NDIR_CO2_Theory.pdf. Accessed April 29, 2010.

18. E. T. Heggs, E. K. Price, and S. R. Proffitt. CO_2 measurements for TOC analysis using static pressure reading of an NDIR. U.S. patent application. Available at www.faqs.org/patents/app/20080198381. Accessed April 29, 2010.

19. L. Lawson. *Using a Single Calibration Curve to Analyze a Wide Range of TOC Samples*. Teledyne Tekmar, Mason, OH. Available at http://www.schmidlin-lab.ch/pdf/Tekmar/TOC_F-001.pdf. Accessed April 29, 2010.

20. J. E. Martínez-Rosa. On-line TOC analyzers are underused PAT tools. *BioProcess Int.*, 2(11):30–38, 2004.

21. G. Aiken, L. A. Kaplan, and J. Weishaar. Assessment of relative accuracy in the determination of organic matter concentrations in aquatic systems. *J. Environ. Monit.*, 4:70–74, 2002.

22. B. Wallace, M. Purcell, and J. Furlong. Total organic carbon analysis as a precursor to disinfection byproducts in potable water: oxidation technique considerations. *J. Environ. Monit.*, 4:35–42, 2002.

23. S. K. Bansal, et al. *Qualification of Analytical Instruments for Use in the Pharmaceutical Industry: A Scientific Approach*. American Association of Pharmaceutical Scientists, Arlington, VA, 2004.

24. *GAMP Good Practice Guide: Validation of Laboratory Computerized Systems*. International Society for Pharmaceutical Engineering, Tampa, FL, 2005.

25. *U.S. Pharmacopeia*, Chapter ⟨643⟩, Total Organic Carbon. USP 31–NF 26. USP, Rockville, MD, 2008.

26. *European Pharmacopoeia* 4, Method 2.2.44, Total Organic Carbon in Water for Pharmaceutical Use, European Pharmacopoeial Commission, London July 1, 1999.

27. *2.59 Test for Total Organic Carbon. General Tests. Japanese Pharmacopeia XV*, April 1, 2006. Available at http://jpdb.nihs.go.jp/jp15e/. Accessed April 29, 2010.

28. *U.S. Pharmacopeia*, Chapter ⟨1325⟩, Validation of Compendial Procedures. USP 31–NF 26. USP, Rockville, MD, 2008.

29. Porter. D. V. Qualification, validation, and verification. *Pharm. Technol.*, Apr. 2007. Available at pharmtech.findpharma.com/pharmtech/QC%2FQA/Qualification-Validation-and-Verification/ArticleStandard/Article/detail/415111. Accessed April 29, 2010.

30. *Validation of Analytical Procedures: Text and Methodology*. CPMP/ICH/381/95. European Medicines Agency, London, June 1995.

31. *U.S. Pharmacopeia*, Chapter ⟨1058⟩, Analytical Equipment Qualification. USP 31–NF 26. USP, Rockville, MD, 2008.

32. S. Kojima. *Lost in Translation with On-line TOC*. Japan Pharmaceuticals and Medical Devices Agency, Tokyo, 2007.

33. R. A. Godec. Science based performance comparison of on-line TOC analyzers. GE, Water & Process Technologies Analytical Instruments. May, 2006. Available at www.pharmamanufacturing.com/whitepapers/2006/043.html. Accessed April 29, 2010.

34. *Standard Methods for the Examination of Water and Wastewater*, 19th ed. suppl. American Public Health Association, American Water Works Association, and Water Environment Federation, Washington, DC, 1996.

35. M. A. Burns, R. Clifford, and J. Strait. *Determining the Correct Concentrations of TOC in Purified Water and Water for Injection*. Shimadzu Scientific Instruments, Columbia, MD, 1998. Available at www2.shimadzu.com/apps/appnotes/app22.pdf. Accessed April 29, 2010.

14

INSTRUMENT PERFORMANCE VERIFICATION: MICROPIPETTES

GEORGE RODRIGUES AND RICHARD CURTIS

Artel, Inc.

1 INTRODUCTION

Handheld micropipettes are among the most common pieces of calibrated equipment in the typical life sciences laboratory. More than 1 million of these instruments are sold worldwide each year, with about 90% of them being operated manually. A growing number of motorized electronic micropipettes are also sold (approaching 100,000 per year) and the number of features available and the complexity of these instruments are increasing. With a life cycle that can exceed 20 years in some applications, the total number of active-duty micropipettes in the world is impressive. The inventories of active-duty micropipettes at a large pharmaceutical campus can exceed 10,000 pieces, each of which must be controlled, maintained, calibrated, and verified on an appropriate schedule.

The primary use of a micropipette is to aspirate and dispense a defined volume of liquid accurately, and there are many applications for these quantitative uses. Most often, the volume of liquid is measured in units of microliters (cubic millimeters) or sometimes in milliliters (cubic centimeters). Macropipettes such as volumetric glass transfer pipettes are usually denominated in units of milliliters, whereas micropipettes are typically specified in units of microliters.

In addition to quantitative uses, micropipettes are sometimes used for nonquantitative purposes. These include sample collection, removing supernatant,

Practical Approaches to Method Validation and Essential Instrument Qualification,
Edited by Chung Chow Chan, Herman Lam, and Xue Ming Zhang
Copyright © 2010 John Wiley & Sons, Inc.

and mixing contents in a tube. Validation of these nonquantitative uses is outside the scope of this chapter. Throughout the chapter the term *macropipette* is used to describe glass and plastic volumetric devices, and the terms *micropipette* and *pipette* are used to mean modern piston-operated volumetric devices.

2 SCOPE OF THE CHAPTER

In this chapter we address the need for, and methods of, verification of handheld piston-operated mechanical action pipettes, including manually operated, motorized, air-displacement, and positive-displacement devices. Not within the scope of this chapter are other macropipette devices, typically constructed of glass or single-use disposable plastic pipettes: single-volume transfer pipettes such as ASTM (American Society of Testing and Materials) class A glass, graduated glass or disposable plastic pipettes, and simple unmarked transfer devices such as Pasteur pipettes.

2.1 Background Information and Pipette Description

Micropipettes can be quite diverse in design and functional features. However, the most commonly encountered micropipette is the air-displacement, manual-action pipette shown in Figure 1. This device seems simple in operation. The thumb-operated plunger causes linear motion of a piston, which pushes air out of the device. When the thumb is released, a spring returns the piston to the top of its stroke, and air is moved back into the body of the pipette. It is this displacement of air that couples the piston to the liquid being aspirated or dispensed by the pipette. Most modern air-displacement pipettes are also equipped with a blowout. This is an additional "extra travel" at the end of the piston stroke that permits the operator to expel the last droplet of liquid that remains in the tip after dispensing.

Some pipettes are built as fixed-volume devices which are used to deliver one and only one volume of liquid. These pipettes will be marked to indicate the fixed or nominal volume. In the early history of the micropipette the fixed-volume version predominated, but the majority of pipettes sold today are of the variable-volume type. Adjustable pipettes will have some sort of mechanism (e.g., a threaded screw) that permits the operator to change the volume. An odometer-style mechanical readout or electronic display shows the volume setpoint. Adjusting the volume is accomplished by turning the threaded screw, which alters the limits of travel for the piston.

Pipettes are adjusted during calibration so that the volume of liquid delivered by the pipette matches the volume indicated. The adjustment mechanism can be thought of as a zero or offset adjustment which adds or subtracts a constant number of microliters from all volume settings in the pipette. Typically, there is no second adjustment or span setting, so a variable-volume pipette must be adjusted at one volume and then checked at other volumes. Because most pipettes lack a span adjustment, they cannot be corrected for nonlinearity, so pipettes

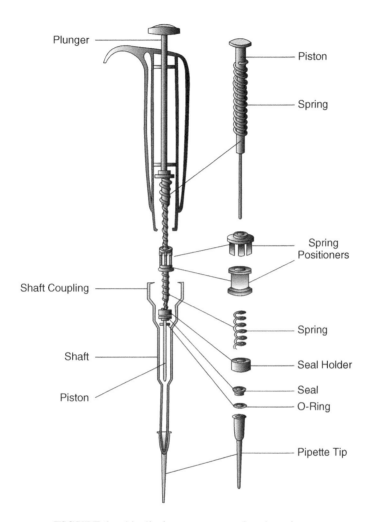

FIGURE 1 Air-displacement manual action pipette.

that display a significant nonlinearity (a variation in accuracy that changes with volume setting) must be repaired to correct the nonlinearity.

Air-displacement pipettes are typically calibrated to deliver accurately in the forward mode. Figure 2 shows a forward-mode pipetting cycle for an air-displacement pipette with blowout. The pipette is prepared for aspiration by depressing the plunger to the first stop and then, while holding the plunger at the first stop, the pipette tip is placed into the liquid in a source vessel. While keeping the end of the pipette at a constant immersion depth below the surface of the liquid, the plunger is released smoothly and liquid is aspirated into the tip. After filling, the pipette and tip are withdrawn from the source vessel and the tip should contain a volume of liquid equal to the pipette's indicated volume.

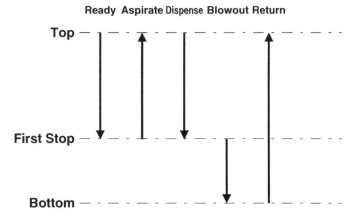

FIGURE 2   Forward-mode pipetting with blowout.

To dispense the liquid contained, the pipette tip is positioned into a receiving vessel and the plunger is depressed to the first stop, then to the bottom stop, where the blowout volume of air works to expel the last droplet from the pipette tip. Returning the plunger to the top stop completes the forward-mode pipetting cycle.

The alternative to forward-mode operation is reverse mode, and some operators have been trained to use this technique. Any pipette that has a blowout can be operated in either the forward or reverse mode. As shown in Figure 3, reverse-mode pipetting begins by pressing the plunger all the way to the bottom stop, then aspirating a full volume, which contains both the setpoint volume and the volume of the blowout. Delivery is made by depressing the plunger to the first stop. In reverse-mode pipetting, a volume of liquid will remain in the tip after dispensing. This volume is approximately equal to the blowout volume of the

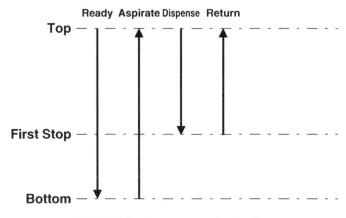

FIGURE 3   Reverse a mode pipetting.

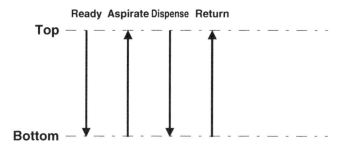

FIGURE 4 Pipetting with no blowout.

pipette and may be discarded when changing tips or can be left in the tip for subsequent reverse-mode pipettings.

It is important to note that pipettes are almost always calibrated to deliver correctly in the forward mode. When using the reverse mode and delivering liquid by "touching off" against glass, pipettes that have been calibrated in forward mode will tend to overdeliver (deliver slightly more liquid) when operated in the reverse mode. For accurate laboratory work the pipetting mode should be specified in work instructions, and each pipette should be calibrated or verified according to the mode in which it will be used.

Some air-displacement pipettes are constructed and sold without a blowout. The operation of these pipettes is simplified as shown in Figure 4. When using a pipette that does not contain a blowout, there is only one mode of operation, and depending on the fluid type and the materials used in construction of the receiving vessel (i.e., glass, plastic, etc.), it is common that a small droplet will remain in the pipette tip after use.

Single-Channel Manual-Action Air-Displacement Pipettes   Because the air-displacement pipette uses air as a mechanical coupling between the piston and the fluid being pipetted, environmental factors that affect the properties of this "captive air" can influence the pipette and cause it to deliver too much or too little fluid. A mathematical description of this effect is included in this chapter, and Table 1 is a summary of how the volume delivered changes under various conditions.

Additional information on these effects is available [1]. When variable-volume air-displacement pipettes are used near their maximum volume, these effects are reduced. Conversely, using pipettes at their minimum volume setting (typically, 10% of maximum setting) increases the ratio of captive air to liquid and renders the pipette more susceptible to environmental influences.

To avoid cross-contamination between samples, modern pipettes use a replaceable plastic tip that is fitted to the end of the pipette, and this tip is the only part of an air-displacement pipette that should come into direct contact with the liquid. When a fresh tip is fitted to a pipette, it is common laboratory practice to prewet the tip prior to dispensing. Prewetting is usually accomplished by

TABLE 1   Effect of Pipetting Conditions on Volume Delivered

Pipetting Conditions	Effect on Volume
Decreased barometric pressure	Decreases
Dry air environment	Decreases
Pipetting hot liquids	Decreases
High-vapor-pressure liquids	Decreases
Pipetting cold liquids	Increases
Very humid	Increases
Reverse-mode pipetting	Increases
Prewetting of tip	Increases

aspirating liquid and dispensing it back into the source vessel. Up to five cycles of prewetting are recommended prior to beginning quantitative pipetting [2].

Prewetting helps the air volume inside the pipette to come into equilibrium with the vapor pressure of the liquid being pipetted. This reduces the tendency of liquid to evaporate into the pipette tip during the aspiration cycle. Prewetting also conditions the interior surface of the pipette tip and makes a series of pipette deliveries more consistent. Some laboratory technicians are accustomed to using the prewetting technique, and some procedures call for prewetting whereas others do not. It is worth noting that most pipettes are calibrated using procedures that include prewetting, which increases the delivered volume. Pipettes that have been calibrated with prewetting but are used without prewetting may be delivering less than the volume expected. This is another example of why it is good practice to calibrate or verify pipettes in the laboratory environment and in accordance with the technique that is actually employed by the user.

The quality of the disposable plastic tips used with the pipette is another variable in the pipetting process. Figure 5 shows photomicrographs of two different tips. The tip on the left is a high-quality tip that is free of flashing and has uniform wall thickness. The tip on the right contains flashing inside the tip that may retain liquid droplets and will also affect the hydrodynamics of the pipetting process. Poor-quality tips will be less precise and accurate. For the best accuracy and precision in pipetting, racked tips from a reputable manufacturer should be used. When other tips are used (such as for particular specialty applications or for cost considerations) the performance of the pipette with the alternative tips should be verified.

In contrast to air-displacement pipettes, positive-displacement pipettes are designed to operate with the piston in direct contact with the liquid being handled. Positive-displacement pipettes come in two basic types: repeater pipettes and disposable-tip nonrepeating pipettes. Positive-displacement pipettes are similar in operation to an air-displacement pipette, but a plastic piston is located inside the plastic tip and is designed to remain in contact with the fluid being pipetted. The tips may be disposable or reusable, depending on the model.

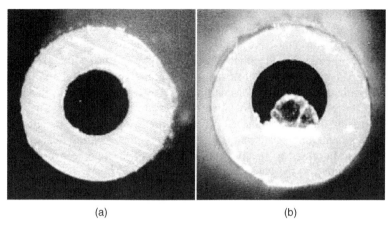

(a)   (b)

FIGURE 5 Working end of two pipette tips: (a) the tip is of good quality; (b) the tip has an opening that is off center and contains interior flashing that will degrade pipette performance. (Courtesy of H. Ulrich.)

Because they do not have an internal captive air volume, positive-displacement pipettes are less sensitive to environmental factors than are their air-displacement counterparts. However, they have a few attributes that limit their acceptance. The consumable tips for positive-displacement pipettes are significantly more expensive than their air-displacement counterparts and under the best environmental conditions when handling a nonviscous aqueous liquid a positive-displacement pipette tends to be less precise than an air-displacement model.

The repeating positive-displacement repeater pipette is designed to dispense multiple aliquots of the same solution. To use a repeater pipette, sufficient solution is first aspirated, then liquid is dispensed by successive depressions of the plunger or thumb lever. Figure 6, shows the motion of the piston in repeater

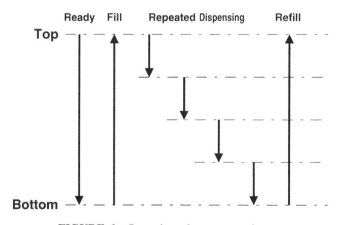

FIGURE 6 Operation of a repeater pipette.

pipetting. Some models of repeater pipettes will deliver inaccurately with the first and last quantities dispensed from a filled tip. Newer models with improved design are available that have reduced or eliminated this source of error. Users of repeater pipettes should either discard the first and last aliquots when using repeater pipettes, or verify that their instrument delivers the first and last "shots" accurately under their particular conditions of use.

Motorized electronically controlled pipettes are also available. These pipettes are constructed on the same basic operating principle as manual pipettes but contain a battery and an electric motor to operate the piston plunger. They also have an electronic interface to receive commands from the operator and control the motor. In addition to the basic pipetting modes (forward and reverse), electronic pipettes may have additional modes, such as repeat, dilution, and mix functions. To use electronic pipettes correctly, the instruction manual should be followed carefully.

Multichannel pipettes are able to handle multiple aliquots at the same time. The multiple tips are usually arranged in a single line, with 8- and 12-channel models being the most common. These pipettes are most often used to increase speed and reduce repetition when working in 96- and 384-well microtiter plates. Verification of multichannel pipettes should be performed in the same way as for single-channel pipettes, with each channel being evaluated for accuracy and precision.

Most pipettes are sold in a calibrated condition as shipped from the factory, but there are many good reasons that the accuracy and precision of pipettes should be verified in the user's own laboratory. Pipettes are sensitive to local environment, and calibration needs to be verified at the temperature, humidity, and barometric pressure in which the pipette is actually used. Pipette performance also depends on the particular choice of pipetting technique (forward or reverse mode, prewetting or not prewetting, etc.) and on the design and quality of the tips that are used. So pipettes should be verified using the technique that will be used routinely in the laboratory. Finally, pipette performance is very much dependent on operator skill, so in-lab verification provides an opportunity to ensure that the laboratory skill level is checked at the same time that the pipette is verified.

Physical and Mathematical Description of an Air-Displacement Pipette   By definition, the liquid aspirated into an air-displacement pipette is separated from the plunger by a cushion of air, whose purpose it is to prevent carryover of sample materials from one aspiration to the next as well as to prevent the plunger and seals from being degraded by contact with the sample material. This cushion of air acts to couple the pipette's plunger to the liquid and allows the use of a low-cost plastic tip that can be disposed of instead of being washed, thus giving good ease of use and overall economy as well as assuring lack of carryover from one delivery to the next. This design has proven popular over the years; however, it can present unexpected sources of error for the unwary, especially when attempting to make highly accurate deliveries. Typically, someone concerned with validating or calibrating an air-displacement pipette is the most affected, since they are concerned with highly accurate and repeatable deliveries.

The air cushion (also known as *captive air* or *dead air*) is trapped within the pipette as soon as the tip of the pipette is immersed into the source liquid. The number of molecules of argon, oxygen, and nitrogen in that trapped air remain fixed as the liquid is aspirated by withdrawal of the plunger. To first approximation, the volume of air is expected to remain constant, so when the plunger is withdrawn by a swept volume V_s, we expect that the volume of liquid aspirated, V_l, will be equal, or $V_l = V_s$. In reality, the volume of captive air does not remain constant, as we will seen below, leading to an error in the amount of liquid aspirated.

The captive air trapped in the pipette closely obeys the ideal gas law:

$$P_a V_a = n_a R T_a \tag{1}$$

If the number of moles of gas, n_a, the temperature of that gas, T_a (absolute kelvin) and its pressure, P_a, all remain constant during the aspiration, its volume, V_a, will also remain constant. In practice, none of those quantities remain constant. For example, if the pipette is not in temperature equilibrium with the ambient air, the trapped air's temperature will be changing to come into equilibrium with the pipette. If the liquid being aspirated is at yet a third temperature, the trapped air is trying to come into equilibrium with it as well. As we will see below, a seeming miniscule change in the temperature of the trapped air by $0.1°C$ during aspiration will lead to enough error so that the pipette performs outside the manufacturer's specification.

The number of moles of gas trapped in the captive air space would seem to be constant. The amount of argon, oxygen, and nitrogen is indeed fixed; however, if the liquid can evaporate from the upper surface of the meniscus, those evaporated molecules of whatever is being aspirated go into that captive air space, leading to an increase in the number of moles of gas trapped there. Again, it would seem to be an insignificant factor, but as we will see below, the amount of error caused by this can easily exceed the manufacturer's specification for the device being calibrated.

Even the pressure of the trapped air will change during aspiration. The most obvious cause is the weight of the column of liquid that has been raised above the surface of liquid in the source vessel. A partial vacuum, relative to pressure outside the tip, is required to hold the column of liquid in the tip. This change in pressure leads to less liquid being aspirated into the tip than expected. This effect is greatest when pipetting relatively large volumes (e.g., 1000 μL). At smaller volumes, the surface tension of the liquid and its contact angle against the plastic of the tip material will tend to alter the pressure of the trapped air, again causing an error in results.

Typically, the effect of changes in pressure are compensated when the device is calibrated, minimizing error in subsequent liquid delivery using that pipette. However, this calibration is only valid for a given barometric pressure. Since barometric pressure changes with altitude above sea level, pipettes should be calibrated at the elevation where they are to be used.The effects of evaporation within the tip, and of temperature disequilibrium, can be minimized as well by carefully adhering to recommended calibration protocols.

By analyzing the ideal gas law equation [Eq. (1)] we can estimate the magnitude of the errors likely to occur due to each of the three causes discussed above. Consider the case in which the temperature of the trapped air changes from T_a at the start of aspiration (when the tip is inserted into the liquid in the source vessel) to T_f at the end (when the tip is removed from the source vessel), but neither P_a nor n_a change from the initial conditions. The volume of liquid aspirated, V_l, differs from the volume swept by the movement of the plunger, V_s, by the change in trapped air volume:

$$V_l = V_s + (V_a - V_f) \tag{2}$$

Using the ideal gas law gives us

$$V_l = V_s + \left(\frac{n_a R T_a}{P_a} - \frac{n_f R T_f}{P_f}\right) \tag{3}$$

In this case we are assuming that neither P_a nor n_a change, giving

$$V_l = V_s + \left(\frac{n_a R T_a}{P_a} - \frac{n_a R T_f}{P_a}\right)$$

$$= V_s + \frac{n_a R}{P_a}(T_a - T_f) \tag{4}$$

The relative error in liquid delivery due to a temperature discrepancy, e_T, is the difference between the aspirated liquid volume V_l and the swept volume V_s divided by the delivered liquid volume V_l:

$$e_T = \frac{V_l - V_s}{V_l}$$

$$= \frac{n_a R}{P_a} \frac{T_a - T_f}{V_l}$$

$$= \frac{V_a}{V_l} \frac{T_a - T_f}{T_a}$$

$$= \frac{V_a}{V_l} \frac{\Delta T}{T_a} \tag{5}$$

The error introduced by a change in temperature during aspiration is proportional to the relative temperature change times a factor equal to the ratio of captive air volume divided by liquid volume. This factor is a troublemaker, as it can be very large.

To take an example, if a certain brand and model of 20-µL pipette is used to deliver 2 µL, the ratio of captive air to liquid delivered is 49 (measured value). If

ΔT is $0.1°C$ and the absolute temperature is 293 K, the error in delivery volume due to temperature dis-equilibrium is

$$e_T = 49 \left(\frac{0.1}{293} \right) = 1.67\% \tag{6}$$

This example illustrates the significant effect that a seemingly insignificant lack of temperature homogeneity can have on results. This is why protocols for calibration of pipettes typically call for at least 2 hours of temperature equilibrium before calibration and temperature stability of the ambient within $0.5°C$. Failure to observe these conditions is a frequent cause of calibration or validation failures. It must be noted that since pipettes in a working laboratory are almost never used under ideal conditions of thermal equilibrium, the actual delivery volume from a pipette under typical laboratory conditions is likely to be in error from this cause.

In a similar way, the error in liquid delivery due to evaporation of liquid into the captive air space e_n and the error due to changes in pressure e_p are

$$e_n = \frac{V_a}{V_l} \frac{\Delta n}{n_a} \tag{7}$$

$$e_p = \frac{V_a}{V_l} \frac{\Delta P}{P_f} \tag{8}$$

These results also contain the ratio of air to liquid volumes, V_a / V_l. Some typical values of this ratio are given in Table 2 for different pipette sizes and settings of delivery volume. Each brand and model of pipette has a different captive air volume, a fact that should be taken into account in designing assay protocols. If a protocol calls for delivering 2 μL of liquid from a container at $60°C$, it makes a big difference whether the pipette used is a 2-μL pipette set for 2 μL or a 20-μL pipette set for 2 μL. It is likely that there would be a 5 or 10% difference in volume delivered, even though both pipettes were well calibrated and functioning correctly. The protocol should at least specify the size of pipette to be used, and preferably even the make and model, to minimize errors of this sort.

TABLE 2 Typical Values for Captive Air Ratios

Pipette Nominal (μL)	Volume Setting (μL)	Captive Air (μL)	Ratio
1000	1000	1500	1.5
1000	100	1500	15
20	20	100	5
20	2	100	50
2	2	40	20
2	0.2	40	200

The captive air volume may also vary depending on the make and model of disposable tip being used, so the pipette should either be calibrated with the same tips being used in the laboratory, or the laboratory should verify that changing the tip model does not significantly alter accuracy or precision.

3 VERIFICATION PRACTICES: VOLUME SETTINGS, NUMBER OF REPLICATES, AND TIPS

Performance verification is accomplished by using the micropipette to dispense a series of aliquots into an appropriate volume-measuring system. Standard practice is to dispense 10 replicates for the evaluation of both accuracy and precision, or four replicates when performing a "quick check" test for accuracy only [3]. Micropipettes of the fixed-volume type are evaluated at their nominal volume, while variable-volume pipettes should be verified at both the nominal volume (100% of full scale) and minimum settable volume (typically 10% of nominal, but can range from 2 to 25% of nominal volume).

To improve accuracy and precision when using a variable-volume pipette at the lower limit of its range, some laboratories will place artificial restrictions on the minimum permissible setting (e.g., 20% of nominal). In this situation, pipettes should be verified at 100% of nominal and at the lower limit of the restriction. Pipettes tested in this way should be marked (e.g., with a restricted-use sticker) to indicate the calibrated and allowable range of operation.

Because pipettes are sensitive to environmental conditions, operator technique, and user skill, it is recommended that pipettes be verified in the laboratory environment in which they are used and that standard laboratory technique be used during the verification. In addition, pipette performance depends on the quality of the disposable tip, so pipettes should be verified using the particular model of tip that is used in each laboratory. The methods most commonly employed for measurements of microliter volumes are the gravimetric and photometric methods.

3.1 Gravimetry

The gravimetric method uses a balance to weigh liquid volumes. The method seems simple and involves dispensing individual aliquots into a receiving vessel and recording the weight change. The balance reports a weight gain, and that weight is converted to mass and then to volume using conversion factors which may be found in tables, calculated from formulas, or produced by software packages. The gravimetric method is described in detail in consensus standards such as ASTM E 1154 and the International Standards Organisation's ISO 8655-6. The methods in these standards are similar, but there are some differences in the details of the procedures. Both methods give similar results at volumes above 100 µL, while some bias between methods can be observed at smaller volumes [4].

Gravimetry has several advantages, including the wide availability of weighing devices in most laboratories and a negligible cost of liquids (usually deionized water) used in the testing. In addition, gravimetry is a well-accepted technology. Gravimetry is frequently the method of choice for measuring device performance when handling larger volumes. For example, a 1000-μL aliquot weighs approximately 1g and can be weighed quite reliably on a modern laboratory analytical balance. However, as volumes decrease in the microliter range, weighing becomes more challenging, for several reasons. First, measuring microliter volumes requires more specialized balances (producing measurement results to five or six decimal places on the gram scale). Such balances are delicate, require a stable platform to limit vibration, and are not as portable as the less sensitive models used for measuring larger liquid volumes. Illustrating the need for sensitivity, both ASTM E 1154 and ISO 8655-6 standards require that volumes of 10 μL or smaller be measured on a six-place (microgram) balance.

Because microgram balances take some time to settle, gravimetric measurements of small volumes can be time consuming. In addition, gravimetry is affected by a variety of environmental conditions, including evaporation, static electricity, and vibration; and as volumes become smaller, these error sources become more significant. Underscoring the need for attention to detail at small volumes, modern air-displacement pipettes can deliver volumes as small as 0.1 or 0.2 μL, volumes so small that they evaporate completely in less than 1 minute. Evaporation traps of varying effectiveness are available to help reduce evaporation and stabilize the sample, and it is customary for pipette calibration laboratories to work under high humidity to further reduce evaporation. However, laboratory verification of pipettes should be conducted under normal humidity conditions, and regardless of efforts to reduce evaporation, it is important to measure the actual evaporation rate and correct for the resulting volume variation.

Electrostatic effects also cause some uncertainty with gravimetric methods because plastic pipette tips are typically used to transfer liquids. Static electricity, which is imparted to the balance pan, draft shield, or evaporation trap, induce a force that changes the apparent weight. When working with small volumes, the error due to static electricity can be significant. Vibration must also be controlled for, and this often requires making measurements on a solid-marble weighing table.

Because gravimetric measurements calculate volume by converting weight to mass and then to volume, accurate measurement is contingent on knowing the density of the fluid being pipetted. The density of water varies with temperature and at room temperature is always less than 1 g/mL, its commonly accepted value. Since balances are calibrated using metal weights (usually stainless steel), there is also an air buoyancy correction that must be made when weighing less dense materials, such as water.

These details need to be accounted for if very precise measurements are required. Consider a device with accuracy specifications of better than 0.6%, which is a typical specification for high-accuracy pipetting of 1000 μL. Failure to correct for air buoyancy in the weighing and liquid density of the aliquot when pipetting water can lead to error in the 0.3 to 0.5% range, which is nearly as large as the acceptable error for the entire piece of equipment. Fortunately, the calculations needed for the gravimetric method are spelled out in detail in the ASTM and ISO standards, and software is available to automate these calculations.

In summary, gravimetric calibration is best suited for measuring the performance of pipettes at larger liquid volumes, usually above 200 to 1000 μL (the precise lower limit for effective use of gravimetry depends on the tightness of the tolerance to be met and the quality of the measuring equipment and procedure employed).

3.2   Photometry

The photometric method is described in outline form in the body of ISO 8655, part 7 [5], and has become popular due to the availability of commercial systems provided by MLA and Artel. The photometric method is based on using the pipette to deliver an unknown volume of a dye containing a known concentration of a chromophore dye molecule. The absorbance change in the receiving vessel is measured and a mass balance calculation is performed to determine the volume of the aliquot delivered. When dyes are chosen and calibrated properly, the photometric method is capable of a standard uncertainty of measurement of 1% or better, regardless of volume. Photometric calibration requires a photometer and stable dyes that absorb light in the visible or ultraviolet range.

Single-Dye Photometry   The simplest form of volume measurement photometry is single-dye photometry. In this approach, a dye solution is delivered into a cuvette, a measuring cell, or a clear-bottomed microtiter plate and is mixed with a known volume of colorless diluent. A beam of light at a specified wavelength is passed through the solution, and the photometer measures the quantity of light that passes through. The amount of light that is absorbed is proportional to the concentration of dye present, and a mass balance equation is used to calculate aliquot volume. An example of single-dye photometry is given in Annex B of ISO 8655-7.

The photometric method produces excellent measurements of precision and is generally less sensitive than gravimetric methods to environmental conditions. Another benefit of photometric calibration methods is the ability to provide precision information about each channel in a multichannel device. Absorbance dyes which are readily available and commonly used for this application include tartazine and potassium dichromate. There is also a commercially available single-dye method for single-channel pipettes sold by MLA, which is commonly used in the clinical laboratory industry [6]. However, the authors are unaware of any validated use of the MLA system within the FDA-regulated pharmaceutical or in vitro diagnostics industry.

ISO 8655-7 recognizes the use of single-dye photometry for liquid-handling device calibration. However, according to this standard, photometric methods should be accompanied by an uncertainty analysis that describes the measurement uncertainty. This analysis may include such error contributions as accuracy of the photometer and reagents, dye instability, deviation from ideal Beer's law behavior, and the like. To account for the dyes as a source of error, data on the stability of the dye, either from the manufacturer or developed in-house through a stability or validation study, is important. Because light is passed through the sample and an optical wall, the optical quality of the microtiter plate or cuvette used in the method can affect the accuracy and precision of the measurement, and laboratories must also account for this.

Photometric methods must be properly standardized to obtain quantitative results for accuracy measurements. The traceability of the method depends on many factors, including how carefully the standardization is carried out. For traceable photometric readings, a standard curve must be developed by using a known liquid delivery device (calibrated pipette) or by weighing volumes. This process can be time consuming and tedious. In addition, this sort of standardization *assumes* that the liquid-handling device used to develop the standard curve is reliable, and this adds a level of uncertainty.

In summary, single-dye photometric calibration is well suited to measuring precision, particularly when handling volumes too small to be weighed on a balance. Accuracy measurements can also be made provided that the method is properly standardized and an uncertainty analysis yields acceptable performance.

Ratiometric Photometry The ratiometric photometric calibration method is a refinement of photometry designed to overcome the accuracy and traceability limitations of traditional single-dye photometric volume measurements. Ratiometric photometry employs two standardized dyes, and its measurement process produces absorbance readings in pairs that can be combined into absorbance ratio readings. An example of the ratiometric method is given in Annex A of ISO 8655-7.

The primary benefit of the ratiometric approach is improved accuracy and robustness of measurement compared to nonratiometric methods. Absorbance ratios can be measured more accurately than individual absorbances, leading to a higher degree of accuracy and precision in ratiometric methods versus traditional single-dye photometric methods. The underlying reason for this is that the absorbance of photometric calibration standards drifts over time, whereas ratios exhibit greater stability.

The ratiometric method offers enhanced capability in small-volume measurements. Ratiometric photometry provides accuracy as well as precision measurements and can do so to a traceable standard because the dyes function as an internal standard. Measuring the second dye in comparison to the first dye provides a nearly automatic compensation for the most common photometric error sources.

Systems based on ratiometric photometry provide information about each individual channel in multichannel devices and good reproducibility plate to plate. However, for ratiometric photometry to produce benefits, it must use well-characterized plates and carefully calibrated solutions of good stability. In addition, to function properly, ratiometric photometric methods require use of specially formulated dyes to produce accurate absorbance ratios. Finally, this technology is not highly applicable to measuring larger volumes, as other technologies can produce equally accurate measurements more cost-effectively.

In summary, ratiometric photometry calibrations are best applied when measuring small liquid volumes for protocols requiring traceability and a high degree of accuracy and precision.

4   PARAMETERS: ACCURACY, PRECISION, AND UNCERTAINTY

Pipettes should be verified for both accuracy and precision. Usual practice is to perform a comprehensive verification for accuracy and precision based on the mean and standard deviation of 10 replicate dispenses at three different volume settings representing the low, middle, and high points in the adjustable range (typically, 10, 50, and 100% of the nominal volume). This comprehensive verification should be performed quarterly unless otherwise justified [7].

A quick check for accuracy can only be done using the mean value of four replicates at two volume settings, corresponding to the minimum and maximum settable volumes of the pipette. For pipette verification, usual practice is to calculate relative inaccuracy and relative precision as a percentage of target volume [8]. Recommended frequency is quarterly for the comprehensive test and monthly for the quick check. A functional check should be performed daily when used.

Uncertainty is a concept that combines the performance of the "unit under test" (the pipette in our case) mathematically with the "measuring system" (either a gravimetric or a photometric system) to arrive at a numerical estimate of the uncertainty in a pipette calibration. An estimate of uncertainty is required for pipette calibrations, which are intended to comply with ISO 17025 and related standards.

The recommended practice for uncertainty estimation in pipette calibration is to calculate the "expanded uncertainty of the mean," which is based on the expanded uncertainty of the measuring system, plus the standard deviation of the mean, which is approximately equal to the measured standard deviation during pipette calibration divided by the square root of the number of replicates. More detail on uncertainty calculations is available in the ISO *Guide to the Expression of Uncertainty in Measurement* [9], and a specific example for pipette calibrations may be found in an ISO technical report [10]. Pipette calibration software that automates the calculation and reporting of uncertainty is also available.

4.1   Specifications

Appropriate tolerances for pipettes depend on the requirements of the tasks for which they are used. Because pipettes are dependent on user skill, the operator's choice of technique, and local environmental conditions, it should not be expected that the published manufacturer's tolerance specifications will always be achievable in the working laboratory environment. It is recommended practice to establish scientifically justified laboratory process tolerances which are appropriate to the test being performed and achievable in the laboratory. Guidance on setting achievable tolerances is available [11].

In contrast, manufacturer's published specifications for pipette performance are generally established based on data collected in a highly controlled environment, usually operated at elevated humidity (which improves pipetting consistency) and using highly trained technicians with specialized skills. Furthermore, there are no industrywide standards or guidelines for how pipette specifications are established. Pipette manufacturers vary in the degree of safety margin (or metrological coverage) that is included in their published specifications, and most often this coverage is not disclosed in the specification. Consequently, the statistical probability that a properly functioning pipette will perform within manufacturer's tolerances varies depending on manufacturer specification policy and local laboratory conditions.

4.2 Good Laboratory Practices

All pipettes are subject to some degree of error, due to changes in environmental variables such as temperature, humidity, and barometric pressure. As was shown earlier, the air displacement design is particularly sensitive to environment.

Temperature and Relative Humidity   Humidity and ambient temperature within a laboratory can change throughout the course of a year, can vary significantly between laboratories within the same building, and can even differ in various sections of the same laboratory. It is also important to account for variations in temperature and humidity when quality control, research, and manufacturing projects are outsourced to contracting laboratories or methods are transferred to another building or location within the company. Although costly, the ideal solution for laboratories is close control of humidity and temperature. When humidity changes must be tolerated, pipette performance can be improved by prewetting. Avoiding using pipettes near the lower end of the adjustable range will also improve performance when temperature is high or humidity is low.

Barometric Pressure   It is best to have pipettes calibrated at the barometric pressure (elevation) where they are being used. For this reason, the performance of pipettes that have been calibrated at a different elevation should be verified before being placed into service at another location. Using pipettes near the maximum end of their adjustable range will decrease the impact of barometric pressure changes. Normal variations of barometric pressure at a single location (such as changes in barometric pressure due to weather) are small enough that pipette performance will not be altered significantly.

Thermal Disequilibrium   Thermal disequilibrium is a dynamic phenomenon, so the impact on pipetting is difficult to predict in a quantitative way. In general, pipettes underdeliver warm fluid and overdeliver cold fluid; the magnitude of volume variation varies depending on details of the pipetting process [12]. The longer a pipette is in a warm microenvironment (such as pipetting from a warm flask), the warmer the captive air volume becomes. This leads to greater air expansion and a greater impact on volume. Therefore, the magnitude of error caused by thermal disequilibrium depends on several protocol-specific details, such as pipetting speed, type of sample container, and volume of captive air in the pipette tip.

Actions that minimize risk of volume variation due to thermal disequilibrium include equilibrating fluids to the temperature of the environment and liquid-handling device, minimizing captive air volume by pipetting as close to the maximum volume as possible, and minimizing exposure of the pipette tip to the warm or cold liquid and microenvironment. Methods that require pipetting fluids that are not in thermal equilibrium should be validated to demonstrate that the associated pipetting error is insignificant.

Operator Technique   Operator technique encompasses the particular combination of actions that are used in operating a pipette [13]. Subtle differences in choice of technique can have a significant impact on pipetting results [14]. Forward mode vs. reverse mode and prewetting vs. not prewetting are two common variants in laboratory technique that exert a measurable effect on delivered volume. Best practice is to calibrate pipettes using the same technique that will be used in the laboratory. The laboratory should validate protocols that require using a technique that differs from that used in calibration,

Operator Skill   Operator skill relates to the consistency with which an operator executes a chosen pipetting technique. Studies have shown that operator skill is the single largest source of pipetting error in most laboratories, and the usual formula of training, education, and experience is inadequate to ensure that laboratory technicians have attained sufficient skill. For pipette use, the best practice is to ensure operator skill through documented quantitative evidence. This can be done by having the operator demonstrate acceptable precision and accuracy using a known good pipette. It is recommended that operators first attain and demonstrate good precision by establishing a numerical coefficient of variation specification (e.g., 2% or better) which all new technicians are required to meet. Accuracy should then be assessed by comparing the mean values of the different operators to demonstrate that each has attained comparable accuracy. Qualification on pipetting should be done on initial hire, with an annual requalification. Materials are available to help train operators [15].

Silent Failures   Pipettes that have failed to perform within acceptable limits of accuracy or precision but are still aspirating and dispensing some liquid are known as silent failures because the operator does not recognize that the pipette is malfunctioning. Silent failures can be minimized by a program of interim verification at appropriate calendar intervals. A laboratory can estimate the rate of silent failure by tracking the number of in-use pipettes that are found to be outside acceptable tolerances during routine calibration.

Good laboratory practice requires that pipettes sent for calibration or maintenance be tested in their "as-found" condition (prior to repair, cleaning, or maintenance). Best practice is to keep as-found failure rates as low as is economically feasible. As-found failure rates above 5% are generally regarded as suboptimal. Some pharmaceutical laboratories maintain as-found failure rates below 2%.

Pipette Storage   The proper way to store a pipette is upright in a stand that is designed to accommodate the pipette. During pipette use, liquids are sometimes accidentally aspirated beyond the pipette tip and into the pipette shaft. It is also possible for condensate to form inside the shaft when pipetting hot or volatile liquids. Upright storage helps these liquids to drain away from the piston and toward the bottom of the pipette. When pipettes are left laying on their side, it increases the chances that liquids inside the pipette will contact the piston, and this increases the risk that steel pistons will corrode or that residues will be deposited on the piston and cause the seal to fail prematurely. Proper storage racks also protect pipette ends from scuffing on the sealing surface that contacts the disposable tip.

Disposable Plastic Tips   Best practice is to treat pipettes and tips as a system. Generic tips may be of inferior quality, may not fit properly, or may change the dead air ratio so that the pipette no longer performs accurately. Pipettes and tips should be verified together. Any changes in make or model of pipette tips used in the lab should be validated or verified. Small plastic tips packed loosely in bulk bags can become damaged and deformed at the fine-pointed end that contacts the liquid. This sort of tip deformation degrades pipetting accuracy.

Tip Mounting Force   Pipettes and tips are designed to fit together and form an airtight seal with a modest amount of force. Most operators use more force than necessary when installing tips. Tips should never be "hammered" on by repeated tapping into the box. Excessive force causes premature wear on the pipette sealing surface. Some newer pipettes have been designed with spring-loaded shafts or other features that provide the operator with biofeedback, indicating that proper force has been applied. These pipettes illustrate the slight force that is required to mount tips properly. Typically, less than 1 kg or 2 lb of force is all that is required to mount a good-quality pipette tip.

5 SUMMARY

Piston-operated micropipettes are ubiquitous and capable of excellent accuracy and precision. However, achieving optimal performance in the environment of a typical laboratory requires thorough training and careful attention to detail.

REFERENCES

1. Artel Extreme Pipetting Website, www.artel-usa.com/extreme.

2. *Piston-Operated Volumetric Apparatus*, Part 6, *Gravimetric Methods for the Determination of Measurement Error*. ISO 8655-6-2002. ISO, Geneva, Switzerland, 2002.

3. *Standard Specification for Piston or Plunger Operated Volumetric Apparatus*. ASTM E 1154-1989 (re-approved 2003). ASTM International, West Conshohocken, PA, 2003.

4. G. Rodrigues. *Bias, Uncertainty and Transferability in Standard Methods of Pipette Calibration*. Artel, Westbrook, ME, 2003.

5. *Piston-Operated Volumetric Apparatus*, Part 7, *Non-gravimetric Methods for the Assessment of Equipment Performance*. ISO 8655-7-2005. ISO, Geneva, Switzerland, 2005.

6. Pipette volume calibration kits. www.mlasystems.com/resevoirs.asp. Accessed Jan. 3, 2008.

7. *Calibration Frequency for Pipettes*. Lab Report Issue 6. Artel, Westbrook, ME, 2003.

8. *Defining Accuracy and Precision*. Lab Report Issue 4. Artel, Westbrook, ME, 2001.

9. *Uncertainty of Measurement*, Part 3, *Guide to the Expression of Uncertainty in Measurement* (GUM:1995). ISO/IEC Guide 98-3:2008. ISO, Geneva, Switzerland, 2008.

10. *Determination of Uncertainty for Volume Measurements Made Using the Gravimetric Method*. ISO/TR 20461:2000. ISO, Geneva, Switzerland, 2000.

11. *Setting Tolerances for Pipettes in the Laboratory*. Lab Report Issue 5. Artel, Westbrook, ME, 2005.

12. S. Ylatupa. *Optimizing Your Pipetting Performance*. Biohit Liquid Handling Application Notes. Available at www.biohit.com/pdf/app1.pdf. Accessed Dec. 20, 2007.

13. S. Ylatupa. *Choosing a Pipetting Technique Affects the Results of Your Analysis*. Biohit Liquid Handling Application Notes. Available at www.biohit.com/pdf/app2.pdf. Accessed Dec. 20, 2007.

14. *Impact of Pipetting Technique*. Lab Report Issue 2. Artel, Westbrook, ME, 2001.

15. *Ten Tips to Improve Pipetting Technique*. Artel, Westbrook, ME, 1996.

15

INSTRUMENT QUALIFICATION AND PERFORMANCE VERIFICATION FOR AUTOMATED LIQUID-HANDLING SYSTEMS

JOHN THOMAS BRADSHAW AND KEITH J. ALBERT

Artel, Inc.

1 INTRODUCTION

Automated liquid-handling systems are used in many life science laboratories to transfer liquids and perform a wide array of protocols and assays. These systems essentially remove any human impact on pipetting variability and are therefore employed to increase throughput, reproducibility, and repeatability of volume transfer steps. As throughput increases and assay volumes decrease, there are more demands for accuracy and precision of volume transfer tasks, which can include aspirating, dispensing, mixing, washing, and diluting steps. Most, if not all, assays performed in life science laboratories are volume dependent. The concentrations of chemical and biological components in an assay, as well as the associated dilution protocols, are dependent on the amount of volume transferred. If the target volume is not delivered to the assay accurately, the concentrations of species in solution will be unexpectedly different. If the amount of volume transferred is different for each replicate assay, the experimental results cannot be trusted. Moreover, the volume differences might go unnoticed and data analysis

Practical Approaches to Method Validation and Essential Instrument Qualification,
Edited by Chung Chow Chan, Herman Lam, and Xue Ming Zhang
Copyright © 2010 John Wiley & Sons, Inc.

may be carried out without any knowledge of a volume transfer discrepancy. It is absolutely imperative to quantify the amount of volume delivered by the liquid handler. A volume verification method, which can be used to quantify the amount of volume transferred, is an essential component enabling proper liquid-handler qualification and performance verification. A volume verification method can be used to quantify most, if not all, critical volume transfers so that the behavior of each liquid handler's automated task is understood. Knowing the behavior of a liquid handler will undoubtedly facilitate instrument qualification and performance verification. Furthermore, when a volume verification method is employed as a diagnostic tool, the need for maintenance or redeployment [1] becomes clear and can be conducted at the appropriate time, thereby reducing downstream troubleshooting and economic loss. There are multiple ways to measure the accuracy and precision of volume transfer steps, and in this chapter we discuss the many types of volume verification methods used for such tasks. The overriding themes discussed herein emphasize the importance of using a robust volume verification method for liquid-handler qualification and performance verification. Some of the many volume verification methods in use in laboratories today are discussed together with their advantages and disadvantages in understanding automated liquid-handler behavior.

1.1 Importance of Verifying the Performance of Liquid Handlers

A volume verification method is defined herein as a scientifically driven process aimed at measuring the accuracy and precision of volumes transferred with a pipette or an automated liquid handler; see Figure 1 for a schematic defining

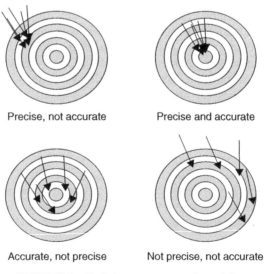

Precise, not accurate   Precise and accurate

Accurate, not precise   Not precise, not accurate

FIGURE 1   Defining accuracy and precision.

accuracy and precision. Mathematical definitions for accuracy and precision are discussed later in the chapter. Along with the ability to evaluate the performance of a liquid handler, the volume verification method can be used to compare the performance of multiple liquid handlers or that of automated methods (scripts), to identify a liquid handler's dispense drift over time, to compare tip-to-tip reproducibility, or for statistical comparison of individual and sequential quantities dispensed. The importance of a robust and reliable volume verification method cannot be underestimated, and such processes could enable instrument qualification or facilitate assay optimization where there is currently a significant bottleneck [2]. A rapid, versatile volume verification method also allows automated method parameters to be tweaked and modified on the fly during developmental stages for quick optimization and performance verification prior to use. Ideally, an evaluation method should be implemented analytically with documented procedures. Additionally, the method should be relatively quick and easy to integrate into a laboratory to minimize instrument downtime *and* required resources (e.g., labor, reagents).

One area of critical importance of a volume verification method is troubleshooting and diagnostics. A recent study showed that a liquid handler's system fluid was diluting the sample volume, and these findings were discovered only after comparing two volume verification methods side by side [3]. A volume verification method can also be used to facilitate the transfer of benchtop assays to an automated platform to show improvements in robustness and stability and to monitor assay variability [4]. By knowing exactly how each liquid handler is transferring liquids, it may become very apparent when performance optimizations, calibration, service, and/or maintenance should be implemented or required. Also, regardless of the frequency of liquid-handler calibration intervals, a volume verification method should be used for interim checks to replace "faith-based" performance monitoring, which would help operators know exactly when liquid handlers fail specific tolerance limits. Performing frequent quick volume spot checks also allows the user to have confidence in automated methods and potentially avoids the initial loss, or unnecessary destruction, of rare or expensive reagents on a day-to-day basis.

For many situations involving the qualification of an automated liquid handler, it may be important to ask questions about each volume transfer step, such as:

- What is the goal of the step?
- What are the tolerances, or specifications, for the assay or system for each specific volume transfer?
- How are these tolerances measured?
- Is the performance of the liquid handler good enough for a given assay?
- Is the accuracy and/or precision critical?
- Can the dispensing technique be optimized to facilitate the instrument qualification process?

- Is there a volume verification method better suited for the particular automated protocol?
- If the target volume is not exact, is it better to under- or overdispense?

Automated liquid handlers are generally used to increase the productivity and repeatability of volume transfers, but as discussed below, they are still prone to error. It is therefore important to understand how some errors can be recognized and avoided to maintain liquid-handling quality assurance. It is difficult to manage and minimize error if there is no means to identify error in the first place. The more frequently liquid-handler checks are performed, the sooner malfunctioning liquid handlers will be detected, fixed, and brought back online. Before the various volume verification methods are described, some of the more common types of liquid-handling errors affecting performance evaluation of liquid transfer are discussed.

1.2 Automated Liquid Handlers as Sources of Error

Although automated liquid handlers remove the human variable from pipetting, these systems are still prone to multiple types of error, especially if the interface, programming, and materials are not defined or employed correctly. In the rest of this section we discuss some common volume transfer errors encountered with liquid-handler use, which can readily affect instrument qualification and performance verification of a liquid handler.

Liquid-Handler Errors: Tip Types and Contamination The types of tips employed on the liquid handler are *very* critical to the accuracy and precision of each volume transfer. When using disposable tips, it is highly recommended to use vendor-approved tips rather than generic tips. Generic tips may cause unseen errors in volume transfer because tip material, properties, wettability, shape, and fit are all important factors for volume transfer repeatability. For example, the cheaper, generic tips may have variable characteristics that affect liquid delivery, such as differences in virgin plastic content or sizing of the upper diameter and/or tip orifice, or they might contain "flash," which is residual plastic residue. Generic tips might not fit well on the liquid handler, and they may have variable wetting and delivery properties. All vendors of pipette and liquid-handler equipment recommend that approved tips are employed for each pipettor to guarantee the best results in liquid delivery. Universal disposable pipette tips do not exist. Without using approved tip types, accuracy and precision cannot be guaranteed, and in some cases, the liquid handlers may be blamed incorrectly for variable performance when generic tips are the root cause of error. Some liquid handlers employ fixed or permanent tips (including pin tools), which help avoid the recurring consumable cost required for disposable tips. When using fixed tips, however, there must be rigorous and efficient tip washing protocols in place to minimize error in liquid transfer. Otherwise, unwanted residual sample may be carried over and cause contamination in subsequent transfer steps.

Another source of error that may occur when using liquid-handler equipment is contamination. For example, the liquid-handler gantry/head moves many times during an automation task. It will move to a predetermined deck location, aspirate reagent, move, dispense reagent or aspirate another reagent, move to a different location, dispense, repeat, eject or wash tips, and so on. It is possible that contamination can occur while the head is moving across the liquid handler's workspace, where droplets can fall from the tips, especially when slippery or organic reagents are employed. Users should evaluate their systems and tips to ensure that droplets are not remaining after a sample is dispensed. Some users minimize these possibilities by using an airgap after a reagent aspiration to minimize the chances of anything "falling out" of a tip. Additionally, to minimize the chances of contamination caused by random reagent splatter onto the liquid handler's workspace, technicians should use caution when and where tips are ejected during the automated task.

Liquid-Handler Errors: System Fluid Interaction and Various Liquid Types As with many types of manual pipettes [5,6], many liquid handlers use air displacement to move and transfer volumes. One of the biggest errors in liquid handling when using air-displacement pipettors is variable temperature [7], so it is always recommended to work at equilibrium temperature. A recent conference presentation discussed error when using warm and cold liquids when using air-displacement pipettes [8]. Instead of air displacement, some liquid handlers incorporate a fluid displacement pipetting approach which requires movement of a system fluid within a length of tubing. This fluid provides a noncompressible environment and allows for reagents to be aspirated and dispensed by moving the system fluid within the lines. The fluid might range from water to various organic solvents or mixtures, but it is extremely important that the system fluid not degrade the parts to which it makes contact because this could cause performance errors over time. Even deionized water can be somewhat corrosive when in contact with certain grades of stainless steel. Most important, it must be ensured that the system fluid does not mix or come in direct contact with reagents being transferred, because the reagent will be diluted with each successive assay and will lead to experimental inaccuracies, contamination, and/or carryover. This dilution effect and its discovery by using two different volume verification methods side by side were recently reported [3]. To avoid system fluid interaction it is usually recommended that the automation method parameters include sufficient airgaps to separate the sample fluid from the important reagents that will be used in the assay.

Liquid handlers are certainly capable of handling a wide array of reagent types, but it is commonly known that performance parameters can vary between different types of liquids. If a system is validated or calibrated for to dispense water accurately but is used to dispense a different type of solution, the performance characteristics could be quite different for the test liquid. The ability to verify dispensing performance for assay-specific reagents (e.g., complex and/or nonaqueous solutions) can be an advantage of some volume verification methods because the methods can help an operator define and validate liquid-class parameters within a liquid handler's software package.

Liquid-Handler Errors: Sequential Dispensing Inaccuracies   In some liquid-handling protocols, a large volume of reagent is aspirated and dispensed sequentially across a microtiter plate. Although this method can save time, there are often errors associated with variable accuracy. The user should ensure that upon dispensing, the tips do not touch liquid in the wells to avoid contamination or unwanted reagent dilution. It is usually recommended that this protocol be a "dry dispense" (dispensing into a dry well) or dispensing in a noncontact fashion above the wells. If a liquid-handler method employs a sequential transfer, a volume verification method should be used to validate that the amount of volume is dispensed equally in each step. With sequential dispensing it is often common for the first and/or last quantity dispensed to contain a slightly different amount of reagent. In these situations, the first or last shot effect is usually avoided by dispensing the first and/or last quantity dispensed back into the sample reservoir or to the waste area.

Liquid-Handler Errors: Serial Dilution Transfers   Many laboratories perform some type of dilution testing, such as detection limits, dose response, percent inhibition, toxicity, and drug efficacy. A serial dilution is a systematic assay where an important reagent is reduced in concentration sequentially across a microtiter plate. Automated liquid handlers are employed routinely to perform serial dilution protocols, but these methods are more complicated to validate (e.g., the users might not have the skill or necessary volume verification methodologies to verify that the volume transfer is accurate and that each well is mixed efficiently before the next transfer takes place). The concentration of target reagent will be *very* different from the assumed (theoretical) concentration across the plate if the reagents in the wells are not mixed efficiently before each transfer. The experimental results may be flawed and the technician may have no indication that inefficient mixing is to blame. The authors know of one commercial volume verification platform that can be used to validate some serial dilution applications and applied to validate the efficiency of mixing methods [9–14].

Liquid-Handler Errors: Pipetting Methods and Method Parameters   One of the first steps in helping to minimize error in liquid handling with automated systems is to choose the proper pipetting technique for an assay. Two common pipetting techniques are forward mode and reverse mode. Forward mode is a very common technique in which the entire aspirated reagent in the tip is discharged and is suitable for aqueous reagents with or without small amounts of proteins or surfactants. In the reverse-mode pipetting technique, more reagent is aspirated into the tip than is dispensed. For example, if 3 μL of serum is required, the pipettor might be programmed to aspirate 5 μL of serum and then dispense the 3 μL into the well of interest, followed by dispensing the remaining reagent either back into the reagent reservoir or into waste. The reverse-mode method is usually suitable for viscous or foaming liquids.

Along with pipetting techniques to consider, some errors with automated liquid handling may occur when variables within the liquid handler software are defined incorrectly. For example, the user should ensure that procedural variables (e.g., aspirating dispensing heights and rates, liquid-class settings, target volumes requested, pauses), consumable types [e.g., reagent reservoir size(s), microplate types/footprints], and workspace layouts (e,g., position and location of hardware and consumables) are defined properly for performance evaluation. It is also important to try to maintain a tip depth of about 2 to 3 mm below the surface of the reagent when aspirating liquid. If the reagent continues to be removed from the reservoir, and tip heights are not compensated for those differences, pipetting errors may occur. For some liquid handlers, liquid "sensing," or conductive, tips are employed to indicate the surface depth of the liquid in the reservoir or plate. When these tips are lowered into bubbly and/or frothy reagents, the system may falsely identify liquid being present. Also, sensitivity of the fluid sensing measurement is highly dependent on the ionic strength of the reagents used.

Economic Impact of Error in High-Throughput Campaigns If a volume verification method is not employed or implemented properly to evaluate the performance of liquid handlers, there may be severe economic consequences. If liquid handlers are not transferring the desired amount of reagent, it is likely that *unseen* error will propagate increasingly within a process. Even small discrepancies in the amount of reagent transferred may compromise results, possibly leading to poor quality, useless data, and downstream costs associated with remedial actions. The economic impact of allocating resources for a liquid-handling process, which is based on potentially false information, may be severe. Moreover, if the liquid delivery systems are overdelivering target volumes of rare and expensive reagents, there will be a significant economic impact, due to the rapid overuse of those precious materials. An average high-throughput screening laboratory may screen about 1 to 1.5 million wells in each screening event, with an average screening frequency of about 20 to 25 times per year. With an approximate range of $0.10 per well, the total cost is approximately $3.75 million per year (1.5 million wells × 25screenings × $0.10/well).

If the liquid handlers continuously, overdispense critical reagents so that the average cost per well increases by only 10%, the average additional annual cost would be $375,000 and the company could risk the faster depletion of those rare compounds. With more critical reagent in each assay and depending on the screening type, there could be more prevalence of false positives, and those particular compounds would be used in subsequent screenings. False positives are detrimental to the process by consuming laboratory time, resources, and materials during screening before they are "tested out" of the process. On the other hand, if the liquid handlers were unknowingly underdispensing critical reagents, the cost per well may decrease but there could be an increase in false negatives, which can be severely detrimental to the integrity of the entire screening process. To the screener, a false negative is no different than a nonperformer, and these compounds are not used in subsequent screenings. The next blockbuster pharmaceutical drug may go unnoticed, and in some respects it could cost the company billions in future revenues. The underlying emphasis is that by not assessing the continuous performance of the liquid handlers with readily available volume verification methods, a laboratory might be faced with huge economic costs associated with over- or underdispensing important reagents.

2 COMMONALITIES BETWEEN VOLUME VERIFICATION METHODS FOR PERFORMANCE EVALUATION

Regardless of the type of volume verification method employed to evaluate liquid-handler performance, it is most important that the methodology be scientifically based, analytically implemented, and executed properly. If the volume verification method is not well thought out or properly implemented, a false sense of liquid-handler performance may result. For instance, in some photometric-based volume verification methods, solution mixing and the types of microtiter plates are critical to the measurement. Volume verification methods that average data across all pipetting tips (more common with gravimetric methods) might not observe individual tip misfires because the individual tip data can be washed out during signal averaging. Again, these and other advantages and disadvantages are discussed below for each volume verification method.

2.1 Environmental Considerations

Volume verification methods used for instrument qualification should be performed in the same environment (e.g., temperature, pressure, humidity) in which the liquid handler operates. There are times when instrument qualification cannot be performed in the same environment as the liquid handler, and for these situations, there should be some caution. For example, the liquid-handler dispensing head may be sent to the factory for recalibration, or there is a chance that an analytical balance (for gravimetry) or spectrophotometer (for photometric methods) are not in the immediate vicinity of the liquid handler. Transport of components and materials/parts handling may introduce more error in the instrument qualification process. When handling microtiter plates with reagents, there is a possibility of evaporation and loss of reagent, which may also introduce error into the measurement that should not be attributed to the performance of the liquid handler. The best-case scenario is that the liquid-handler qualification and performance evaluation is implemented in close proximity to the liquid handler, using a standard delivery approach, while under the same environmental conditions that are present when the liquid handler is employed routinely.

2.2 Measuring Accuracy and Precision

For qualifying liquid handlers and/or assessing their performance, it is critical to measure both the accuracy and precision of each target volume transfer. To determine the accuracy and precision, the mean and standard deviation values

TABLE 1 Summary of Equations for Determining Accuracy and Precision

Mean Volume	$\overline{V} = \dfrac{\sum_1^n V_i}{n}$
Standard deviation	$\text{std} = \sqrt{\dfrac{\sum_1^n (V_i - \overline{V})^2}{n - 1}}$
Coefficient of variation	$\text{CV} = \dfrac{\text{std} \times 100}{\overline{V}}$
Relative inaccuracy[a]	$\text{inacc} = \dfrac{(\overline{V} - V^P) \times 100}{V^P}$

[a] V^P is the target volume.

must first be determined. Table 1 is a summary of the equations discussed in this chapter. Accuracy can be defined as the degree of veracity (e.g., the measurement's closeness to a reference point or standard value). In the analogy presented in Figure 1, the arrows closest to the bull's-eye in the center of the archery target are more accurate than the arrows farther away from the bull's-eye. With regard to accuracy, the difference between the mean of the measurements and the reference value can be determined. Often, these values are expressed as a relative inaccuracy. For example, if the liquid handler is supposed to transfer a 2-μL target volume but, instead, transfers a mean volume of 1.97 μL, the volume transfer is 98.5% accurate. In this case, the relative inaccuracy of the transfer is -1.5%. The relative inaccuracy value helps the user rapidly identify measurements that fall in or out of the tolerance criteria for the specific volume transferred.

Precision is the closeness of independent measurements of a quantity made under the same conditions and tells us how well a measurement can be made without reference to a true value. With the target analogy, the spread of the arrow cluster, or group, defines the precision. For example, if all the arrows were grouped tightly, the measurement is precise. Precision is usually defined in terms of the standard deviation or the coefficient of variation (CV). In the 2-μL example above, if five volume transfers were measured to be 2.04, 2.06, 1.91, 1.94, and 1.90 μL, the mean is 1.97 μL, the standard deviation is 0.07 μL, and the coefficient of variation is 3.8%. As with the relative inaccuracy value, the coefficient of variation measurement is widely used in the liquid-handling community when discussing the precision of a volume transfer. The CV values also allow rapid identification of measurements that are out of tolerance for the specific volume transfer task (e.g., if the tolerance for a 2-μL transfer is 5%, the example above would be within the accepted criteria). Volume transfer measurements may be accurate and imprecise, precise but inaccurate, both accurate and precise, or inaccurate and imprecise.

To reiterate, it is important to measure *both* the accuracy and precision of the volume transfer for instrument qualification. The accuracy will allow you to determine if the volume amount is correct, which really means that the concentration of each component in the assay can be calculated correctly (assuming that the starting concentration of critical reagent is known). The precision value in the measurement will determine the variability between component concentration levels within the group of volume transfers. There is always a chance that individual tips within a multichannel liquid handler can perform with some variability. Defining this variability, or determining which channels "misbehave," allows assay results to be interpreted properly and the results may help predict the possibility of required system maintenance. For example, most tips in an eight-tip liquid handler behave the same, but one or two might behave poorly. In the proof-of-point example in Table 2, six of the tips show good accuracy and good precision values, tip five is inaccurate but precise, and tip seven has good accuracy but is imprecise. With data such as these it is obvious that a volume verification method is most valuable when it can be used to measure performance data on a tip-by-tip basis. It should be noted, however, that it is not possible to achieve reliable accuracy in individual measurements without precision: If the arrows are not tightly clustered, each arrow cannot be close to the bull's-eye. The case in point with automated liquid handlers is that it is of the utmost importance to make sure that all tips are behaving the same (precision), and once the variability, if any, is defined, the liquid-handler software can be adjusted to deliver the correct target volume (accuracy).

2.3 Number of Replicates

Guidelines for the number of replicate quantities dispensed per tip of an auto-mated liquid handler are not as specific as those for handheld single-channel pipettes. For single-tip handheld pipettes, several regulatory agencies, including ASTM [15] and the International Organization of Standardization (ISO) [16], offer guidance and recommendations. For example, ASTM (American Society of Testing and Materials) E 1154 [15] specifies taking four replicates monthly for a "quick check" and 10 replicates quarterly for a comprehensive volume verifi-cation. This recommendation is generalized and does not take into account the needs of individual laboratories. When a performance evaluation is conducted for an automated liquid handler, at least three replicates per tip should be col-lected. The authors recommend that four to 10 replicates be collected per tip on a monthly to quarterly basis until a system's baseline performance is established. At this point, the frequency of performance verification (volume checks) could be reevaluated for shorter or longer durations, depending on the reliability of the liquid handler under test.

3 VOLUME VERIFICATION METHODS

A number of volume verification methods are used for liquid-handler qualification and performance verification [3,4,9–11,16–28]. A majority of laboratories implement either gravimetric, photometric (e.g., absorbance based), or fluorometric methods, or a combination thereof. Our goal in this section is

TABLE 2 Measured Volume Transfers per Well with Statistics for Each Tip[a]

Well ID in 96-Well Plate	Column 1	Column 2	Column 3
Row A	2.01	1.99	2.03
Row B	1.97	1.95	1.99
Row C	1.95	2.02	2.01
Row D	1.95	1.94	1.91
Row E	1.86	1.84	1.85
Row F	2.03	2.10	2.00
Row G	2.20	1.80	1.91
Row H	2.00	1.97	1.99

Individual tip statistics:

Tip ID	Mean (Volume μL)	Inaccuracy (%)	Standard Deviation (μL)	CV (%)
1	2.01	0.50	0.02	1.00
2	1.97	−1.50	0.02	1.02
3	1.99	−0.33	0.04	1.90
4	1.93	−3.33	0.02	1.08
5	1.85	−7.50	0.01	0.54
6	2.04	2.17	0.05	2.51
7	1.97	−1.50	0.21	10.49
8	1.99	−0.67	0.02	0.77

[a]Three repetitions of a 2-μL target volume per tip into a 96-well plate; in μL.

to describe the different volume verification methods most often employed and their associated advantages and disadvantages for assessing volume transfer performance of liquid handlers.

3.1 Gravimetric-Based Volume Verification Methods

Because gravimetry is used so widely for many laboratory processes, and because lab practitioners are generally familiar with using a balance, it is no surprise that this method has been implemented for volume measurement purposes. The key advantage to gravimetric analysis is its simplicity. The simplest application of a gravimetric measurement of a volume of liquid involves dispensing the liquid into a receptacle stationed on the balance pan and measuring the weight of the delivered volume. The weight, or mass, of the delivered solution can then be converted to a volume by dividing by the density of solution:

$$V = \frac{m}{\rho} \tag{1}$$

where mass, m, is measured in units of grams and density, ρ, is measured in units of grams per cm^3 (i.e., mL). Repeating this measurement process multiple times thus allows for statistical assessment of the accuracy and precision with which the delivery device is performing.

Although such an approach appears straightforward on the surface, proper use of gravimetry for testing volume delivery is a bit more complex than is often recognized. For example, the simplistic approach defined above would be a reasonable approach for large volumes (e.g., >100 μL) delivered from single-channel instruments. However, when considering multichannel devices (e.g., handheld multichannel pipettes or multichannel robots), this approach may not be capable of providing as much information as described in the scenario above. If a 96-tip pipetting head is used, for example, the inability to drive each pipetting channel individually often requires dispensing all 96 tips into a microtiter plate, which is then weighed on the balance. The weight measured represents the total volume dispensed by all 96 tips. Thus, dividing the total weight by 96 gives the average weight, or volume, that was dispensed by a tip. Clearly, such an approach cannot be used to assess the performance of each individual channel simultaneously, and may not be able to pinpoint performance problems of a given pipetting device unless rigid tolerances are set for the allow-able standard deviation across all wells measured. This same scenario will exist for all pipetting devices possessing more than one pipetting channel unless each channel can be individually controlled. In some cases the 96 tips can dispense liquid to be weighed into 96 individual containers, such as test tubes, with the 96 tubes weighed individually on the balance. This process can take hours to days, especially if at least three replicates are used for statistical determination of performance values per tip.

There are a few other facets of proper application of gravimetry that are often overlooked by the common practitioner but which may result in errors in the final volume measurements. Such facets include:

- Assumptions about the true density of the delivered solution(s)
- Correcting the solution density properly for environmental conditions such as temperature, barometric pressure, and relative humidity
- Accounting for solution evaporation during the measurement process
- Reducing static electricity
- Correcting mass measurements for air buoyancy factors resulting from exist-ing environmental conditions

The effect that each area could have on gravimetric measurement of solution volume is mentioned briefly here, but a fuller discussion of the impact of each can be found in Chapter 14 of this book.

Although every practitioner would probably admit that using the solution density of water ($\rho = 1.0$ g/mL) for the gravimetric assessment of liquid delivery is only an approximation, few truly understand how such an assumption might affect the end result. Using the solution density of water when measuring a 20 mM KHP buffer solution will result in a -0.3% error in the volume calculated.

Although such an error may be acceptable in some circumstances, more accurate gravimetry requires use of the actual density of the solutions being used.

Another assumption that is commonly used when determining volume via gravimetry, but is rarely recognized, is that of constant density for a given solution. Although many practitioners probably understand that they are using an assumed density of water for their test solutions, few understand that the density of any solution is related directly to the environmental conditions of its surroundings and thus assume incorrectly that their solution's density is constant under ambient lab conditions. The density of any solution will change with the temperature of that solution and with the barometric pressure and humidity of its surroundings. The effect that such environmental changes will have on the density of any solution can be accounted for by tracking the change that will occur for the density of water. In other words, the impact of temperature, pressure, and humidity on the density of pure water is well known and can be adjusted through a quantity called the *z-factor* [15,16]. For a given set of ambient environmental conditions, a z-factor can be calculated, and the solution density relative to water can be determined by

$$V = \frac{m}{R}\, z \qquad (2)$$

where R is the relative density of the solution compared to that of pure water ($R = \rho_{solution}/\rho_{water}$), and the z-factor corrects for changes in the density of water resulting from environmental conditions. The z-factor, given in units of mL/g, also accounts for the effect that these environmental conditions impart on the mass measured by the balance. These same environmental conditions affect the "buoyancy" of any object being weighed, which results in an error in the measurement of the true mass of the object. Thus, by applying a z-factor correction to the measured mass of a solution of interest, whose density relative to water has been determined at a controlled temperature, the environmental effects can be removed from the determined volume.

Those who have used the balance to measure liquids have probably observed the typical negative drift of the measured mass due to the evaporation of water. The opposite effect, wherein the solution weight can increase due to the absorption of water, can also be observed for hygroscopic solutions such as dimethyl sulfoxide (DMSO) or concentrated sulfuric acid. Clearly, such a drift in the gravimetric measurement will lead to errors in the volume result and must be accounted for if the best accuracy is desired. Additionally, static electricity can create problems for gravimetric measurements, particularly under low-humidity conditions.

When using gravimetric methods of volume verification, the user should also consider that skill, patience, and time are required as well as draft shields and proper balance resolution for the target volumes of interest. Balances are accurate for larger target volumes ($>20\ \mu$L), but are usually not as suitable for low volumes ($<20\ \mu$L). The balance should also be cleaned routinely and used on a proper, stable table, which usually means that they cannot be brought to the liquid handler for volume verification (i.e., the sample must be transported to the balance, thereby increasing the probability of error).

3.2   Spectroscopic-Based Volume Verification Methods

There are two predominant spectroscopic-based volume verification techniques: one fluorescence-based and the other absorbance-based. In both cases, a dye solution is used to generate a measurable "light" signal that can be attributed or used in some fashion to determine the precision and/or accuracy of a volume transfer within a microtiter plate. In many respects, the two methods can be carried out in very similar fashion. They both require a plate reader, which includes a light source, a wavelength-selection device (e.g., bandpass filter, monochromator), and a detector(s). Both techniques may also require that one or more solutions be prepared (with and without dye). For either the fluorescence or absorbance method, the dye solution is dispensed with the liquid handler into a microtiter plate. In many cases, a buffered solution is added to the microtiter plate to provide a total working volume in each well of the plate.

Considerations When Using Spectroscopic Methods   In each photometric method there are a significant number of procedural variables, or events, which can considerably contribute to the error in the performance evaluation. There must be skill and patience in the weighing of dye powder, filtering insoluble components from the solutions, and determining the stability and/or shelf life of the prepared solutions. The methods should employ the necessary controls, such as using an appropriate baseline subtraction (e.g., zero measurement) and accounting for plate reader drift. Also, when using a single-dye fluorescence or absorbance method, the volume of buffer used to fill the well to a total working volume needs to be known for most plate types; otherwise, the uncertainty in the buffer volume cannot be separated from the uncertainty in the sample solution volume. In essence, these particular methods require the accurate and precise transfer of both target volume and buffer volume. Otherwise, an incorrect dilution of the dye in each well will lead to an increase or decrease in light measurements, which will undoubtedly lead to calculation of a different target volume for the liquid handler. This effect is a result of the commonly occurring tapered sidewall of most standard plate types. For example, a common round-well, 96-well microtiter plate generally possesses a sidewall taper angle of $1.3°$, whereas a common square-well 384-well microtiter plate can have a sidewall taper angle of $>7°$. A dual-dye absorbance-based method, which is protected from the dilution effect observed in single-dye methods, is discussed below.

The amount of signal generated by the diluted dye in the microtiter well must be in a measurable range for the plate reader (e.g., the light signal cannot be too weak or too intense). In many situations, only the precision of the volume transfer can be calculated, because the accuracy measurement usually requires additional steps (see below). The precision in the measurement is determined by measuring the standard deviation of the optical signal over replicate transfers of the dye solution.

When using fluorescence- or absorbance-based volume verification methods in microplates, the types and quality of the plates can have a strong effect on the performance assessment, especially if plates and/or lots of plates are interchanged during the evaluation process. Additionally, it is of the utmost importance that dye solutions are mixed efficiently in each well for the verification method to be executed properly. Proper sample mixing cannot be stressed enough, and without it the verification method may not be useful in monitoring dispensing performance for a liquid handler. In fact, without proper mixing, the performance for the liquid handler could be misinterpreted, especially when dispensing into 384-well and higher-density plate formats, where the mixing step becomes even more critical and challenging [14]. The perceived performance of a liquid handler could lead to misguided data interpretations if the mixing step is not performed correctly. For many of the reasons noted above, the applicability of standard photometric volume verification methods between laboratories or sites becomes a daunting task.

By comparison, absorbance measurements typically do not have the degree of sensitivity inherent to fluorescence, but absorbance dyes are often more stable over longer periods of time. Also, absorbance methods can be traced with known certainty to internationally recognized standards, which allows for much easier standardization between methods conducted on different days, or in different labs, or with different lots of dye solutions.

Absorbance-Based Methods Absorbance-based volume verification methods are commonly employed for liquid handler performance evaluation because most laboratories have a spectrophotometer, which could be in the form of an ultraviolet–visible cuvette-based system or a vertical beam microtiter plate reader. The latter instrument is the most popular for assessing liquid-handler performance for target volumes in microtiter plates. It is required that the plates have an optically clear flat bottom so that the vertical light beam of the plate reader can have an unobstructed path from the light source through the solution to the detector, thereby providing reliable light transmittance measurements through the plate material. The Beer–Lambert law, or simply, Beer's law, is used to define the relationship among the measured absorbance (A) of a chromophore at wavelength λ, the molar absorptivity (ε) of the chromophore at wavelength λ, the concentration (C) of the chromophore, and the pathlength (l) of light through the sample:

$$A_\lambda = \varepsilon_\lambda C l \tag{3}$$

This relation is common known to, and used by, many scientists. Arguably the most common use is to determine the unknown concentration of a sample in solution by collecting the absorbance with a horizontal beam spectrophotometer in a fixed-pathlength cuvette. For such use, Beer's law is fairly simple and is easily applied. However, closer inspection demonstrates that this relation becomes more complex when used for vertical beam spectrophotometers, as in the case of microtiter plate readers. For this scenario the pathlength is no longer fixed but is a function of the well shape and total volume of solution in the well. Although this acts to complicate the Beer's law relation for vertical beam measurements, it also opens the door to using Beer's law to more easily determine solution volume. For such volume determinations, a solution with known dye concentration is used, so it becomes convenient to combine the concentration (C) and molar absorptivity (ε):

$$\varepsilon_\lambda C = a_\lambda \tag{4}$$

where a_λ is the absorbance per unit pathlength for a defined solution.

As mentioned previously, the precision of the volume delivery can readily be determined by measuring the standard deviation of the absorbance units (e.g., optical density values) and calculating the coefficient of variation (CV) values from well to well and/or tip to tip. In the example shown in Figure 2, an eight-tip liquid handler was used to dispense a dye solution into a 96-well microtiter plate over 12 replicates for each tip. The baseline-corrected absorbance values shown in Figure 2a were used to calculate the average absorbance values, standard deviation values, and CV values by sequential dispense and tip numbers. As stated, the precision can be determined easily by simply preparing a dye solution such as food coloring in an aqueous buffer. Examples of common absorbance dyes include tartrazine and Ponceau S. A buffered solution is recommended for all such work because most common dyes are weak acid compounds that have pH-dependent molar absorptivities. When working in microtiter plates, it is important to use this same buffer, without the dye, as a blank measurement in the same plate that will be used for the determination. Using the same buffer for a blank measurement ensures that any light scatter due to meniscus curvature will be subtracted properly from the sample absorbance measurements.

Whereas the precision of volume delivery can be determined using a dye with a relatively unknown concentration as long as it provides a measurable absorbance signal, the determination of accuracy is not as easily achieved. With that said, measuring the absorbance of light of a volume transfer can be related to accuracy values if a standard curve is generated or if a gravimetric method is used as a complementary method (see below).

A few remaining areas that must be considered when using an absorbance method for determining accuracy of volume delivery include the effect that the diluent volume will have on the sample volume calculated, as well as common errors associated with deviations of the chosen dye from Beer's law [30,31]. When using a vertical beam format, dispensing too much diluent will lead to a corresponding decrease in dye concentration. However, overdispensing will also lead to a longer pathlength of light. For a perfectly cylindrical well, and for a solution with a perfectly flat meniscus, the degree of dye dilution is inversely proportional (1 : 1) to the change in pathlength. Thus, considering the effect these changes would have on the absorbance as predicted by Beer's law, it is readily seen that the two effects cancel each other and the absorbance remains unchanged for perfectly cylindrical wells and for chromophores that obey Beer's law perfectly.

This relation also holds for any well cross-sectional geometry (e.g., square well, square well with rounded corners) as long as the sidewalls are not tapered and the bottom is flat. In reality, such plates do not exist. Even commonly used

(a)

	1	2	3	4	5	6	7	8	9	10	11	12
A	0.557	0.558	0.559	0.56	0.559	0.56	0.559	0.558	0.558	0.558	0.558	0.558
B	0.558	0.559	0.559	0.559	0.56	0.559	0.559	0.559	0.56	0.559	0.558	0.557
C	0.558	0.56	0.559	0.561	0.56	0.56	0.559	0.56	0.56	0.559	0.558	0.557
D	0.558	0.559	0.559	0.56	0.559	0.557	0.558	0.56	0.559	0.557	0.556	0.556
E	0.558	0.558	0.558	0.56	0.56	0.558	0.559	0.56	0.56	0.56	0.557	0.557
F	0.558	0.56	0.561	0.561	0.563	0.561	0.56	0.56	0.56	0.558	0.558	0.558
G	0.558	0.559	0.56	0.56	0.56	0.559	0.559	0.56	0.559	0.559	0.558	0.558
H	0.557	0.556	0.558	0.558	0.557	0.557	0.557	0.557	0.557	0.557	0.557	0.557

(b)

	Column 1, Dispense 1	Column 2, Dispense 2	Column 3, Dispense 3	Column 4, Dispense 4	Column 5, Dispense 5	Column 6, Dispense 6	Column 7, Dispense 7	Column 8, Dispense 8	Column 9, Dispense 9	Column 10, Dispense 10	Column 11, Dispense 11	Column 12, Dispense 12
Average (au) =	0.558	0.558	0.559	0.559	0.560	0.559	0.559	0.559	0.559	0.558	0.558	0.557
Standard Deviation (au) =	0.0005	0.0011	0.0013	0.0010	0.0017	0.0015	0.0009	0.0012	0.0011	0.0011	0.0005	0.0007
CV (%) =	0.083%	0.202%	0.233%	0.185%	0.298%	0.261%	0.159%	0.208%	0.201%	0.190%	0.096%	0.127%

(c)

	Average (au)	Standard Deviation (au)	CV (%)
Row A, Tip 1	0.558	0.0010	0.176%
Row B, Tip 2	0.559	0.0009	0.155%
Row C, Tip 3	0.559	0.0011	0.204%
Row D, Tip 4	0.558	0.0011	0.200%
Row E, Tip 5	0.559	0.0011	0.194%
Row F, Tip 6	0.560	0.0015	0.265%
Row G, Tip 7	0.559	0.0007	0.132%
Row H, Tip 8	0.557	0.0005	0.092%

FIGURE 2 (a) Baseline-subtracted raw absorbance data for an eight-tip liquid handler dispensing 12 replicates into a 96-well microtiter plate. Using the raw data, the average absorbance values, standard deviation values, and CV values were calculated per sequential dispense number (b) and per tip number (c).

363

nontapered plates have shallowly tapered sidewalls, which is a necessary aide in removing the plates from the molding pins used to form the wells during the plate molding process. Because of this slight taper, and because solutions rarely form perfectly flat menisci, errors in diluent volume will ultimately carry through into the sample volume determination. In most cases, the degree of error associated with uncontrolled diluent volume is a small percentage of the final results; however, this depends on the size of the sidewall taper angle and the degree of meniscus curvature for the solution used. Additionally, a common assumption made when using Beer's law relates to the linear absorbance response of a dye over a given concentration range. In most cases such an assumption is close, but not exact. The best accuracy thus requires a quantitative measure of the degree of nonlinearity in absorbance over the working concentration range for the dye chosen.

Fluorescence-Based Methods As with absorbance, fluorescence spectroscopy is also known and used quite widely in many laboratories. Fluorescence methods are typically well known and provide some unique advantages inherent to this spectroscopic format. Because fluorescence measurements are collected against a significantly lower background signal than exists in absorbance measurements, sensitivity enhancement occurs [29–31]. This sensitivity often leads to a limit of detection of fluorescent dyes that is two to three orders of magnitude better (i.e., a lower limit of detection) than that of an equivalent absorbance measurement. In general, a fluorescent dye can display a linear dynamic operating range four to five orders of magnitude of concentration change, whereas most absorbance dyes only provide a measurable linear concentration change over two to three orders of magnitude. In practice, this means that one fluorescent dye solution will cover a testable volume range that might require two or three different concentrations of absorbance dye to adequately measure the same range.

A fluorescence-based volume measurement assay would proceed in a fashion similar to that of an absorbance method. A premade fluorescent dye solution is dispensed into the wells of a microtiter plate by the liquid handler under evaluation. Some common fluorescent dyes used for volume measurement assays include fluorescein and rhodamine. A diluent, often in the form of a buffered salt solution, is used to fill each well to a defined working volume. The solutions in the plate are mixed, and the fluorescence intensity is measured in each well. The fluorescence intensity, F, is related to the concentration [30] of the fluorescent dye in the plate wells by

$$F = 2.3K'\varepsilon\, bC P_0 \tag{5}$$

where the constant K' is related to the quantum efficiency of the fluorescent dye, ε, b, and C are the Beer's law terms defined above, and P_0 is the power of the light beam incident on the solution. By comparison to Beer's law, the fluorescence intensity is related to the fluorescent dye concentration, *but* it is also related to the quantum efficiency of the dye and the incident power of the lamp. These two additional components can be problematic, as described below.

It should be noted that this equation will only hold for dye concentrations with absorbance values below about 0.05 OD (i.e., $A < 0.05$).

Many of the same details described above for the absorbance approach also apply to fluorescence-based methods used for measuring the precision of the volume delivery. For example, the precision of any tip of a liquid handler can be determined easily by measuring the standard deviation of the fluorescence intensity measured across multiple wells of solution dispensed by each tip. As mentioned, a buffered solution is often used as the diluent because many fluorescent dyes demonstrate a pH-dependent fluorescence. A proper fluorescence method will also require adequate background subtraction using a plate filled with the same diluent that will be used in the volume determination assay.

As with absorbance methods, using a fluorescent dye solution for precision testing of a liquid handler is quite straightforward and typically easy to implement. However, determining the accuracy of liquid delivery using fluorescence is typically more problematic than is observed for its absorbance-based counterpart. Accuracy determination requires more control over not only the dye solution but also the fluorescence plate reader used. The well-known variability inherent to fluorescent dyes caused by quenching and photobleaching, which affect the quantum efficiency of the dyes, can make an accurate volume determination difficult to achieve. Also, because the fluorescence intensity measured is tied to the intensity of incident light from the plate reader, as demonstrated in Eq. (5), the measured response is strongly tied to the particular plate reader used (e.g., fluctuations in lamp intensity can be problematic in the performance verification process).

These two areas pose significant challenges to determining the accuracy of volume delivery when using a fluorescence assay. Despite these challenges, methods exist for overcoming both. For example, an accuracy determination using fluorescent dyes requires development of a standard curve (see below), which is no different than absorbance-based methods. However, because the fluorescence intensity is related to the quantum efficiency of the fluorescent dyes and to the incident power of the light beam emitted by the fluorometer, small deviations in either of these areas requires reestablishment of the standard curve in order to maintain a sufficient level of accuracy in the final volume results. Thus, accuracy determination using fluorescence methods generally requires more work than absorbance methods.

One additional challenge encountered when using the fluorescence approach for accuracy determination is that the measured fluorescence intensity is dependent on the dye concentration. Although the same is true for absorbance methods, the concentration dependency is offset by an inverse change in pathlength, and thus becomes a minor contributor to the error when measuring absorbance in the vertical-beam format (e.g., a plate reader). However, the fluorescent intensity measured is not defined by a pathlength of light, as is the case for absorbance measurements. Instead, fluorescence measurements are collected from an emission volume, which is a function of the collection optics and slit width of the particular instrument used as well as the concentration of the fluorescent dye. This emission volume may or may not extend all the way to the bottom of the well, which in turn may depend on whether the particular fluorescence plate reader is top- or bottom-reading. If the volume over which fluorescence is collected does not extend all the way to the bottom of the well, what happens if you add more diluent? In this case, the solution moves "higher" up in the well because the total volume increases. Also, because the fluorescent dye has been diluted, some change in the emission volume may also occur, but such a change is difficult to predict. Thus, in most cases the fluorescence intensity measured is directly affected by the accuracy of diluent addition. To achieve optimum accuracy when using a fluorescence approach, the accuracy of diluent addition also needs to be known and/or controlled, which poses a different challenge.

As with absorbance-based methods, the fluorescence approach has some limitations on the plate types that can be used. Since many fluorescence-based microtiter plate readers allow for excitation and emission from the top of the plate, a flat, optically clear plate bottom is not needed for fluorescence assays. However, one important feature of the plate used is that it must be made from a material that exhibits a low amount of autofluorescence. Plates that produce a large autofluorescence background significantly reduce the limit of detection achievable by the fluorescent dye used, thereby defeating one of the key advantages of this approach. This problem is widely known and has led to the development of plate types that are designed specifically for fluorescence testing.

Although fluorescence-based methods can be used for determining the accuracy of liquid delivery, it is most often employed as a precision-only test. Because of the effort required to use fluorescence to determine liquid delivery accuracy, it is most commonly combined with other tests which are more suitable for accuracy determination, such as gravimetry. We discuss such approaches below.

Use of a Standard Curve for Accuracy Assessment A standard curve can be generated and used to measure the accuracy of a liquid handler on a tip-by-tip basis. To use a standard curve method, a calibrated pipette or syringe is used by a skilled operator to dispense the desired target volumes of dye solution into a microplate. The buffer, or diluent, also has to be added to the target volume in the microplate. If the buffer is not dispensed accurately and reproducibly, especially for single-dye volume verification methods, the calibration curve can be seriously flawed because the dye could be diluted differently in wells where they should be the same. A dual-dye approach is protected against the quantitative addition of buffer to the target volume, as discussed below. The measured optical signal is plotted versus the calibrated volume of the pipette or syringe (Figure 3a). From these data, a "best fit" standard curve equation is generated over the volume range of interest and is subsequently used in determining the accuracy of the unknown transferred target volume from the liquid handler under evaluation (Figure 3b). If an accurate standard curve is prepared, the measured light signals of a target volume dispensed by a liquid handler under evaluation can be used in the performance verification of said liquid handler. Ideally, the standard curve is linear over the volume range of interest and the linear fit produces an r^2 value close to 1, indicating a good fit. Although the curve in Figure 3 shows a slight deviation

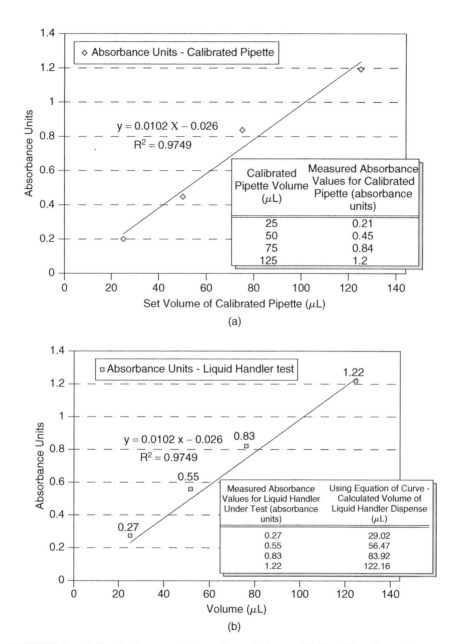

FIGURE 3 (a) Standard curve with a calibrated pipette; (b) liquid handler under test.

from an ideal standard curve (with $r^2 = 0.9749$), the figure is used simply to demonstrate the process for preparing and implementing the standard curve.

Care and skill must be practiced in the generation of the standard curve (e.g., in the preparation of the dye solution (weighing dye powder, filtering, protecting from light, labeling) as well as the use of the calibrated pipettes). Because there are *substantial* differences from operator to operator when using calibrated pipettes [5,6], the standard curve generated by one technician with a calibrated pipette could be quite different from a standard curve generated by a different technician for the same pipette and same solution. The same might also hold true for different standard solutions prepared by different technicians where the dye solutions and/or quality of the solutions and the associated documentation (e.g., SOPs or lab notebooks) could be different from person to person or lab to lab. Also adding to the variability is the amount of buffer added, as alluded to above. Additionally, it might be in the laboratory's best interest to have the dye solutions and standard curve prepared fresh on the same day as the instrument qualification.

Volume verification using a single absorbance-based dye solution will be limited because of the smaller dynamic range of absorbance-based spectrophotometers, which is why some absorbance methods require multiple dye solutions (and multiple standard curves) to cover the volume range of interest. For example, one standard solution might be used to measure a volume range between 1 and 10 μL over an absorbance range of 0.2 to 2 absorbance units for the plate reader. A second standard solution might have a similar absorbance range (0.2 to 2.0) but cover a different volume range, such as 10 to 100 μL. Also, preparing a standard curve for very low volumes (e.g., >1 μL) requires the use of more accurate techniques, due to the rapidly declining accuracy of pipettes or syringes in this low volume range. One example of a more accurate technique would be to prepare a gravimetrically measured large volume dilution of the sample solution in buffer. In this way, the small volume can be scaled to a more accurately measurable volume, and an equivalent dilution can be made by adding the appropriately scaled volume of buffer. The spectroscopic measurements collected will then provide a more accurate standard curve for the optical response at the low volume range. For fluorescence methods, there is normal drift of a plate reader and the quenching and photobleaching of fluorescent dyes. Due to these characteristics of fluorescent methods, the standard curve developed for a given test solution may only be applicable (or reliable) for a very short time (e.g., typically a few days).

Dual-Dye Ratiometric Photometry Method   Although the advantages and disadvantages of absorbance-based methods for volume determination have been described in detail above, there is one form of this method that bears more discussion. By expanding to two the number of dyes in solution, some specific advantages can be gained as compared to single-dye absorbance, single-dye fluorescence, and even gravimetric methods, as described herein. The dual-dye photometric approach has been used to develop two commercially available products:

the Artel PCS Pipette Calibration System, which is used primarily for single-channel pipette calibration, and the Artel MVS Multichannel Verification System, which is used to calibrate multichannel pipettes as well as automated liquid handlers. Both of these products are described in detail elsewhere [3,8–13,17], but the dual-dye absorbance approach on which they are based is described briefly here to provide insight into some of the key advantages of the dual-dye method.

The basis of the dual-dye method resides in using two different types of solution: (1) a sample solution that contains known concentrations of two dyes (dyes 1 and 2), and (2) a diluent solution that contains a known concentration of dye 2 but *at the same concentration as in the sample solution*. The operating principle of this method involves dispensing the target volume of sample solution containing both dyes 1 and 2 into the wells of a microtiter plate. The two different dyes have nonoverlapping absorbance peaks. This sample solution is dispensed by the liquid-handling device that is being tested. The diluent is also dispensed, either before or after the sample dispense, into each well to a defined total working volume (the diluent need not be dispensed by the same device under test). After the solutions are mixed, the absorbance is collected for both dyes in the mixed solutions in the microtiter wells.

The photometric measurements collected for each well containing the mixed sample and diluent solutions are used to calculate the aliquot volume in three steps: (1) calculation of the total solution depth or pathlength (based on the Beer–Lambert law), (2) calculation of the total volume of solution in each microtiter well, and (3) calculation of the volume of sample solution dispensed by the liquid delivery device under test.

Because the diluent contains the same known concentration of dye 2 as does the sample solution, no dilution of this dye occurs when diluent is added to the sample, and Beer's law [Eq. (3)] can be used to determine the pathlength of light (l) through the total volume of solution in each well. By measuring the absorbance of dye 2 ($A_{\lambda 2}$) and incorporating the known absorbance per unit pathlength of dye 2 ($a_{\lambda 2}$), the sample solution depth is determined by

$$l = \frac{A_{\lambda 2}}{a_{\lambda 2}} \tag{6}$$

Once the sample solution depth is known, the geometrical equation for the volume of the wells in the microtiter plate can be used to determine the total volume of sample solution in the wells. For common 96-well microtiter plates, the total volume can be determined using the geometry for a truncated cone. For a truncated cone shape, calculation of the total volume, V_T, is based on the total solution depth, l, as determined from Eq. (6), and the taper angle, θ, and diameter, D, of the microtiter plate wells, as shown in

$$V_T = \pi l \frac{D^2}{4} + \pi D l^2 \frac{\tan\theta}{2} + \pi l^3 \frac{\tan^2\theta}{3} \tag{7}$$

The reader will note that Eq. (7) is a standard geometrical relation used for calculating the volume of a section of a truncated cone with a base diameter D, a height l, and a sidewall taper θ. The calculation of the sample volume, V_S, is based on the total solution volume, V_T, the measured absorbance of dye 2, $A_{\lambda 2}$, the measured absorbance of dye 1, $A_{\lambda 1}$, the quantities of the absorbance per unit pathlength of dye 2, $a_{\lambda 2}$, and the absorbance per unit pathlength of dye 1, $a_{\lambda 1}$, as given by

$$V_s = V_T \left(\frac{a_{\lambda 2}}{(a_{\lambda 1})} \frac{A_{\lambda 1}}{(A_{\lambda 2})} \right) \tag{8}$$

Although it may not be directly apparent, Eq. (8) has been derived from the Beer's law relationships for two different dyes. This relationship is established through use of the fixed concentration of dye 2 in both the diluent and the sample solution [9]. By controlling this concentration, the calculation required for the volume of sample solution becomes independent from calculation of the total volume.

As can be surmised, one of the key advantages of the dual-dye photometric method is that the sample volume can be determined independently from the diluent volume. Because of the dual-dye ratiometric calculations that are conducted with this method, errors in diluent volume are accounted for.

Combination of Methods for Volume Verification   There are several methods described in this chapter that are available to lab practitioners for measuring the performance of liquid delivery devices. As described above, some methods are more readily applicable for either accuracy measurements or precision measurements, but few methods are easily employed to collect both accuracy and precision information (one exception would be the dual-dye photometric method described above). For example, when using a microtiter plate-based assay, gravimetry can be used to determine the average volume of solution in each well of the plate. However, a determination of the precision is not possible using this approach unless the replicate dispenses are weighed individually. By comparison, an absorbance- or fluorescence-based approach readily allows for precision determinations but requires additional work to make accuracy measurements.

In response to some of the shortcomings of the individual methods described above, some practices have been developed that employ a combination of methods. One such combination that has been used for testing the performance of pipetting heads with ganged tips, such as are used on most 96-tip pipetting robots, includes a combination of gravimetry and fluorometry. The gravimetric method used does not involve pipetting a full plate. Instead, the solution dispensed from some (typically, one to four tips) of the tips is carefully collected and weighed on the balance. This process allows for a determination of how accurately each of the few measured tips is performing. Following this test, a fluorescent dye is dispensed by all 96 tips into a 96-well plate and the fluorescence intensity is measured. The relative standard deviation in measured fluorescence intensity between all wells provides a measure of the precision between each of the 96 tips. Also, although only the accuracy of a few of the tips was measured gravimetrically, a relative accuracy of each tip can be determined from the fluorescence data collected. To do this, the gravimetric volume collected for one of the tips is linked to the fluorescence intensity measured for the volume dispensed by that tip. The measured fluorescence values for all other wells are then normalized to that gravimetrically determined volume, thus allowing a relative accuracy determination for each tip used.

Another example of a combination of methods links a tedious, careful gravimetric accuracy determination, performed annually or semiannually, with a more easily implemented absorbance method used for interim (e.g., daily, weekly, or monthly) accuracy and precision testing. Yet another example combines a gravimetric determination used for relatively large volumes (e.g., 100 μL to 5 mL), an absorbance method for lower volumes (e.g., 30 nL to 100 μL), and a fluorescence method for very low volumes (e.g., 1 to 30 nL). In each case, the advantages of each of the methods described are exploited, to provide an overall solution to volume delivery measurements.

Other Volume Verification Methods   There are other volume verification methods that are not discussed here but are discussed in the literature. For the optical determination of total volume in containers, such as the wells of microplates, a dual-wavelength process that determines a total solution pathlength based on measurements of two solvent absorbance bands is employed in commercial plate readers [32,33]. Another method used to determine total solution volume is based on acoustic measurement between a transducer head and the solution in a well, thereby determining the solution height. Yet another method employs titration to determine the uncertainty of the volume measurements [26].

4 IMPORTANCE OF STANDARDIZATION

The need to ensure quality in a laboratory process has become increasingly important [34–36]. As the scientific community becomes more dependent on regulatory compliance, documentation, and system validations, volume verification methods that are traceable to international standards may, in fact, become the standardized approach for quantifying dispensed volume. Traceable verification methods may allow for the creation of standard operating procedures for liquid handlers in laboratories that follow good manufacturing practices (GMPs) or similar guidelines. Traceable verification methods may also help when developing a performance qualification for an isolated volume transfer task or for a group of tasks. Volume transfer for critical target reagent screening should be standardized to compare all liquid handlers within a laboratory, and the verification method should try to mimic the assay transfer task (e.g., follow the same automated parameters as the assay). Ideally, the volume verification method used for liquid transfer performance evaluation would have measurable results that are linked, or traceable, to an international standard unit of measurement. Such standards are defined in the international system of units and maintained by national standards organizations such as the National Institute of Standards and Technology. A standard method of instrument qualification would help laboratories with liquid-handling quality assurance. By using a standardized method, liquid handlers with slight differences in performance could be identified rapidly, which would be very important if the systems were used in parallel (e.g., in assay scale-up, transfer, or for preparing reproducible samples). Comparisons between liquid handlers using a standard volume verification method would be ideal when equipment is new, if the equipment is transferred from another laboratory, if the equipment has been dormant and is being brought back online, or if the equipment is being compared directly to a standard or calibrated liquid handler.

An evaluation method should be standardized, fast, easy to implement, and should minimize instrument downtime and required resources. Volume transfer for critical target reagent screening should be compared for all devices within a process, especially for liquid handlers that are performing similar or identical tasks. If liquid handlers in Boston and London are performing the same tasks for the same company, those systems should be evaluated directly using a standardized procedure on the basis tip-by-tip accuracy and precision. Additionally, a standard volume verification method would also offer the opportunity to understand liquid-handler device behavior for quality control purposes, trending patterns, diagnostic troubleshooting, method transfer, factory and site acceptance testing, and employee training.

5 SUMMARY

If automated liquid handlers are not dispensing the exact, or desired, volume for critical reagents, it is likely that unseen error can propagate increasingly as a process continues. Without knowing the exact volume transferred at each step from each liquid handler, for instance, true concentrations of species in solution may be unknown and results could be interpreted falsely. Even slight discrepancies in delivered volume can compromise results, leading to poor-quality (or useless) data and downstream costs associated with remedial actions. The economic impact of allocating resources for research or production efforts, which is based on potentially false results, may be severe. Moreover, if the liquid-delivery systems are overdelivering target volumes of expensive and/or rare reagents, there may be a significant economic impact, due to the loss of precious materials. The resulting downstream economic loss suggests that a method of performance evaluation should be implemented continuously to verify that all liquid handlers are dispensing critical volumes accurately. As process control within a laboratory continues to be emphasized, a volume verification method should be implemented so that liquid-handler behavior is known and optimized to deliver the desired target volumes for all levels of assay development. A robust and reliable volume verification method is an essential tool for knowing an assay's exact volume and component concentrations, which is critical for assay interpretation, optimization, avoiding unnecessary downstream costs, and achieving liquid delivery quality assurance. To qualify instruments and evaluate performance, as well as reduce and minimize errors from liquid-handling systems, it is in a laboratory's best interest to quickly identify those systems that are failing by implementing regular calibration programs and verification checks for volume transfer accuracy and precision. Instrument qualification and performance verification should be obtained through liquid-delivery quality assurance programs that focus on accurate and precise transfer and measurement of target volumes.

REFERENCES

1. N. Benn, F. Turlais, V. Clark, M. Jones, and S. Clulow. An automated metrics system to measure and improve the success of laboratory automation and implementation. *J. Assoc. Lab. Autom.*, 11:16–22, 2006.

2. M. Atekar et al. Assay optimization: a statistical design of experiments approach. *J. Assoc. Lab. Autom.*, 11:33–41, 2006.

3. H. Dong, Z. Ouyang, J. Liu, and M. Jemal. The use of a dual dye photometric calibration method to identify possible sample dilution from an automated multichannel liquid-handling system. *J. Assoc. Lab. Autom.*, 11:60–64, 2006.

4. P. B. Taylor et al. A standard operating procedure for assessing liquid handler performance in high-throughput screening. *J. Biomol. Screen.*, 7:554–569, 2002.

5. R. Curtis. Performance verification of manual action pipettes: 1. *Am. Clin. Lab.*, 12(7): 8–10, 1994.

6. R. Curtis. Performance verification of manual action pipettes: 2. *Am. Clin. Lab.*, 12(9): 16–17, 1994.

7. D. N. Joyce, J. P. P. Tyler. Accuracy, precision, and temperature dependence of disposable tip pipettes. *Med. Lab. Tech.*, 30:331–334, 1973.

8. D. R. Rumery, A. B. Davis, A. B. Carle, and G. R. Rodrigues. Errors associated with pipetting warm and cold liquids. Poster presented at AACC Annual Meeting and Clinical Lab Expo, San Diego, CA, July 15–19, 2007.

9. J. T. Bradshaw, T. Knaide, A. Rogers, and R. H. Curtis. Multichannel verification system (MVS): a dual-dye ratiometric photometry system for performance verification of multichannel liquid delivery devices. *J. Assoc. Lab. Autom.*, 10:35–42, 2005.

10. K. J. Albert, J. T. Bradshaw, T. R. Knaide, and A. L. Rogers. Verifying liquid handler performance for complex or non-aqueous reagents: a new approach. *J. Assoc. Lab. Autom.*, 11:172–180, 2006.

11. T. R. Knaide, J. T. Bradshaw, A. L. Rogers, C. McNally, B. W. Spaulding, and R. H. Curtis. Rapid volume verification in high-density microtiter plates using dual-dye photometry. *J. Assoc. Lab. Autom.*, 11:319–322, 2006.

12. K. J. Albert, and J. T. Bradshaw. Importance of integrating a volume verification method for liquid handlers: applications in learning performance behavior. *J. Assoc. Lab. Autom.*, 12:172–180, 2007.

13. J. T. Bradshaw, R. H. Curtis, T. R. Knaide, and B. W. Spaulding. Determining dilution accuracy in microtiter plate assays using a quantitative dual-wavelength absorbance method. *J. Assoc. Lab. Autom.*, 12:260–266, 2007.

14. B. W. Spaulding, J. T. Bradshaw, and A. Rogers. A method to evaluate mixing efficiency in 384-well plates. Poster presented at SBS 12th Annual Conference, Seattle, WA, Sept. 17–21, 2006. Also, B. W. Spaulding, L. Borrmann, J. T. Bradshaw, and W. Wente. Photometric measurement of mixing efficiency using the Eppendorf Mixmate mixer. Poster presented at SBS 13th Annual Conference, Montreal, Quebec, Canada, Mar. 15–18, 2007. Also, B. W. Spaulding, J. T. Bradshaw, P. Chang, and I. Feygin. Optimization of the Techelan TEOS orbital shaker using a dual-dye photometric protocol. Poster presented at Lab Automation 2007, Palm Springs, CA, Jan. 27–31, 2007.

15. *Standard Specification for Piston or Plunger Operated Volumetric Apparatus*. ASTM E1154-1989. ASTM, West Conshohocken, PA, 2008.

16. *Piston Operated Volumetric Instruments*, Parts 1–7. ISO 8655. ISO, Geneva, Switzerland, 2002.

17. R. H. Curtis, and A. E. Rundell, inventors; Artel, assignee. Reagent system for calibration of pipettes and other volumetric measuring devices. U.S. patent 5,492,673, Feb. 20, 1993.

18. E. L. McGown, K. Schroeder, and D. G. Hafeman. Verification of multi-channel liquid dispenser performance in the 4–30 μL range by using optical pathlength measurements in microplates. *Clin. Chem.*, 44:2206–2208, 1998.

19. E. L. McGown, and D. G. Hafeman. Multi-channel pipettor performance verified by measuring pathlength of reagent dispersed into a microplate. *Anal. Biochem.*, 258:155–157, 1998.

20. J. Peterson and J. Nguyen. Comparison of absorbance and fluorescence methods for determining liquid dispensing precision. *J. Assoc. Lab. Autom.*, 10:82–87, 2005.

21. J. P. Clark and H. Shull. Gravimetric and spectrophotometric errors impact on pipette calibration certainty. *Calibr. Lab.*, pp. 31–38, Jan.–Mar. 2003.

22. R. Pavlis. Surprising statistics on pipet performance. *Am. Lab.*, pp. 8–9, Mar. 2004.

23. P. B. Taylor. Optimizing assays for automated platforms: experimental design and automation accelerate the development of drug assays. *Modern Drug Discov.*, 5(12): 37–39, 2003.

24. M. J. Felton. Liquid handling: dispensing reliability. *Anal. Chem.*, 75:397A–399A, 2003.

25. M. Connors and R. Curtis. Pipetting error: a real problem with a simple solution: 2. *Am. Lab News*, 31(25): 12, 1999.

26. J. Peters. Determination of uncertainty for volume measurements made using the titration method. *Am. Lab.*, pp. 14–22, Oct. 2004.

27. H. Rhode, M. Schultze, S. Renard, P. Zimmerman, T. Moore, G. A. Comme, and A. Horn. An improved method for checking HTS/μ HTS liquid-handling systems. *J. Biomol. Screen.*, 9:726–733, 2004.

28. R. H. Curtis, inventor; Artel, assignee. Photometric calibration of liquid volumes. U.S. patent 6,741,365. May 25, 2004.

29. J. D. Ingle and S. R. Crouch. *Spectrochemical Analysis*. Prentice Hall, Englewood Cliffs, NJ, 1988.

30. D. A. Skoog, F. J. Holler, and T. A. Nieman. *Principles of Instrumental Analysis*, 5th ed. Brooks/Cole Thomson Learning, Belmont, CA, 1998.

31. J. R. Lakowicz. *Principles of Fluorescence Spectroscopy*, 2nd ed. Kluwer Academic/Plenum Publishers, New York, 1999.

32. D. G. Hafeman and C. T. Chow, inventors; Molecular Devices Corporation, assignee. Determination of light absorption pathlength in a vertical-beam photometer. U.S. patent 5,959,738. Sept. 28, 1999.

33. P. Hale. *Converting Spectrophotometric Assays to Biotek's Microplate Readers: Determination of Microplate Path Lengths*. Application Note. BioTek Instruments, Winooski, VT.

34. E. Fowler. Analytical technologies for real-time monitoring of biopharmaceutical manufacturing processes. *Am. Lab.*, pp. 30–34, Feb. 2006.

35. *Guidance for Industry: Process Analytical Technology, a Framework for Innovative Pharmaceutical Development, Manufacturing and Quality Assurance*. FDA, Washington, DC, 2004.

36. *Q8 Pharmaceutical Development*. International Conference on Harmonization Draft Guidance. FDA, Washington, DC, 2004.

16

PERFORMANCE QUALIFICATION AND VERIFICATION IN POWDER X-RAY DIFFRACTION

ANICETA SKOWRON

Activation Laboratories Ltd.

1 INTRODUCTION

In this chapter we briefly outline the methodology and steps for performance qualification and routine performance verification of a powder x-ray diffraction instrument. The basic principles of x-ray diffraction are outlined first, followed by a brief description of the essential instrument components for diffraction measurement, main sources of experimental errors, and finally, the description of essential performance qualification and verification steps and procedures. Powder x-ray diffraction is already a mature technique and is applied widely in the pharmaceutical industry as both a research and development (R&D) tool and for quality control tasks [1]. Typically, pharmaceutical applications concentrate on identification and characterization of polymorphic forms of active pharmaceutical ingredients, monitoring the stability of the chosen crystal form and on the characterization of polymorphic-phase transformations, including transitions between metastable amorphous and crystalline forms.

As with many other techniques, the specifics of the application dictate the measurement methodology, required precision, and critical variables and errors. When powder x-ray diffraction is used for R&D purposes full crystallographic

Practical Approaches to Method Validation and Essential Instrument Qualification,
Edited by Chung Chow Chan, Herman Lam, and Xue Ming Zhang
Copyright © 2010 John Wiley & Sons, Inc.

characterization of the sample is usually aimed for, and consequently, the instrumental and experimental demands are the highest. In the pharmaceutical quality control context, the usual applications concentrate on the confirmation of polymorphic form identity, purity, and stability—most often by comparison with known standards. Although at times some of the instrumental attributes, such as resolution, may not be as critical as for the R&D and structural study application, for quality control there is particular stress on adequate, monitored, and well-documented performance qualification and verification of the diffraction instrument.

2 BASICS OF X-RAY DIFFRACTION

The field of crystallography is facilitated by the interaction (i.e., diffraction) of waves with periodic objects such as crystals. A crystal, defined broadly as a periodic arrangement of atoms (or molecules) in three dimensions, is shown schematically in Figure 1 as a two-dimensional array of points. The crystal can also be viewed as a collection of imaginary atomic planes extending in various directions and separated by characteristic distances d. X-ray wave is sensitive to such regular atomic planes and diffracts from them; the angle of diffraction depends on the separation of the planes.

Diffraction of an x-ray beam from a crystal was first described in 1913 by English physicists Sir W. Lawrence Bragg and his father, Sir W. H. Bragg, to explain why the cleavage faces of crystals appear to reflect x-ray beams at certain angles of incidence. Lawrence Bragg formulated a general relationship governing the diffraction from crystals, known since as Bragg's law:

$$n\lambda = 2d \sin \theta \tag{1}$$

where λ is the wavelength of the x-ray radiation, d the separation of the diffracting atomic planes in the crystal, θ the angle at which the diffracted x-ray beam is observed, and n is an integer number. Figure 2 illustrates Bragg's law.

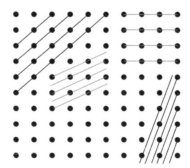

FIGURE 1 Two-dimensional crystal lattice viewed as a collection of sets of equidistant atomic planes.

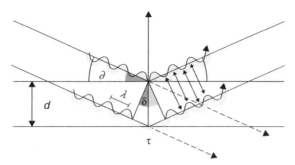

FIGURE 2 Diffraction of an x-ray beam from atomic planes separated by a distance d. The reader may find an interactive illustration of Bragg's law at www.journey. sunysb.edu/ProjectJava/Bragg/home.html.

Figure 3 shows how successive peaks originate from diffraction of the x-ray beam from different sets of atomic planes: The smaller the diffraction angle, the larger the distance between the planes. The powder diffraction pattern observed is a collection of the diffracted peaks originating from all the accessible crystal planes in a powder sample. Since a particular crystal form can be considered as a unique collection of atomic planes, separated by their unique interatomic distances, the resulting diffraction pattern is a unique sequence of peaks and, consequently, can serve as the crystal fingerprint [2].

A diffraction pattern contains information about both peak positions and peak intensities. The peak positions, expressed as diffraction angles, 2θ, or equivalently, as distances between the reflecting planes, d, identify the crystal form and can be used for characterization of its unit cell and symmetry. The determination of the angular peak positions and their intensities is the core x-ray diffraction measurement. The peak intensities reflect the total scattering from each set of planes in the crystal and are directly related to their atomic makeup. However, while the accuracy of the measurement of peak positions for a well-aligned instrument in a given laboratory can be $\pm 0.01°$ 2θ, the determination of correct peak intensities is much more difficult since the intensities can be modified severely by the sample form, most notably by preferred orientation. Nevertheless, if care is taken in the sample preparation step and suitable corrections are applied, the peak intensities can be used for the refinement, or determination, of the atomic positions in the unit cell of the crystal structure [3].

2.1 Powder X-ray Diffraction Measurement

Most commercial x-ray diffraction systems employ Bragg–Brentano parafocusing geometry, developed in 1950s [4], in which a divergent beam of radiation, properly collimated, strikes a flat sample at an angle θ and, after diffracting from it, leaves the sample at an angle 2θ to the incident beam (and θ to the sample surface), as illustrated in Figure 4 [5]. After leaving the sample, the diffracted beam is guided to a detector by passing through a collimator,

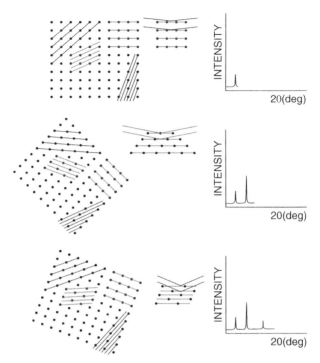

FIGURE 3   Each diffraction peak originates from diffraction off a different set of atomic planes. The diffraction from the set of planes with the largest separation results in the peak at the lowest angle 2θ.

receiving slits, and a monochromator. The x-ray source and the detector, each with its own beam collimation assembly, are placed on the circumference of a circle (goniometer circle) and the sample is placed at the center of the circle. Depending on the configuration adopted in each particular system, two of the three elements—x-ray source, sample, or the detector—are rotated with appropriate relative speeds such that the radiation is always diffracted by the sample at an angle of 2θ to the incident beam, and the detector is rotated to be at that particular angle to record and measure the diffracted radiation. By changing the angle continuously among the x-ray source, the sample, and the detector, at a controlled rate and between preset limits, an x-ray diffraction pattern is obtained.

Care must be taken to align a diffractometer properly for optimal performance. It is most important to set the goniometer zero to obtain collinearity of the x-ray source, the sample surface, and the detector. The second most important setting is the speed of the relative θ/2θ (or θ/θ) movement of the detector and sample (or source): Errors in these adjustments will introduce errors in the measured 2θ positions of the peaks. With the great many models of diffractometers available commercially, and the details of the alignment specific to each instrument, the reader is referred to the instrument manual and/or vendor technical help for

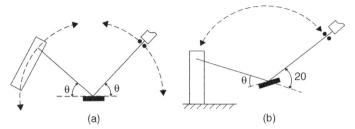

FIGURE 4 Bragg–Brentano parafocusing diffraction geometry configurations. In the $\theta/2\theta$ configuration (a) the sample is stationary while the x-ray source and the detector rotate. In the θ/θ configuration (b) the x-ray source is stationary while the sample and the detector rotate. (From [5].)

detailed alignment procedures and tools, and to x-ray diffraction literature for general information on diffraction geometry and physics [2]. Here we focus on the postalignment performance qualification checks and describe the steps for routine performance verification.

2.2 Main Sources of Errors in Powder X-ray Diffraction

Although the proper alignment of a powder diffractometer profoundly affects the diffraction results, there are many other sources of errors that also affect the quality of x-ray diffraction measurement. Table 1 summarizes the more common types. In addition to those related to instrument alignment, there are also errors due to basic instrument geometry, specimen preparation, presentation, and some incurred during data acquisition and processing [6]. As indicated in the table, the performance of an x-ray diffraction instrument can be evaluated by eliminating, or keeping constant, all the systematic errors other than those related to the instrument alignment and basic instrument geometry. Consequently, to eliminate the variability in sample preparation, permanently mounted standards are recommended for the instrument performance checks.

3 PERFORMANCE QUALIFICATION

After an initial installation, alignment, and operation qualification of a newly installed powder x-ray diffractometer, its performance qualification must be carried out to verify that the alignment is adequate for sufficiently accurate measurements. A similar performance qualification should be carried out after each major realignment of the instrument. To diagnose the instrument alignment, a diffraction pattern can be obtained using a suitable reference standard that will allow three performance qualification parameters of the diffraction pattern to be examined.

TABLE 1   Main Sources of Errors in Powder X-Ray Diffraction

Error Source	Effect on Diffraction Pattern	Comment
Instrument Alignment		
Error in the zero 2θ position	Horizontal shift of the entire diffraction pattern; all peak positions changed by an angular increment	All the instrument misalignment errors are the object of performance checks.
Missetting of the 2 : 1 adjustment (the receiving slit should rotate at exactly twice the angular speed of the sample)	Contraction, or stretching, of the entire diffraction pattern; incorrect peak positions	
Other instrument misalignment (axial alignment, slit alignment)	Asymmetrical peak shape; general low peak intensity; incorrect relative intensity of the peaks; decreased resolution	
Basic Instrument Geometry		
Flat specimen error	Influences peak shape and slightly peak position	These errors cannot be avoided; they are inherent to the geometry.
Axial x-ray beam divergence		
Sample Presentation and Form		
Sample displacement due to instrument misalignment	Change in peak positions and decrease in intensity	This error is minimized in the diffractometer alignment step.
Sample displacement due to packing into the sample holder	Angle-dependent shift of the entire diffraction pattern; all peak positions changed by varying angular increment	
	Decrease in peak intensity with increasing displacement	
Sample transparency due to sample porosity or a low mass absorption coefficient	Asymmetry of the peaks—tailing toward lower angles	The sample errors are minimal for performance check using a suitable, permanently mounted standard.
Preferred orientation of crystallites; the sample is not truly random	Incorrect relative peak intensities in comparison with a truly random sample	

TABLE 1 (*Continued*)

Error Source	Effect on Diffraction Pattern	Comment
Increase in the crystal defect density introduced by excessive grinding of the sample	Peak broadening	
Data Acquisition and Processing		
Counting statistics—poor selection of scanning rate	Influences the precision of intensity data	These errors can be avoided by validation of scanning parameters.
Poor selection of step width in step scanning		
Peak distortion due to Kα_1 and Kα_2 wavelengths[a]	Peak displacement to higher angles	Most XRD manufactures provide Kα_2 peak stripping software.

Source: Data from [6–8].

[a] A copper source emits x-ray radiation that has two close wavelengths: CuKα_1 (1.54056 Å) and CuKα_2 (1.54439 Å). Unless a special monochromator is used, both wavelengths contribute to the diffraction pattern. Most processing software allows mathematical subtraction from the diffraction pattern of the contribution from Kα_2.

3.1 Peak Positions: Instrument Linearity

When the diffraction peak positions are measured using a reference standard with well-known lattice parameters, the peak positions expected may be calculated using Bragg's law and compared with the observed peak positions. An agreement between the peak positions expected and measured over the typical angular range of 20 to 120°, 2θ, constitutes the first essential check on the performance of the diffraction instrument. There are several choices of reference standards certified for the measurement of peak positions, as listed in Table 2. Apart from the National Institute of Standards and Technology's (NIST's) SRM 1976 sintered alumina plate, most of the other standards, such as silicon or lanthanum hexaborate, are supplied as powders and require preparation for permanent mounting in the x-ray holder to minimize the variability associated with sample preparation. For this purpose the powders may be embedded in an epoxy resin, the resulting plate cemented into the sample holder, and its surface ground to an approximate flatness of ±10 μm.

   In some diffractometers the incident x-ray beam passes through a variable-rather than a fixed-width divergence slit. If a variable divergence slit is used, it is also useful to check the low-angle alignment by measuring the position

TABLE 2 Common Powder X-ray Diffraction Standards

Standard Material	Traceability	Physical Form	Certified Parameters
Si	NIST SRM 640d	Powder	Line positions and line shape
LaB_6	NIST SRM 660a	Powder	Line positions and line shape
Al_2O_3	NIST SRM 1976a	Sintered plate	Relative intensities and line positions
Al_2O_3	NIST 676a	Powder	Relative intensities and line positions
ZnO, TiO_2, Cr_2O_3, CeO_2	NIST SRM674b	Powders	Relative intensities and line positions
Mica	NIST SRM 675	Powder and solid	Line positions, low 2θ ($\sim 8.8°$)
Silver behenate	ICDD round robin	Powder	Line positions, very low 2θ (1.5 to 20°)
Quartz–novaculite	Secondary standard	Polycrystalline plate	Performance evaluation: instrumental response and resolution

of a low-angle (large d-spacing) peak [9]. Certified standards for low-angle calibration are somewhat sparse. NIST supplies mica standard for calibration of peak positions above 8.8°. For lower angles, silver behenate, $C_{22}H_{44}O_2$.Ag, with d_{001}-58.37′ can be used for calibration between 1.5 and 20°, 2θ. Although silver behenate has been characterized in a round-robin study coordinated by the International Center for Diffraction Data [10] and is available commercially [11], it has not yet been certified by NIST.

After measuring the peak positions using a chosen calibration standard, it is useful to plot versus 2θ the differences between the peak positions observed and expected for the entire angular range [6]. The example in Figure 5 shows a typical calibration curve for an α-alumina reference standard measured over the range 20 to 90°, 2θ. The horizontal line at $\Delta 2\theta = 0.0$ represents the "theoretical curve" that would be obtained in the absence of all errors and distortions. The "practical curve" is obtained by fitting a least-squares polynomial through the points observed. The practical curve includes the peak shifts due to all inherent instrumental aberrations but does not incorporate the specimen displacement, which will vary from one specimen to another. Although the calibration curve can be predicted theoretically, it is best to determine this curve experimentally after each instrumental alignment as a part of instrument performance qualification.

It is also useful to define an acceptable range of spread of the data points to the calibration curve. Dashed lines in Figure 5 show acceptance limits of $\pm 0.04°$, but the value chosen in a particular laboratory will depend on the accuracy being sought.

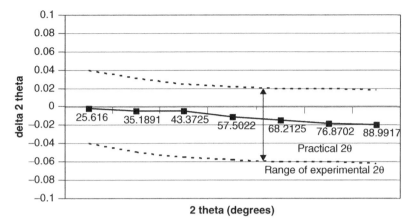

FIGURE 5 An experimental calibration curve for α-alumina is shown with squares. An example of acceptable range of spread for $\Delta 2\theta$ is indicated by dashed lines. (After [6].)

3.2 Peak Intensities: Instrument Sensitivity

Instrument sensitivity is determined by the quality of the peak intensity measurement. Alumina sintered plate NIST SRM 1976 is the most widely used peak intensity standard, with 14 diffraction peaks selected as the intensity reference. By plotting the ratios of measured relative intensities to the certified values vs. the 2θ angle, the diffractometer sensitivity curve, as shown in Figure 6, will be obtained. Published results from intensity measurements in round-robin tests [12] using NIST SRM 1976 alumina plate indicate that 5 to 15% intensity variations

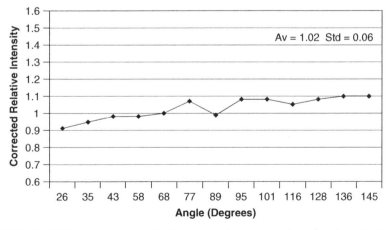

FIGURE 6 Diffractometer sensitivity curve from a well-performing instrument. (From [6].)

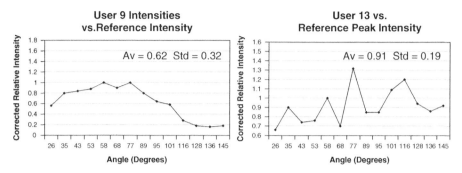

FIGURE 7   Sensitivity curves from poorly performing diffractometers. (From [6].)

within the 2θ range may be found for specific instruments and diffraction optics. If larger differences are observed, such as those shown in Figure 7, then a problem with instrument alignment or functioning of some accessory may be indicated. The performance of variable divergence slits have been shown to be a sensitive spot that could be diagnosed using a sensitivity curve.

3.3   Peak Width: Instrument Resolution

The instrument resolution is defined as its ability to resolve closely spaced peaks. Resolution of a powder diffractometer can be expressed as the angular peak width, HWHM (half width at half maximum). To demonstrate an instrument's resolution, samples with well-defined crystallite size, such as Si or LaB_6 standards, are typically used by diffractometer manufacturers. The curve in Figure 8 shows an example of resolution of a modern state-of-the-art diffractometer measured using LaB_6 and expressed as HWHM. Also shown the dependence of the resolution on the diffraction angle 2θ.

An alternative and historically earlier method for expressing instrument resolution uses the quartz quintuplet. The diffraction pattern from a plate of polycrystalline α-quartz in the range 67 to 69°, 2θ (Figure 9) shows the quartz quintuplet. Three sets of diffracting planes—(212), (203) and (301)—give peaks in this range. Since each peak is a doublet arising from the two wavelengths of the x-ray radiation, CuKα₁ (1.54056 Å) and CuKα₂ (1.54439 Å), a total of six peaks should be present in this angular range. In practice, however, two of the six peaks are superimposed, and the total number of peaks is five, giving rise to the characteristic five-peak pattern known as a quartz quintuplet or five fingers of quartz.

The resolution is defined as the ratio H_1/H_2, where H_1 is defined as the intensity of the (212) $K\alpha_2$ line and H_2 is the average value of the three lowest "troughs" in the five-finger pattern. In the example above the resolution is $3.35/1.57 = 2.15$. The average HWHM of the peaks shown in Figure 9 is 0.09°, which would place the instrument on which the reflections were measured as having only a moderate resolution.

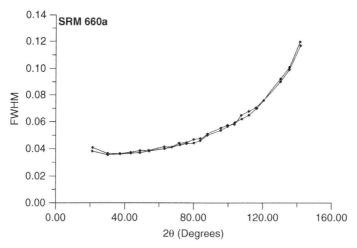

FIGURE 8 Example of instrumental resolution and its angular dependence for a modern state-of-the art diffraction instrument. (From www.bruker-axs.de/uploads/tx₁inkselectorforpdfpool/LynxEye_DOC-S88-EXS027ᵥ3₁ow.pdf.)

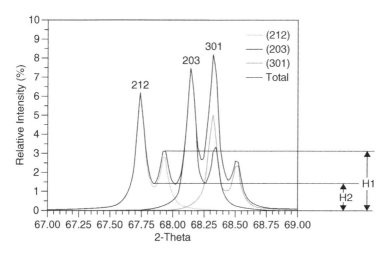

FIGURE 9 Diffraction pattern from a plate of polycrystalline α-quartz in the range 67 to 69°, 2θ. The quartz quintuplet is used as a measure of instrument performance. (From [6].)

4 PERFORMANCE VERIFICATION: CALIBRATION PRACTICE

Most practitioners in the field of x-ray diffractometry recommend weekly analysis of one or two peaks for position, intensity, and width at both low and high 2θ angles. Such a practice will assure confidence in the instrument performance and help to detect developing problems since the set of these six measurements constitutes a fairly powerful diagnostic tool [6]. A deterioration of alignment will probably result in one or more of the following changes:

1. An overall loss of intensity
2. An angle-dependent loss of intensity
3. A systematic shift in all 2θ values
4. An angle-dependent shift in 2θ values
5. Loss of resolution (i.e., distortion in peak shape)

Table 3 shows how the changes in the peak positions, intensities, and widths of two reflections, one at a low and one at a high 2θ angle, may be linked with instrument misalignment. Details will vary with the instrument, and the manual and/or vendor should be consulted for help if necessary.

The suitability of a particular standard for performance verification depends somewhat on the focus of each laboratory, i.e., the predominant questions being asked. Generally, there are no hard and fast rules, but the best guidance is dictated by a balance between the value of each measurement and the number of standards that have to be acquired.

In the context of the pharmaceutical industry, there are perhaps two major types of questions: quality control questions, concerned primarily with polymorphic purity, and more general questions, typically asked by R&D staff, such as details of phase transformations or structure solution. In the context of applications to pharmaceutical quality control, the stronger emphasis tends to be on the accurate measurement of peak positions rather than on accurate measurement of peak intensity. This is caused by the fact that pharmaceutical materials tend to give x-ray diffraction patterns with intensities strongly dependent on details of

TABLE 3   Calibration Data as an Indicator of Misalignment

Low θ_1	High θ_2	Intensity 1	Intensity 2	Resolution[a]	Probable Cause
Same	High	Same	Same	Same	2 : 1 error[b]
High/low	High/low	Same	Same	Same	Zero error or sample displacement
Same	Same	Same	Low	Same	Divergence slit
Same	Same	Low	Low	Same	Monochromator or PHS[c]
Same	Same	Low	Low	Low	X-ray tube aging
Same	Same	Low	Low	Low	Wrong receiving slit

Source: Adapted from [6].
[a]Either with one selected reflection or the resolution value obtained from the quartz quintuplet.
[b]Relative speed of sample and detector rotation.
[c]PHS, pulse height selector.

particle size and morphology (e.g. display strong preferred orientation). Also, the detailed check of the peak shape is usually not critical, and it suffices to ensure the resolution (i.e., the HWHM of a selected peak) is maintained. For the calibration practice to reflect the common application, it might be sufficient to use a standard such as alumina plate, which has the additional benefit that it does not require embedding in an epoxy.

On the other hand, if a laboratory is concerned with pharmaceutical research and development and desires to perform structure solutions or quantitative analysis, then apart from calibration for the peak positions, detailed calibration of instrument sensitivity (i.e., accuracy of intensity measurement) and check of peak shape become essential [3]. In this case, use of two or three standards might be necessary: Si or LaB_6 for the peak shape, Al_2O_3 for peak intensities, and either one for the peak's position. If very low angles are involved, or a variable divergence slit is used, an additional low-angle standard may also be appropriate.

REFERENCES

1. *U.S. Pharmacopeia*, General Chapter <941>, X-Ray Diffraction. USP, Rockville, MD, 2006.

2. B. D. Cullity. *Elements of X-Ray Diffraction*, 2nd ed. Addison-Wesley, Reading, MA, 1977.

3. R. A. Young, Ed. *The Rietveld Method*. Oxford University Press, New York, 1981.

4. W. Parris, E. A. Hamacher, and K. Lowitzsch. The Norelco x-ray diffractometer. *Phillips Tech. Rev.*, 16:123–133, 1954.

5. R. Jenkins. Instrumentation. In *Modern Powder Diffraction*, D. L. Bish and J. E. Post, Eds. Mineralogical Society of America, Washington, DC, 1989, p. 22.

6. R. Jenkins and R. L. Snyder. *Introduction to X-Ray Powder Diffractometry*. Wiley, New York, 1996.

7. V. E. Burhke, R. Jenkins, and D. K. Smith. *Practical Guide to the Preparation of Specimens for X-Ray Fluorescence and X-Ray Diffraction Analysis*. Wiley-VCH, Weinheim, Germany, 1997.

8. F. H. Chung and D. K. Smith. *Industrial Applications of X-Ray Diffraction*, 1999.

9. R. Jenkins, T. Hom, C. Villamizar, and W. N. Schreiner. Calibration of the powder diffractometer at low values of 2θ. *Adv. X-Ray Anal.*, 25:289–294, 1982.

10. T. N. Blanton et al. JCPDS-ICDD round robin study of silver behenate: a possible low-angle x-ray diffraction calibration standard. *Powder Diffraction*, 10:91–95, 1995.

11. Gem Dugout, State College, PA. www.thegemdougout.com.

12. R. Jenkins and W. N. Schreiner. Intensity round-robin report. In *JCPDS-ICDD Methods and Practices Manual*, Sec. 13:2. ICDD, Newton Square, PA, 1989.

INDEX

21 CFR Part 11, 81
3C's, 259
3M's, 261, 292
3R's, 261, 274, 282
3S's, 290
4Q, 8, 9
4Q's, 112
4Q's, system, 272

acceptable limit, 98, 99, 101, 102, 105
acceptance
  criteria, 41, 98, 99, 104, 105, 107, 172
  test, 83
accuracy, 14, 16, 23, 30, 37, 39
  definition, 261
  GC, 204, 205, 208, 209, 210, 211, 212, 213,
    218, 220
  of liquid delivery, 342, 348, 351, 355 (table),
    362, 365, 368
  TOC, 309, 310, 312, 313, 314, 317
acetaminophen, 251, 253
acetic acid, 321
acidification, 302, 303
acoustic measurement, 371
active
  pharmaceutical ingredients (APIs), 93
  residues, 95
admixtures, 282

adulterated, 93
agglomerates, 256, 283
agglomeration, 264, 265, 277, 285, 289, 293
AIQ (analytical instrument qualification), 4, 8
air
  displacement pipette, 328ff, 334
  draft, 162, 164
  jet mill, 283
  vs. fluid displacement pipettors, 351
AIST, 267
allergenics, 95
ambient climate, 162
American Association of Pharmaceutical
    Scientists (AAPS), 310, 315, 316, 317
analytical
  method, 94, 95, 98, 100, 101, 104, 106
  procedure, 24
anthracene, 242, 243, 245, 246, 249, 250
antigen, 96
APPIE, 267
Artel MVS® (Multichannel Verification
    System), 369
Artel PCS® (Pipette Calibration System), 369
aspirin, 279
assay
  optimization, 349
  assay variability, 349

Practical Approaches to Method Validation and Essential Instrument Qualification,
Edited by Chung Chow Chan, Herman Lam, and Xue Ming Zhang
Copyright © 2010 John Wiley & Sons, Inc.

ASTM (American Society for Testing and Materials), 188, 189, 196, 214, 215, 238, 267, 269, 270
ASTM E11–04, 274
ASTM E29, 271
ASTM E56, 271
atomic planes, 378, 380
attrition, 281, 284
automatic cleaning and disinfecting, 97

back to back cleaning, 107
background, 304, 307, 308, 315, 322, 323
  (laser diffraction), 290, 293
bacteria, 97, 301, 302, 308
balance, xii, 155, 156
  analytical, 157, 170
  micro, 157
  top loading, 157
balling, 256
barometric pressure, 343
batch size, 105, 106, 107
BCR, 267, 268, 293
Becke Lines, 291
Beer-Lambert law, 361, 364, 369
before, during, and after sonication (BDAS), 282, 285
bicarbonate, 300, 303
bioburden, 98, 101
biofilm, 302
biological drugs, 94, 96, 101
blank, 219, 223, 224, 306, 308, 310, 311, 314, 321, 323
BP, 278
bracketing, 23, 96
Bragg's Law, 378, 379
Bragg-Brentano geometry, 379
Breakfast cereal problem, 262
British Pharmacopoeia, 177, 178, 188, 189, 190, 191, 195
Brownian motion, 268
BS 5295, 269
BSI, 255
bubbles (laser diffraction), 294
bulk size, 286, 287

caffeine, 239, 240, 242, 251, 253
calcium sulfate, 287, 288
calibration, 204, 205, 212, 306, 319, 320, 322, 323
  balance, 158
  of pipettes, 334, 338ff
campaign cleaning validation, 107, 108
capsules, 109
captive air in pipette, 331
carbides, 321

carbon dioxide (CO_2), 300, 302, 303, 304, 305, 306, 307, 308
carbonate (CO_3^{2-}), 300, 303, 304, 312, 320
carbonic acid (H_2CO_3), 303, 305, 306
Cargille immersion fluids, 291
carrier, 211, 212, 213, 224
carryover, 101, 104, 105, 106, 107, 108, 242, 247, 248, 249, 251, 253
cascade impactor, 257
CE, 202
cephalosporins, 95, 106
ceramics, 289
cGMP, 93, 95
change and configuration management, 67
change control, 148
CHARN, 278
chelating, 32
chemistry, manufacturing, and control (CMC), 48, 71, 72
chemometric, 55, 59
ChemSketch, 291
chloroform, 321
chromatographic detection, xii
chromatography data system (CDS), 115, 117
class 1 weight, 158
Class 100, 278
class 2 weight, 159
class 3 weight, 159
class 4 weight, 159
cleaning, xi
  agent residues, 95
  in place (CIP), 97
  procedures, 94, 96, 97, 98, 99, 104, 108
  validation, 93, 94, 95, 96, 98, 99, 100, 101, 102, 103, 104, 105, 106, 107, 108, 110
  validation protocol, 98, 107
cleanliness, 259
collaborative production management, 47
colony-forming units (CFU), 95
column temperature, 43
comminution, 283, 289
common problems, 25, 42, 69, 108, 161, 220, 292, 321
comparability protocol submission, 70, 72
comparative study, 40
compliance, 208, 220, 314
  testing into, 267
concentration, 259, 290
conductivity, 95, 104, 303, 304, 305, 306, 311, 317
conductometric, 304, 305, 321
configuration
  management, 83, 148

specification (CS), 123, 124, 127, 133, 135, 138
configured products, 122
consultant, 126
contamination, 214, 218, 220, 223, 224, 226, 227, 228, 259, 279
  liquid handler, 350
corner loading, 158, 160, 173
cotton-tipped cleaning stick, 100
counter
  air, 278
  condensation nucleus (CNC), 278
  liquid, 259, 278
  particle, 269
covalidation, 40
CRC (Handbook), 291
critical
  area, 98
  process parameters (CPP), 49, 51
  quality attributes (CQA), 48, 51, 67
  reagent, 353
cross-contamination, 96
crystalline solid, 378
curve, 323
custom
  application, 122
  coding, 138
cyanates, 321
cyanides, 321
cytotoxic, 95, 106

data logging, 209, 210
data manager, 58
degradation product, 96
density, 275
derivative ultraviolet spectroscopy, 100
description, 41
design
  qualification (DQ), 5, 6, 112
  space, 52
desolvation, 236, 250, 251
detection, 302, 304, 305, 306, 308, 309, 312, 313, 314, 317, 321
  limit (DL), 14, 30, 33, 37, 101, 308, 312, 313, 314, 317, 321
detector, 207, 212, 213, 214, 215, 216, 217, 218, 222, 224, 226, 227, 228
detergent residue, 94, 95, 99, 100
deuterium lamp, 236, 244
diagnostic tool, volume, 348
differential pressure (DP), 261, 282, 284, 285, 286, 290
diffraction
  angle, 2θ, 378, 379, 381
  peak intensity, 379, 385

peak position, 379, 380, 383
peaks, 380
peak width, 386
diffractometer
  alignment, 381, 382
  calibration practice, 387
  linearity, 383
  resolution, 386
  sensitivity, 385
dilution of dye, 360, 363, 368
direct-conductometric, 305
dispersion, 269, 270, 275, 282, 283, 284, 285, 288, 290, 292, 293
dissolution, 256, 262, 287
dissolved organic carbon (DOC), 300, 309
distribution, volume, 268
documentation, 24
DoE, 279, 281
drift, 214, 215
  tube, 235, 236
dual-dye photometry, 360, 366, 369
Duke Scientific (now ThermoFisher), 267, 268
dwell volume, 42
dynamic range, 232, 236, 238, 239, 242, 246, 248, 251, 253

E. coli, 107
E1458–92, 269
early phase, 13
eccentricity loading, 160
economic impact of volume error, 353, 372
electromagnetic influences, 162, 166
electron
  microscopy, 266
  multiplier voltage (EMV), 228
electronic
  batch record, 116
  lab notebook, 115
  pipette, 334
  record, 143
electrostatic effects, 339
electrozone sensing, 257, 258, 267, 279, 294
emission, 236, 237, 238, 242, 243, 244, 245
endotoxin, 301, 313
enterprise resource planning (ERP), 47
environmental
  conditions, 333, 338, 339, 340, 342
  effect on liquid delivery, 354, 358, 359
  variables, 343
enzymatic defection, 100
equipment surfaces, 93, 97, 100
equivalent columns, 31, 33
Erbium, 245
estimated carryover (ECO), 108

European Medicine Agency (EMEA), 301
European Pharmacopoeia (EP), 177, 178, 179,
    187, 188, 189, 190, 191, 192, 195, 275,
    301, 311, 305, 311, 319, 320
evaporation in pipette, 335
Excel, 282
excipients, 96, 100
excitation, 236, 237, 238, 242, 243, 245
external validation, 86, 87, 160

false negative/positive, 353
FCC (North America EMC), 202
FDA, *see* Food and Drug Administration
Federal Standard 209D/E, 259, 278
fermenters, 96
FID, 213, 216, 226, 227
filler, 109
fingerprint, 75
firmware, 122, 315, 316
fit-for-purpose, 279, 282
fixed volume pipette, 328
flame photometric detector (FPD), 227
flame photometry, 100
flow sensors, 224, 225
flowability, 256
fluidized-bed dryer, 53
fluorescence, 360, 364, 370
foaming and bubbles, 294
focused-beam reflectance method (FBRM)
    particle size analyzer, 55
Food and Drug Administration (FDA), 45,
    61, 70, 255, 257, 260, 268, 274, 279,
    301, 316
formic acid, 321
forward
    mode pipetting, 329
    pressure regulation, 211
    vs. reverse mode pipetting, 352
Fourier transform near infrared (FT-NIR), 53
fragility, 280
friability, 280, 284
Fs values, 29, 34

GAMP (good automatic manufacturing practice),
    5, 62, 67, 113, 122, 152, 310, 316, 317
GAP (good analytical practice), xi
garnet (standard), 274, 293
gas
    absorption, 256
    permeation, 306
    phase sampler, 220, 221, 222
    pycnometer, 275
Gaussian distribution, 264, 276, 279
gauze, 100
GC (gas chromatography), xii, 201
glass, 93

GLP, 4, 121, 146
GMP, 2, 59, 93, 95, 111, 121, 146
gold colloid, 277, 289
Good automated manufacturing practice
    (GAMP), *see* GAMP
Good manufacturing practice (GMP), *see* GMP
gradient, 210, 211
gravimetric method, 338, 354, 357, 362, 370
guidelines, 93, 95, 110
GxP, xii
    assessment, 62
gypsum, 287, 288

halogenated compounds, 321
Handbook of Optics, 291
handling, 307, 308, 321
hardest to clean, 98, 100, 103
hardware, 314, 317
heterogeneity, 264, 266, 279
high temperature catalytic combustion (HTCC),
    303, 304, 305, 323
highly purified water (HPW), 301, 302, 308,
    311, 313
holistic, 202, 204, 205, 206, 207, 214, 215, 218
homogeneity, 261, 262, 265, 276, 280, 285
HPLC, 95, 100, 104, 108
humidity change, 163
hydrophobicity, 32, 33
hygroscopic, 163
hysteresis, 210, 211

ICH (International Conference on
    Harmonization) Q9, 63
ideal gas law equation, 336
identification, 75, 76, 77, 80, 81
identity, 95
image analysis, 258, 270, 272
immunoassay, 101
inactivation, 98
information user, 125
infrared light (IR), 306
infrastructure software, 122
ingredients, 93, 94, 100, 101, 106
injection precision, 205, 206, 216, 217, 221
inlet pressure, 210, 223, 225
inlets, 207, 210, 216, 217, 218, 222, 223,
    226, 228
inner filter effect, 237
installation qualification (IQ), 5, 6, 112, 124,
    128, 136, 272, 274, 295
internal
    calibration, 160
    validation, 86
    validation group, 126
ion mobility, 100

IQ, *see* Installation qualification
IRMM, 304 294
ISO, 263, 270, 278
ISO 13320 (laser diffraction), 263, 265, 266,
    269, 272, 275, 276, 277, 281, 282, 284, 286
ISO 13321 (DLS/PCS), 277
ISO 17025, 342
ISO 3310, 274
ISO TC, 229, 267, 271
ISO TC24/SC4, 267, 270
ISO Working Groups, 270
ISO/ASTM, 356
IT department, 125

jet, 226, 227

laboratory
    information management system, *see* LIMS
    management, 125
    user, 125
laser diffraction, 263, 268, 270, 275, 277, 281,
    283, 284, 286, 293
late development, 14
latex, 275, 276, 277, 278, 286
LD50, 106, 107
legacy system, 144
library validation, 83
life cycle, 5, 11, 28, 111, 121, 123, 126, 152
light
    obscuration (technique), 258, 259, 267
    scattering, 257, 260, 267, 273, 275, 289, 293
LIMS, xi, 111, 114, 115, 122
    environment, 114, 116
    matrix, 117
    project manager, 125
    supplier, 125
linearity, 14, 16, 22, 30, 33, 37, 39, 101, 102,
    158, 160
liquid
    delivery quality assurance, 350, 372
    handler errors, 350, 351, 352
    handler qualification, 348, 354, 356
    handling, xi
LOQ, 16, 24

macropipette, 327
magnetic forces, 168
Malvern application notes, 263, 266
manufacturer testing, 208
manufacturing
    and packaging equipment, 94
    execution system, 47
mass flow, 211, 212, 216
materials, reference, 267
matrix approach, 94, 108
maximum daily dose, 105, 106, 107

mechanical influences, 162, 165
mercury xenon lamp, 244
methanol, 232, 234, 235, 239
method
    development, 12, 13, 279
    revalidation, 27, 36, 38, 40
    transfer, 27, 40
    validation, 12, 28, 40
    verification, 27, 28
methylene chloride, 181, 182, 183, 187
microbial
    load test, 95
    proliferation, 98
micronizing, 291
microscopy, 257, 258, 269, 291
microtiter plate, 352, 358, 360, 370
migrating data, 144
milling, 283, 289
minimal concentration, 97
minimum
    contact time, 97
    dose, 106
    mass (sampling), 263, 264
    number (sampling), 264
mobile phase, 42
modular testing, 128, 202, 204, 205, 206, 207
monodisperse, 269, 284
multiple scattering, 290
multivariate tools, 46

nano materials, 288
NBS, 268
NDA (new drug application), 72
near infrared (NIR), 75, 81, 177, 184, 195
nebulizer, xii, 235, 236, 251
nickel carbonyl, 289
NIR spectrophotometer, 177
NIST, 180, 182, 183, 184, 185, 186, 195, 197,
    198, 267, 268, 276, 277, 281
nitrogen-phosphorus detector (NPD), 226,
    228, 229
noise, 213, 214, 215, 216, 219
nonconfigured products, 122
nonlinear, 237, 246, 248, 251, 253
nonproduct contact surfaces, 95
nonreactive material, 98
nonselective, 232
nonuniform contamination, 108, 109, 110
nonwoven polyester, 100
Noyes-Whitney, 256
NPL, 267

obscuration (laser diffraction), 287, 288, 290
one point calibration, 25
operational qualification (OQ), 5, 6, 7, 112, 124,
    128, 137, 273, 274, 276, 292, 293, 294, 295

operator skill in pipetting, 344
OQ, *see* Operational qualification
oven temperature, 208, 210
overdrying, 97

paints and coatings, 289
paperless laboratory, 118
parameters, 14
parenteral product, 96
particle
    counting, 258, 259, 278
    density, 263
    size analysis, 259
    size distribution, 259, 260, 266, 279
particles, primary, 256
passivating agent, 95
passivation, 95
PAT, *see* Process analytical techniques
pathogen, 96
PCS (photon correlation spectroscopy), 278,
    293, 294
penicillin, 95, 106
performance
    qualification (PQ), 5, 6, 8, 159, 112, 124, 128,
        140, 260, 269, 273, 279, 292, 295
    verification, xi, 8, 177, 178, 180, 191, 192, 195
pertinent parameters, 97
pharmaceutical quality, 2
phase appropriate, 1, 11, 22
photometric
    linearity, 178, 191, 192, 193
    method, 340, 354, 360, 370
    noise, 178, 189, 195
piloting requirements, 132, 133
pipette, xii, 348, 351, 356, 369
    specifications, 342
    storage, 345
pipetting
    conditions, 332
    variability, 347, 353
pipework, 97
planning, 98
polydisperse, 269, 275, 284
polydispersity (PCS), 294
polyester swabs, 100
polymorphs, 291
polysciences, 267
positive displacement pipette, 328
potent steroids, 95, 106
powder, 256, 269, 272, 282, 283, 288, 289
    X-ray diffraction, measurement, 379
    X-ray diffraction, sources of errors, 382
    XRD standards, 384, 389
    XRD standards, alumina plate, 383,
        384, 389

XRD standards, LaB6, 384, 389
XRD standards, mica, 384
XRD standards, quartz, 384, 387, 389
XRD standards, silicon, 383, 384, 389
XRD standards, silver behanate, 384
power on self-test, 208
PQ, *see* Performance qualification
PQRI (Product Quality Research Institute),
    31, 33
precision, 14, 16, 23, 30, 33, 37, 39, 203, 279
    definition, 261
    intermediate, 289
    of liquid delivery, 348 (figure), 355 (Table),
        360, 362, 365, 370
    of pipetting, 342
preservatives, 96
pressure fluctuation, 164
pressure-size titration (PST), 281, 282, 283, 284
preventative maintenance, 219
prewetting of pipette, 332
primary size, 287
problem detection, 203, 206, 207
process
    analytical techniques, 3, 45, 279
    control system, 67
    integration, 97
    map, 119
    train, 94, 108
processing, 95, 96, 97, 98, 108
proper pipetting technique, 352
Pseudomonas aeurginosa, 107
pump-stirrer, 289, 290, 294
purity, 95, 96, 97

QbD, *see* Quality by design
qualification, 56, 202, 203, 207, 208, 209, 215,
    217, 218, 219
quality, 95
    assurance, 125, 138
    by design (QbD), xi, 13, 45, 50, 279
quantitation limit (LOQ and QL), 14, 30, 37,
    39, 101
quartz quintuplet, 387

radiation, 162, 165
Raleigh, 245
Raman, 238, 242, 245
range, 158
rare earth oxides, 180, 181, 182, 183, 184, 185,
    186, 188, 198
ratiometric photometry, 341, 368
readability, 158
real time, 47
recombinant protein, 96
recommissioned equipment, 94
recovery, 101, 102

reference standards, 186, 187, 191
reflectance, 178, 179, 181, 182, 184, 185, 187,
    189, 190, 191, 192, 193, 194, 195, 197,
    198, 199
refractive index, 282, 287, 288, 290
regression coefficient, 101
regulatory compliance, 356, 371
reliability, 289
repeatability, 261, 262, 282, 286, 288
    balance, 158, 160
replicates when pipetting, 338, 342
representative sample, 281, 292
reproducibility, 19, 261, 262, 264, 282, 285,
    289, 292
requalification (RQ), 8
requirements, 81
resampling, 97
residual, 234, 250
response, 213, 215, 216, 217, 227, 228
    linearity, 216, 217
    stability, 178, 191, 192
reverse mode pipetting, 330ff
revision history, 171
ringing, 211
rinse samples, 103
risk
    analysis, 79
    assessment, 126, 127, 129, 133, 202, 230
    based, 1, 2, 11, 13, 46, 50
    -based validation, 111, 152
    management, 3, 49, 113
robustness, 23, 261, 282, 289, 290
roles and responsibilities, 171
root mean square noise, 190, 191
route of synthesis, 27
rubber, 93

safety, 95, 98, 100, 105, 106, 107, 109, 208
Salmonella, 107
sample, 205, 207, 211, 215, 216, 217, 218, 219,
    220, 221, 222, 223, 224, 226, 228
    attributes, 207
    management, 119, 121
    provider, 125
sampling systems, 68
sanitization, 98
saturation, 226
scattering, 236
screens, 260
sedimentation, 257, 268, 270, 274, 293
selectivity, 101
senior management, 125
sensitivity, 101
service level agreement (SLA), 129, 146
shape selectivity, 32

sieves, 274, 291
    problems/solutions, 292
sieving, 257, 268
silanol activity, 31, 32
silent failures of pipettes, 344
silicon dioxide, 226
Silverson mixer, 289
Sinclair-La Mer nebulizer, 269
single-dye absorbance/fluoresence, 360, 366, 368
sink condition, 25
SISPQ (strength, identity, safety, purity, and
    quality), 13
size
    exclusion chromatography (SEC), 257
    primary, 279
smallest therapeutic dose, 106
smooth surface, 98
software, 214
    module specification, 127
    testing, 142
    validation, 113
solid bridging, 287, 289
solubility, 94, 101, 108
soluble materials, 287, 288, 291
solution, saturated, 287
solvents, 99
sonication, 286, 290
SOP, *see* Standard operating procedures
specific surface area, 287
specification, 281
specificity, 14, 20, 30, 33, 37, 39, 101
Spectralon®, 179, 194, 198, 199
spectrophotometer, spectroscopy, 354, 360,
    364, 368
spinning riffler, 266, 281
sponges, 100
SQ (specification qualification), 272
SRM, 277
stability, 210, 256, 257, 261, 280, 282, 283, 285,
    287, 290, 291
    sample, 20, 23
    standard, 20, 23
stabilizer, 289
stagnant water, 98
stainless steel, 93, 101, 102, 105, 108
standard
    error, 262, 263
    materials (list), 297, 298
    operating procedures (SOP), 95, 146, 267,
        273, 274, 280
standards
    concentration , 278
    ISO, 271, 272
    particle size, 266, 275
    size, 293

Staphylococcus aureus, 107
static electricity, 358
sterilization, 98
Stokes equation, 293
strength, 95
stress corrosion milling, 287, 289
sucrose, 233, 234, 239
SUPAC (scale-up and post approval
    changes), 72
supervisory control and data management, 47
supplement submission, 70
surfactants, 99, 290, 293, 294
swab, 95, 100, 101, 102, 103, 104, 106,
    107, 108
  method, 102, 103
swabbing techniques, 103
system suitability, 19, 23, 30

tableting, 256
tailing factor, 32, 33
tanks, cleaning, 97
target volume, 354, 360, 366, 369, 372
technical architecture, 127, 135
Teflon, 93, 101, 102
tefsteel, 101, 102
test design, 143
testable, 133
Texwipe, 100
therapeutic dose, 105, 106
thermal
  accuracy, 204
  conductivity detector (TCD), 228
  disequilibrium, 343
  gradient, 210, 211
  precision, 203
  zones, 226
ThermoFisher, 267, 268
three point calibration, 25
TickIT, 275
tip
  depth, 353
  mounting force, 345
titration, 100
TOC, *see* Total organic carbon
tolerances, 276
total organic carbon (TOC), xii, 103, 299
toxicity, 94, 107, 108
toxicology, 99, 105
trace residual contaminants, 93
traceability matrix, 133
training, 24
  record, 145
transfer waiver, 41
transflectance, 178, 179, 182, 187

transmittance, 178, 179, 180, 182, 183, 185,
    190, 192, 197, 198, 199
Tromp curve, 259
TX662, 100
TX714A, 100
TX758B, 100
TX759B, 100
TX761, 100

UL (device certification), 202
uncertainty in liquid delivery, 342
United States Pharmacopoeia (USP), 95,
    107, 177
universal detector, 250
USA vs. Barr Laboratories, 268
user
  acceptance testing, 124, 128, 129, 140
  requirement specification (URS), 123, 124,
    127, 129, 132, 133, 135
  responsibility, 7
  skill at pipetting, 344
  training, 128, 129
USP <1058>, 8, 112, 113
USP <1119>, 81
USP <1225>, 28
USP <1226>, 28
USP <429>, 272, 275, 276
USP <776>, 258
USP <788>, 259, 278
USP approach, 31
utensils, 95, 96

vaccines, 96
validation, 56, 60
  parameters, 14
  plan, 78, 126, 131, 132
  program, 96, 99
  protocol, abbreviated, 15
  protocol, detail, 15
  reporting, 90
  summary report, 129, 147
van der Waals, 289
variable volume pipette, 328
velocity biasing, 283
vendor
  audit, 127, 130
  responsibility, 7
verifiable, 133
verification, 56, 201, 202, 208, 266, 267, 276,
    292, 295
vertical beam spectrometer, 361, 366
vessels, 97
vibrations, 165
viscosity, 256, 293
visual examination, 95, 105
Viton, 101, 102

volume
  transfer errors, 350
  verification method, 348, 356, 360,
    367, 371

wander, 214, 219
water for injection, 104
water-soluble, 94
wavelength
  accuracy, 178, 179, 180, 181, 182, 183, 184,
    187, 195, 199
  repeatability, 178, 187, 188, 189
wavenumber, 180, 183, 184, 185, 187, 188,
    189, 195

wet suspension, 287
wet-dry comparison, 282, 285, 286
Whitehouse Scientific, 267, 268, 274, 277
workflow, 118, 120, 130
worst-case, 94, 97, 105
written procedure, 95

Xenon lamp, 236, 237, 244
X-ray diffraction, xii

Zantac, 256, 257
z-average, 294
zeta potential, 270, 294

Printed and bound by CPI Group (UK) Ltd, Croydon, CR0 4YY

16/04/2025

14658416-0005